A Third Pot-Pourri

Mrs. C.W. Earle

CAMBRIDGE
UNIVERSITY PRESS

CAMBRIDGE
UNIVERSITY PRESS

University Printing House, Cambridge, CB2 8BS, United Kingdom

Cambridge University Press is part of the University of Cambridge.
It furthers the University's mission by disseminating knowledge in the pursuit of
education, learning and research at the highest international levels of excellence.

www.cambridge.org
Information on this title: www.cambridge.org/9781108076685

© in this compilation Cambridge University Press 2015

This edition first published 1903
This digitally printed version 2015

ISBN 978-1-108-07668-5 Paperback

CAMBRIDGE LIBRARY COLLECTION

Books of enduring scholarly value

Botany and Horticulture

Until the nineteenth century, the investigation of natural phenomena, plants and animals was considered either the preserve of elite scholars or a pastime for the leisured upper classes. As increasing academic rigour and systematisation was brought to the study of 'natural history', its subdisciplines were adopted into university curricula, and learned societies (such as the Royal Horticultural Society, founded in 1804) were established to support research in these areas. A related development was strong enthusiasm for exotic garden plants, which resulted in plant collecting expeditions to every corner of the globe, sometimes with tragic consequences. This series includes accounts of some of those expeditions, detailed reference works on the flora of different regions, and practical advice for amateur and professional gardeners.

A Third Pot-Pourri

Mrs C.W. Earle (1836–1925) was born into the minor aristocracy as Maria Theresa Villiers. After training as an artist, she married Captain C.W. Earle, who inherited family property which enabled a comfortable lifestyle with a town house in London and a small property with a large garden in Surrey. Earle's designs for her garden were much admired by her artistic and literary circle, and she was encouraged to write down her gardening advice. In 1903 she published this work, the third in a very successful series of writings about gardening, cookery, travel and art, but the emphasis in this book is very much on the importance of diet to health, though there are plenty of other topics. The final section of the book contains the last letters home of Mrs Earle's son Sydney, a captain in the Coldstream Guards, who was killed late in 1899 during the Boer War.

Selected books of related interest, also reissued in the
CAMBRIDGE LIBRARY COLLECTION

Amherst, Alicia: *A History of Gardening in England* (1895) [ISBN 9781108062084]

Anonymous: *The Book of Garden Management* (1871) [ISBN 9781108049399]

Blaikie, Thomas: *Diary of a Scotch Gardener at the French Court at the End of the Eighteenth Century* (1931) [ISBN 9781108055611]

Candolle, Alphonse de: *The Origin of Cultivated Plants* (1886) [ISBN 9781108038904]

Drewitt, Frederic Dawtrey: *The Romance of the Apothecaries' Garden at Chelsea* (1928) [ISBN 9781108015875]

Evelyn, John: *Sylva, Or, a Discourse of Forest Trees* (2 vols., fourth edition, 1908) [ISBN 9781108055284]

Farrer, Reginald John: *In a Yorkshire Garden* (1909) [ISBN 9781108037228]

Field, Henry: *Memoirs of the Botanic Garden at Chelsea* (1878) [ISBN 9781108037488]

Forsyth, William: *A Treatise on the Culture and Management of Fruit-Trees* (1802) [ISBN 9781108037471]

Haggard, H. Rider: *A Gardener's Year* (1905) [ISBN 9781108044455]

Hibberd, Shirley: *Rustic Adornments for Homes of Taste* (1856) [ISBN 9781108037174]

Hibberd, Shirley: *The Amateur's Flower Garden* (1871) [ISBN 9781108055345]

Hibberd, Shirley: *The Fern Garden* (1869) [ISBN 9781108037181]

Hibberd, Shirley: *The Rose Book* (1864) [ISBN 9781108045384]

Hogg, Robert: *The British Pomology* (1851) [ISBN 9781108039444]

Hogg, Robert: *The Fruit Manual* (1860) [ISBN 9781108039451]

Hooker, Joseph Dalton: *Kew Gardens* (1858) [ISBN 9781108065450]

Jackson, Benjamin Daydon: *Catalogue of Plants Cultivated in the Garden of John Gerard, in the Years 1596–1599* (1876) [ISBN 9781108037150]

Jekyll, Gertrude: *Home and Garden* (1900) [ISBN 9781108037204]

Jekyll, Gertrude: *Wood and Garden* (1899) [ISBN 9781108037198]

Johnson, George William: *A History of English Gardening, Chronological, Biographical, Literary, and Critical* (1829) [ISBN 9781108037136]

Knight, Thomas Andrew: *A Selection from the Physiological and Horticultural Papers Published in the Transactions of the Royal and Horticultural Societies* (1841) [ISBN 9781108037297]

Lindley, John: *The Theory of Horticulture* (1840) [ISBN 9781108037242]

Loudon, Jane: *Instructions in Gardening for Ladies* (1840) [ISBN 9781108055659]

Mollison, John: *The New Practical Window Gardener* (1877) [ISBN 9781108061704]

Paris, John Ayrton: *A Biographical Sketch of the Late William George Maton M.D.* (1838) [ISBN 9781108038157]

Paxton, Joseph, and Lindley, John: *Paxton's Flower Garden* (3 vols., 1850–3) [ISBN 9781108037280]

Repton, Humphry and Loudon, John Claudius: *The Landscape Gardening and Landscape Architecture of the Late Humphry Repton, Esq.* (1840) [ISBN 9781108066174]

Robinson, William: *The English Flower Garden* (1883) [ISBN 9781108037129]

Robinson, William: *The Subtropical Garden* (1871) [ISBN 9781108037112]

Robinson, William: *The Wild Garden* (1870) [ISBN 9781108037105]

Sedding, John D.: *Garden-Craft Old and New* (1891) [ISBN 9781108037143]

Veitch, James Herbert: *Hortus Veitchii* (1906) [ISBN 9781108037365]

Ward, Nathaniel: *On the Growth of Plants in Closely Glazed Cases* (1842) [ISBN 9781108061131]

For a complete list of titles in the Cambridge Library Collection please visit:
www.cambridge.org/features/CambridgeLibraryCollection/books.htm

A THIRD POT-POURRI

A

THIRD POT-POURRI

BY

MRS. C. W. EARLE

AUTHOR OF 'POT-POURRI FROM A SURREY GARDEN' ETC.

LONDON
SMITH, ELDER & CO., 15 WATERLOO PLACE
1903

Home now your comrades come again,
　　But you come not.
For them life's triumphs still remain;
　　You draw Death's lot.

Oh, lying far from home away,
　　Feel not so far;
For, though all come, my heart does stay
　　There where you are.

E. Fuller Maitland.

June 1902

TO THE MEMORY

OF

SYDNEY EARLE

I DEDICATE THIS BOOK

I WISH to offer my very best thanks to my friend, Miss Adela Curtis, and my niece, Lady Constance Lytton, for the very material help they have given me with this book.

CONTENTS

---◆---

HEALTH

PAGE

Reasons for more about health—A stranger's letter—Encouragement from Dr. Haig—Details of my diet—Reason for early breakfast—Asparagus poison—Arguments of opponents —Dulness of diet—Reason of benefit felt on first going back to mixed foods—Test of underfeeding Dentist story— Opposition of medical profession—Their indifference to diet —The ordinary man at his breakfast-table—Doctors to be educated by the public—Uses of Plasmon—Necessity for mothers and children to learn physiology—Definition of uric acid—Instincts not safe guides—Difficulties of hospitality—Lord Roberts on 'treating'—List of useful books —Home education of girls: two methods—Hindoo love story 1

HEALTH OF OTHERS

Imaginary conversation between two doctors—Advertisements *versus* Dr. Haig—Remedies for depression on beginning diet —The old need not fear change of diet—Tea-drinking and the Chinese—The Bible and meat-eating—Pythagoreans and the bean—'Better Food for Boys,' and the schoolmaster —Difficulties of the diet—Fat and thin women—Mediæval and Greek idea—Individual cases—Maeterlinck's testimony   40

SUPPLEMENT

PAGE

Home started in Buckinghamshire, under Dr. Haig, for instruction and practice in uric-acid-free foods—A second
home in Hampshire for fleshless diet, recommended by
Mr. Eustace Miles, M.A.—Critical paper, by a patient of
Dr. Haig—Dr. Haig's answer to same 76

GOATS

Goats at Naples—Possible solution for milk difficulty in rural
districts—A toothless generation—Ignorance as to nourish-
ing value of separated milk—Mr. Hook on goat-keeping—
Personal experiment—Roast kid and *agneau-de-lait*—
Reasons for prejudice against goats—Suggestions for the
philanthropic—Immunity of goats from tubercular disease
—Day at Guildford—Almonds—The Astolat Press—Mr.
Gates' herd of Toggenburg goats—Feeding of goats—
Chemistry of food to be taught in elementary schools . . 89

WHOLESOME FOOD ON THREE SHILLINGS A WEEK

'Cornhill' budgets—Food reformers and lentils—Taste for
savoury foods—Nervous appetites—Cabinet Minister and
charwoman—The healthy foods—Maeterlinck's appeal
against meat and alcohol—Food values—To feed a family
of four on 12s. a week—Nut milk—A week's *menus*, and
cost—Ditto, with once-a-week cooking—Advantage of living
in country—Goat's milk at a London dairy—Cheapest and
healthiest diet at 2s. 4d. a week—To wean servants from
the beef-beer-tea faith—Possible purpose of meat-eating
phase in evolution—A philanthropist's experiment—
Amateur farmers—A pair of Bushey art students—Receipts 107

EIGHTEEN HUNDRED A YEAR . . 133

FEBRUARY

NOTES FROM NINE MONTHS OF A SCRAPPY JOURNAL

1901–1902

PAGE

Forcing cut branches—*Amygdalus Davidiana*—Early spring flowers under glass—Bulbous irises—Epimediums in pots— Letter to 'Westminster Gazette' on railway carriages and tuberculosis—Congress on tuberculosis in 1901 .   .   . 152

MARCH

Valescure—Tree heath and briar-wood pipes—Fragrant herbs and thyme carpet—Aloes and agaves—Cork-trees—Fréjus— Ruins of the Tuileries: De l'Orme and Bullant—Meissonier's picture of the burnt Tuileries—Cannes—Eucalyptus-trees—La Mortola—Arrival at Florence—Turban ranunculus—Mino da Fiesole—Letter about Florence .   .   . 160

APRIL

Arrival at Naples—Museum—English Society for the Prevention of Cruelty to Animals at Naples—Slaughter-houses in England—Art objects from Pompeii sometimes echoes of modern Japan—Baiæ—Coleridge and Harrison on Gibbon— South of Italy less affected by barbaric invasions than other parts—Aquarium—Goats—Dr. Munthe on the housing of the poor—Mrs. Jameson's picnic—Pompeii: smallness of the houses—Mr. Rolfe on Pompeiian pins and matches— Cenotaphs and war memorials—Pompeiian gardens as models for London—Sorrento and Amalfi—Garibaldi on cremation —'Aurora Leigh'   175

MAY

Apology for more gardening notes—Journey to Ireland—New English Art Club—A modern landscape recalling Claude at his best—Spring in the West of Ireland—Glorification of

PAGE

flat garden by old yuccas—Persian ranunculus—Want of thinning out and pruning a universal fault—An East Coast garden—Cultivation of *Hydrangea paniculata*—'The Wild Geese'—Gardening letter from German friend—Two good spring plants—A sundial—Floating bouquets—The May horticultural show 203

JUNE

Cuttings of double gorse and ericas—A gorse hedge—Gerarde on Solomon's seal—Preserving tulip bulbs after flowering—*Dictamnus fraxinella* in Wiltshire—The globe artichoke as food for man and goats—Peace—Glasnevin—Mr. Linden's garden at Brussels—Old wistaria and bignonias grown as shrubs—How to tell a good soil—Mr. G. F. Wilson's wild garden—How to grow Portugal laurels in boxes—Tamarisks and sea-buckthorns grown inland—The beauties of *Polygonum compactum*—London and the 24th of June—The rose show at Holland House 224

JULY

An account of lately bought gardening books—A lost poem by Milton—Vegetable gardens and rotation of crops—How to easily catalogue a garden—More half-hardy plants suitable for large pots—Carnations at Mr. Douglas'—Spanish rush-broom 244

AUGUST

Cultivation of various plants—Outdoor fig culture—Rhubarb in France—Effects of *Nicotiana sylvestris alba*—Potatoes in succession—Colonial branch of Swanley Horticultural College for Women—'Animal life'—Letter about monkey's food—Hampton Court garden and the old railing—The motor and Bramshill—Building a house—Rose planting—Cooking receipts—Autumn work in a German country-house kitchen—Household receipts 266

SEPTEMBER

PAGE

Visit to Northamptonshire—Peterboro', Fotheringhay, and Kirby Hall—Iris in pans for spring flowering—A last year's autumn letter from Germany—Kew and the smoke curse— Pruning back of shrubby plants to imitate sub-tropical gardening—Japanese anemones in shade—Sunflower seeds as a possible farming industry 305

OCTOBER

Solomon's love of nature— An old letter— Zola and fresh air— Old Harwich inn and curious specimen of *Clematis Vitalba* —Mesembryanthemums for cliff gardens—An old monastery fruit-wall—Three Pergolas—A long-wanted book on trees and shrubs—An old Suffolk breviary—Stories— Wild flowers for garden culture—Wellingtonias on a German hillside— Chrysanthemum culture—Mr. Morley's gift to Cambridge . 326

THE JOURNAL OF A TOUR IN THE NORTH OF EUROPE IN 1825–26 . . . 356

THE LAST LETTERS OF CAPTAIN SYDNEY EARLE, COLDSTREAM GUARDS . . 374

INDEX 419

A THIRD POT-POURRI

HEALTH

Reasons for more about health—A stranger's letter—Encouragement from Dr. Haig—Details of my diet—Reason for early breakfast—Asparagus poison—Arguments of opponents—Dulness of diet—Reason of benefit felt in going back to mixed foods—Test of underfeeding—Dentist story—Opposition of medical profession—Their indifference to diet—The ordinary man at his breakfast-table—Doctors to be educated by the public—Uses of Plasmon—Necessity for mothers and children to learn physiology—Definition of uric acid—Instincts not safe guides—Difficulties of hospitality—Lord Roberts on 'treating'—List of useful books—Home education of girls: two methods—Hindoo love story.

I MUST apologise to the public for the apparent poorness of idea in again repeating my somewhat tiresome title. I heard Mr. Motley, the historian, once say, a title should be 'telling and selling.' A 'Third Pot-Pourri' will very likely turn out to be neither of these, but it seemed to me the most honest title I could think of towards those who were kind enough, not only to read, but to like, my former books. They may find the matter in this book better or worse; the manner is exactly the same as before, and it could hardly be otherwise at my age.

I must, perhaps, also apologise for putting the Health chapters prominently forward at the beginning of this book, and I can only ask those who have no interest in

B

the subject to skip them altogether. They are written for those who asked for them. The chapter headed ' March ' in my second book, ' More Pot-Pourri,' which contained my personal confessions about diet, brought me such a number of touching and appealing letters from people of all sorts in every part of the world, that I cannot help thinking it almost a duty I owe to the readers of that book, to tell them as plainly as I can what I have learnt further about the subject, which for want of a better title we may call Diet, or Food, and its effects on the health of all classes of the community.

A great many people will merely laugh and think it very conceited and ridiculous that I should set up my opinion in matters of health against the great majority of the medical profession; but to anyone who has acquired good health, even late in life, the blessing is so inestimable, that it is only natural to try to help others to attain it. A note received the other day from a complete stranger stimulated me, perhaps, more than any other to feel that the knowledge and experience I have gained in the last three years might really be of some use to a few human beings. In this last of several letters, my un-known correspondent says, ' I am not likely to forget to associate your name with my improvement, and you, on your side, will have the satisfaction of knowing you have been the means of brightening and bettering our family's existence.'

Now, it seems to me that, however ridiculous it may appear to be very much absorbed in any one subject, if taking the trouble to publish a book upon it can call forth such an expression as this, and benefit, say, half-a-dozen families, I am well rewarded. As a further justification of my action in this matter, I should like to quote what T. E. Brown says in one of his delightful ' Letters ' : ' I believe that Jowett, like so many Englishmen, carried the principle of not " pinning his heart upon his sleeve for daws to peck at," so far as to forget that, besides the pecking daws, there are the craving hearts of others . . . craving for the food which, God help us, is not too abundantly spread upon the tables of this world.' Sympathy comes naturally to those who have prosperous circumstances, and I, who enjoy life so abundantly, in spite of age and sorrows, on account of my health, cannot help responding to appeals, from those who suffer, for further information as to the means by which I obtained it.

I am always being asked what I do myself. So far as I can, I will tell this exactly, first briefly stating that my health, which was good three years ago, has been distinctly improving both as regards endurance and nerve-power, and this in spite of heavy trials and sorrows borne a great deal alone, which to a nature like mine, after a life spent as mine has been, is no small additional suffering. Added to this, late in life I have had thrown upon me the entire management of house, garden, servants, stables, hospitality, which means a great strain on memory, especially after a lifetime with a man who shared all this with me, taking on himself the sole responsibility of much of it, and financially directing the whole.

My own conviction is, that though I started by myself on what I consider the right road as regards diet and health, yet without the assistance and support of Dr. Haig I should never have had courage to persevere against all opposition, and so have reached a level of health which has enabled me to withstand all this, and be so much better and stronger than either my mother or most of my aunts and uncles, who, with constitutions strong enough to live to a great old age, did so with much suffering from constant ailments—loss of hearing, sight, and brain-power.

A great many people may think that my improved health
is a matter of imagination, and I am the last to deny that
the mental attitude has an immense effect on the success
of diet; but with due allowance for this, my present
increased mental and physical power is a somewhat
unusual record, considering my family history, which is
one of strong constitution and bad health.

To come now to the details of my diet: At breakfast,
8 A.M., I eat a thick slice of home-made brown bread of
the kind known as 'Graham,' made without yeast [see
receipts in August] (to be bought from Heywood, 42 Queen
Anne Street, London, W., but better made at home), with
butter and marmalade, and a cup of hot separated milk
tinged with coffee. I have reserved to myself the right to
continue this self-indulgence of a small amount of coffee
in my milk, in spite of Dr. Haig's warnings: first, because
I so dislike the taste of milk, and secondly, because it
leaves me something tangible to leave off in case advancing years should make me less well. But I have a
nephew who looks with horror at an aunt whom he used
to think of as a kind of prophetess, who sits down at 8
in the morning in front of a coffee-pot. People often ask
me why I breakfast at 8. My general answer is that I
like it, and that it gives me a nice long morning; but the
real reason why I recommend it to others is, that if food is
taken at all in the morning, it must be taken five hours
before the luncheon time, as I think piling on another
meal before the previous one is digested is one of the
many causes of ill-health in the present day.

At my second meal, 1.30, I eat potatoes and vegetables
that are in season, experiencing no harm from *young* peas
or beans, but finding asparagus quite a poison to me.
Three years ago I wrote to Dr. Haig, saying that I had
been less well, and asking if it could be from asparagus,
as I had been eating it twice a day for ten days during my

full spring supply. He answered that, as far as he knew, asparagus was quite harmless, and that he thought I must have taken a chill. Last year, on the same symptoms reappearing, I wrote again. Dr. Haig replied as follows : ' I write a line at once to tell you what I know will interest you—that the asparagus *is* the cause of all your troubles.' This did not surprise me very much, as I knew that thirty years ago Dr. Garrod, the great gout specialist of that time, used to forbid asparagus to his patients. In winter, for the sake of change, I sometimes eat some well-cooked lentils. At this meal I generally eat salad, with about an ounce of cheese and a good big slice of home-made white bread with butter. If I still feel hungry I eat a milky pudding and some stewed fruit. This is unwise for those who are dyspeptics, as fruit and vegetables are best kept for separate meals (see Dr. Kellogg's ' Science in the Kitchen ') ; fortunately, I have had a good digestion all my life. My great object has always been, within certain health conditions, to keep my feeding as nearly as possible that which will fit in with the non-dietists who surround me. For instance, I always serve potatoes with fish, that I may take something and so save the depressing effect of a person sitting so long at table without eating anything ; and once or twice a month I have been known to take a little bit of fish if I fancied it, especially if I have been lunching or dining out, though I have proved conclusively that so simple a food (according to ordinary ideas) as plain boiled fish, if I eat it two or three days running, has a distinctly injurious effect on the rheumatic pain in my hip.

This sensitiveness to change of food is one of the strongest arguments used by the opponents of diet, and I confess it has some disadvantages ; but this applies to all forms of abstinence, and I would rather suffer occasionally than submit to an habitually low standard of health. The enemies of dieting—and most doctors to whom I have spoken about it are of the number—declare that the great objection to strict dieting is that it weakens the digestion. This, I think, is quite true of the Salisbury diet—namely, meat and hot water, as that gives the digestion next to nothing to do, and dilutes the gastric juices with quantities of hot water; but Dr. Haig's diet of cereals, cheese, milk, salad, raw fruit, and vegetables, is by no means easy of digestion, and the quickness with which I am now made aware of the harmfulness of many things that I used to take with apparent impunity, is in my opinion due, not to a weakened digestion, but to a return of healthy sensitiveness, induced by living for a long time on the natural food of man. I am quite sure if meat were given to horses, cows, or monkeys, though starvation might force them to eat it, they would be made very ill by such a diet. Rightly or wrongly, this seems to me the attitude to take towards the objection raised against what is called the 'weak digestion' of the vegetarian.

We have always been told that dyspeptics live for ever —this only means that nature is severely kind and sets pain as a sentinel to warn them when they have eaten something which they are unable to assimilate, and experience teaches them what to take and what to avoid; whereas the person of strong digestion, warned by no suffering, swallows everything and thinks he may do so with impunity. We all know how healthy children and healthy animals show when anything disagrees with them, and some of us well remember how the old nurses used to say, 'The sick baby always thrives,' meaning the baby whose stomach refused to be overloaded.

At 5 o'clock, for the meal which I still call 'Tea,' with the same truthfulness that I say 'The sun sets' though I know it doesn't, I drink one or two teacups of separated milk and hot water in equal proportions, and eat two or three pieces of toast made from home-made white bread, with butter, jam or honey, or watercress.

The meal at night when I am alone I own I seldom enjoy. I sometimes, besides home-made bread, have melted cheese (see receipts), or macaroni, sometimes rice and onions or other vegetable, with bread and butter and a little dried or fresh fruit, or both. At meals I drink very little indeed, milk being counted as nourishment rather than drink, but if I feel thirsty I take a little water—great thirst I should look upon as a sign of bad health, unless produced by excessive exercise. It is not a necessity, but I constantly drink a tumbler, or half a tumbler, of either moderately hot or cold water on getting up in the morning or on going to bed, or perhaps both.

Many people would say, ' So strict a way of diet would make life unbearable,' but after a time this strictness so changes the taste that the simpler foods are really enjoyed, and I distinctly think, that when people have dieted for several years, the amount of harm done by an occasional relapse is so small that the social convenience of it makes it worth while, so long as it is acknowledged as a concession to weakness and not a thing to be continued. It is what is done every day that matters.

People often tell me they feel so much better when they leave off the diet. This would only be a proof to me that they had not strictly dieted long enough, or had been under-nourished, and that the return to stimulating food does for them what alcohol does for those who already have too much in their system, and is merely a putting back of ultimate cure. I think all who have tried the diet for some time can always regulate it according to their varying requirements, if they will read the books and give the matter a little consideration. I, for instance, am always being told that I underfeed, and Dr. Haig never sees me without expressing his surprise that I am as well as I am, considering that I live a good deal on vegetables and certainly, as a rule, take much below the correct amount of proteids for my age and weight. My under-feeding cannot be serious, for I sleep my six or seven hours, have not lost or gained flesh, and feel perfectly well. I often have tried to add food of a more nourishing kind, such as curd cheese, Plasmon biscuits, milk, &c., but after a few days I generally find it has a tendency to bring on a slight return of rheumatic stiffness. I am inclined to think that the doctors who preach great moderation, whatever the diet, such as Dr. Keith and Dr. Dewey, have a good deal of truth on their side, as, though the proper standard of *strength* will never be attained by the under-fed, still the full allowance of food may go to feed the particular weakness or ailment which people of a certain age are almost sure to have, and will thus prevent them reaching the level of *health* they might have on a lower standard. People seem to assimilate food so differently that, given there is no permanent pallor, especially no very white gums, or sense of fatigue, each one must judge a little for himself what he requires. Did I not suffer less from fatigue than I have ever done in my life, I should try harder to live up to the standard settled by physiologists as necessary to health, and which would doubtless be essential were I younger. I tried some Grape-nuts in the winter and felt a hot Hercules for a few days, but I believe them to be distinctly a gout-making food.

Two or three years ago I had occasion to go to an oculist to see if my spectacles required strengthening. I begged him to test my eyes thoroughly. At the end of the interview I asked him if in every respect they were up to a good average standard for my age. He said most certainly they were, and in a most healthy condition. I then asked whether he would be surprised to hear that I had for some years been what is called a vegetarian. He

immediately assumed a serious medical manner and said,
'Up to now it seems to have done you no harm, but,
please, don't go on with it too long ! ' the manner implying
that terrible things might happen. I smilingly replied that
I promised I would give it up the moment I was less well.
I mention this to encourage people to meet the opposition
which they must expect from all doctors, nurses, aurists,
oculists, and dentists—in fact, all the minds trained on
the lines of the regular accepted medical teaching. The
study of food in relation to health is a branch of medical
science as yet in its infancy, for the best authorities, as
may be seen in the standard text-books on Materia
Medica, own that they know next to nothing of metabolism, or the changes undergone by food in the body.

A great many people tell me that diet involves so
deep a knowledge of physiology that they cannot possibly
undertake it. They cannot risk the responsibility of going
against their doctor. They say to me, 'How can I fight
a man who has given his lifetime to the study of these
things, and who must know so much more about them
than I can, even if I give my best attention to studying
them ? ' But is this the truth ? Has not the doctor been
taught to study drugs for the cure of disease rather than
food as the basis of health ? He never gives diet much
consideration except in the case of over-eating in severe
illness. In giving a mother rules for the health of her
children, doctors will constantly recommend fresh air,
exercise, and above all sufficient nourishment ; but they
rarely give any details as to the best *kind* of nourishment. I have heard of a doctor who recommended a
non-flesh diet to one of his patients, and on expressing
his surprise at finding he had really followed it, said,
' There are at least 120 of my patients who would be
benefited by it, but not one of them would do it.' I
think this is only natural, as the moment the patient was

better he would say, 'Does Dr. —— practise this diet himself and in his own family? If not, why am I to do it now I am better?' Can anyone think that vaccination would be so universally accepted if doctors and their children were not themselves vaccinated?

Only those people who have strength of character enough to take responsibility against public opinion should attempt the simpler food diet. To begin it haphazard with no knowledge and little faith is almost bound to end in failure. The undeniable success of the diet upon myself has caused many people to say, 'There is no doubt this diet suits *you*,' with an emphasis on the 'you,' intended to convey 'what suits you, would be fatal to me.' With these I go no further. The real fact is an immense number of people are very fairly well, and much enjoy the good things of life, including food, between their attacks of illness. They entirely forget the expense of time, strength, and money entailed by these little attacks of colds, bilious headaches, feverishness, &c., the mornings spent in bed, the afternoons on a sofa in a darkened room, the days lost at business from bronchitis and influenza.

If people could once be persuaded that the reduction of luxurious food does mean improved health, I think I should hear less about the extreme self-denial involved in my diet. The man who comes down in the morning and grumbles because he likes neither of the two hot dishes provided for breakfast, would be the last to consider himself either a luxurious liver or an invalid; but, having been convinced by the preaching of years that he must 'keep himself up' by eating well, any change from a three-meal a day meat diet, without considering sandwiches at 5 o'clock tea, strikes him with horror as a low diet which will result in 'running down' and losing the strength of mind and body so necessary for work.

This ' running down ' does indeed happen not infrequently as a result of the old-fashioned unintelligent vegetarianism.

A great obstacle to change of diet is the family doctor. I have known two or three who sadly needed it themselves, and having tried it more or less for a few months, pronounced it a failure because the good result was not instantaneous. The best results never come under eighteen months or two years. I wish to warn people that if they consult their doctor no diet will be tried. The doctors must be educated by the public. The majority of them have no idea of giving up their own food and social enjoyment, though I *have* heard of a few of those very men recommending it to their *wives*! Doctors are good kind men on their own lines, and devoted to their profession as long as it means curing illness by drugs ; but a little reflection teaches us that the members of a learned profession are naturally the very persons least disposed to innovation upon the practices which custom and prescription have rendered sacred in their eyes. A lawyer is not the person to consult upon bold reforms in jurisprudence, and a physician can scarcely be expected to own that diet may cure diseases which resist an armament of phials. Every feeling of a doctor must be against a system which does not profess to be a cure of active disease, but a radical reverse of all the preconceived ideas of maintaining health, and is also a denial of the principles taught by the College of Surgeons. All the same there are hopeful signs that a great change is coming about. Dr. Lionel Beale used to tell young practitioners many years ago that they would often come across cases for which starvation was the only cure, but he said, ' Of course, if a patient came to you and you advised him to starve himself, you would never see him again. But there are many ways of inculcating good advice without shocking the nerves of sensitive people who suppose that abstinence from food for a few hours means death. Tell your patient not to take any solid food for a week. Order him a little beef-tea three times a day. Towards evening he may take with it a biscuit, or a little dry toast. . . . By a little exercise of ingenuity you may suggest various things to take that will satisfy him, but which altogether will not amount to much.'

To-day's post brings me this account of what is called a 'holiday doctor' in a neighbouring village. The patient, a strict vegetarian, had caught measles. She writes: 'We had a sensible young doctor who knew nothing of my ordinary way of living. He gave me no medicine, and recommended me to give up lobster and pork. He was much interested because I had rheumatism in the joints, which he said he had heard of in measles, but never seen. As to medicines, he said he never took any himself. As I said I did not like them, he said he would not offer me any till I asked him for some. "But," he added, "most people *will* have them, even if one only sends them a little coloured water." With high fever I have always before this had headache, but this time I had no headache. The rash *rushed* out, and vanished suddenly the second day. I have had a most comfortable illness, and really actually feel the better for it.' The recommendation not to eat pork and lobster makes me think of a story of forty years ago. An uncle of mine, Mr. Charles Villiers, returning from one of the German baths, told us he had been much amused by the German doctor in his parting instructions, saying, after a good deal of apologetic attempt to soften the blow, 'There is one thing I really *must* beg you not to eat, and that is *bear's* flesh!'

The oft-repeated injunction of doctors to eat chicken and fish after illness has sometimes amusing results amongst the uneducated classes. A soldier who had been off duty for some days with a severe bilious attack came back and said with pride to the inquiring officer that he was much better, and his wife had given him a 'nice dinner of tinned salmon!'

My last word is, if you want to try the diet on *children*, you must have both faith and knowledge enough to fight nine doctors out of ten, although in my experience, with adults, they are quite willing to leave all matters of food and even drink to the patients themselves, merely cautiously changing the wine or recommending none at all. I had one friend, a doctor, who had the honesty to say to me after years of threatening me with every kind of misfortune, 'Mrs. Earle, what would happen to doctors if everybody lived as you do?' He also confessed to me, with great generosity, that in consequence of what I had said, no doubt assisted by the way the profession recommends Plasmon, he had had great success with two inebriate female patients by making them drink dissolved Plasmon, mixed with tea, milk, barley-water, &c. Plasmon should never be used without being first dissolved by boiling it in a little water, and this must be done even when it is introduced into puddings, cakes, biscuits, or any food or drink whatsoever. Plasmon is now being so much used by those who I consider are already suffering from over-feeding, that both Plasmon and Protene, as well as many other concentrated foods, promise to be, in my opinion, a considerable danger in the future.

Sir Henry Thompson, in his most practical and temperately written book, 'Diet in Relation to Age and Activity,' says: 'Respecting the act of eating itself, it is desirable to add a few words here. Not many persons learn the importance of performing it rightly in youth and middle life. Indeed, it ought to be taught among other elementary lessons in physiology at every school in early life, a short course of which would be much more important and far more interesting than some of the other courses which the existing curriculum contains. I mean by this, a simple description of the chief internal organs connected with digestion and how they act. Every child at eight or ten years of age should know what becomes of his bread and butter, and of his meat, when he gets it. I can scarcely conceive a better subject than this for a simple and entertaining talk to a class of these young people, with a diagram on the wall showing the chief organs contained in the chest and abdomen. Another chat about respiration and the circulation of the blood would follow at a later period. The subject is regarded with suspicion by the public, from the imposing effect of the five-syllabled Greek term "Physiology," which suggests the idea that I propose to teach young children " science "! —as if that term, let me remark, whenever it is used, denoted anything more than an " *exact* knowledge respecting the matter in hand." '

How many mothers possess this knowledge which Sir Henry Thompson declares should be familiar to ' every child of eight or ten ' ? Can it be so very abstruse and difficult if children of this age can begin to learn it through the medium of ' a simple and entertaining talk ' ? A gardener who means to be successful is not content to work on hearsay ; he takes care to acquaint himself with the best books on the chemistry of soils, and by careful experiment builds up for himself a first-hand knowledge of the best foods for his plants ; and the same process is followed by the horse-breeder and the farmer of crops and cattle. The human animal alone, most precious and costly of all, is reared on the traditions of nurses and doctors, the mother apparently not thinking it her business to know anything of its anatomy and physiology, or of the chemistry of the food on which its healthy growth must so largely depend. For want of such knowledge, moreover, she is not really capable of judging of the fitness of either doctor or nurse. Many of the women who object that this subject is too difficult for them will spend hours every day in reading current works of fiction, history, biography, travel, politics, &c., in order to establish or keep up a reputation for being 'well-read,' 'cultured,' or whatever the phrase may be which conveys the impression that they can take intelligent, if not brilliant, part in dinner-table conversation on the interests of the hour. If part of the time now spent by women in doing as a matter of course the work which contributes towards 'social success' were to be given to the elements of physiology, hygiene, and the chemistry of food, the health of future generations might be enormously improved. These are not subjects which can give any social brilliance; they merely lay the foundation of physical, mental, and moral well-being in the family and the race. This is my reason for including at the end of this chapter a few of the books which quite clearly explain all that it is desirable to know; Dr. Allinson's books, which are read and understood by thousands of the poorer classes, supply a good deal of instruction in a popular form.

Returning from London, in February this year, where I had caught a cold in the head which never laid me up for an hour, I fancied I was a little less well. A friend sent me the American Dr. E. H. Dewey's book, ' A New Era for Women '—his text being Herbert Spencer's words : ' If there were no eating without hunger, and no drinking without thirst, then would the system be but seldom out of working order.' The book is full of useful information, his panacea being the abolition of breakfast and eating nothing till 12 or 1 o'clock. Out of curiosity I tried this for ten days with the same kind of benefit that I used to feel in old days when I took a tonic that suited me. I did not continue it for the same reasons which prevent my doing several things that I believe would be better for me—the unsociability, and finding that I got too hungry by 1.30, the luncheon hour, which I could not conveniently change. I think a great many people who live on the ordinary food, and have eaten a large dinner the night before, will find benefit from cutting off breakfast altogether, or, at any rate, diminishing it.

My great difficulty is when I pay visits, but as they seldom last for more than a week or ten days, and I have the courage to ask for servants' cheese even of the swellest butler, and as change of air always gives me an appetite, I generally come home feeling better than I went, whether I have underfed or not. This is the case even if I have eaten fish once or twice—boiled fish, one of my remaining temptations, seeming to me one of the most harmless of uric-acid-containing foods. This term, uric acid, I find is bewilderingly mystifying to most of my friends, who seem to think it is a medical glorification of ordinary acidity. So far as I understand it, it is a necessarily component part of our body, but 75 per cent. of modern human beings have an excess of it. Every bit of flesh food, including fish and eggs, contains a certain portion of this substance, of which we have already too much; therefore the very facility with which we digest it adds to its injuriousness, while its tonic properties add to its attractiveness.

One of the questions I am constantly asked is, ' Why may we not follow our instincts, and eat light foods in summer and meat in winter ? ' I should answer : Because I believe that many of our eating and drinking instincts, not coming from that actual healthy hunger which finds dry, good bread and plain water a delicious repast, are on the level of the inebriate's desire for alcohol. Given that we are underfeeding, we need more nourishment—*i.e.*, food containing albumen or proteid—in hot weather than in cold, when we lead, as a rule, much more sedentary lives, and take much less out of ourselves. What many people call 'natural' is often most 'artificial,' and generally wrong. All my generation were brought up to think that beef-tea and port-wine were naturally essential to fight the weakness consequent upon illness and fever ; now, except in very remote country districts, beef-tea is never ordered except as a substitute for brandy—*i.e.*, stimulant without nourishment.

It is no small compliment to what is expected of diet that many people who have been ill or ailing for years under the ordinary regimen, and with the advice of various physicians, when they visit Dr. Haig, and begin his methods, express great anger if not instantly better, and instead of returning to give him the chance of changing his prescription—viz., readjusting the diet— throw up the whole system in disgust as a failure. This is, of course, most unreasonable. The other cause of anger against Dr. Haig is that he changes the details of the dietary he recommends. At first he had to accept, as all doctors do, the conclusions of others as to the constituents of food ; and in consequence he recommended the pulses—peas, beans, and lentils. Further personal experience and first-hand investigation convinced him that these and a few other vegetable substances—as asparagus, tea, coffee, cocoa, and mushrooms—contained a poison differing but slightly from, and in no way less injurious than, the animal poison of uric acid. To me, changes in detail, round a central idea, are the greatest proof of intellectual growth.

A vegetarian friend has related to me the explanation of the well-known story which goes the round of London, and is cited by doctors when asked by patients if they shall try Dr. Haig's diet, or consult him for rheumatism. The story is that when Dr. Haig read a paper on flesh-eating as the cause of rheumatism, before a medical congress some years ago, a member present asked a question which 'floored him'—viz., why rheumatism was prevalent in many countries where no flesh was eaten? The immediate answer did not occur to Dr. Haig, as he had not then carried his researches far enough to know that the pulses—peas, beans, lentils—and other vegetable substances, as tea, coffee, &c., contain even more uric-acid-producing poison than meat itself. The question, however, set him to work on those foods, with the result that he soon discovered the full answer to the question, and published it. This important point seems to have escaped the notice of those members of the profession who content themselves with telling only the first half of the story, which gives an inaccurate and unfair impression. I myself have often been asked why horses get rheumatism. My answer has always been, 'Too many oats'; for years ago, from his study of animals while soldiering in wild countries, my husband used to tell me that excess of cereal foods, with their high percentages of acid salts, caused rheumatism in horses unless well balanced by fresh green foods.

I have found that my difficulties in wishing to provide my guests with what they like to eat have been immensely increased of late years. As long as one's friends are mixed feeders, a great variety of diet is necessary. The best solution has seemed to me that the invalids should have their meals apart, and that in mixed company there should be three side-tables. On one of these, soup, fish, meat, and vegetables; on another, farinaceous foods and stewed fruits, Plasmon and other biscuits; and on another, fresh raw fruits and nuts. The great thing is to reduce courses and to serve dinners more as luncheons or suppers, everything being brought in at the same time This helps to disguise what people eat, and what they don't eat. Cheese is always offered with the salad to those who do not eat meat. I have no wish to be extravagant and give dainties which no one will eat, and a whole dinner of soup, fish, two meats and a sweet, for one or two people, while the rest eat only some vegetarian food, is rather obviously uncomfortable not only for the hostess but for the guest—and I confess a general tone of abstinence, and desire to prove indifference to food and drink, have greatly increased in my house in the last few years. This is in a measure due, no doubt, to people's kindness of heart and friendliness towards an old woman with a crank; for I know my sons have often been condoled with on the excessive pity it is that their mother has gone mad on the food question, the speaker putting it down, no doubt, to nervous depression from starvation, or failure from old age.

I, myself, have now considerable dread that mere abstemiousness should lead to underfeeding, which all seem agreed upon as the greatest danger for the young; and if mixed feeders ask my advice, I say, ' Knock off wrong foods at breakfast as much as you like, but keep your other meals very much what you have been accustomed to until you have time to attend properly to the subject, and decide what is really the best diet for you.' Sportsmen and others who lead an habitually outdoor life can eat most things, and a middle-aged nephew, who has always led a healthy country life, said to me lately, ' My system is never to allow my stomach to dictate to me. The other day I took a glass of beer which, as you know, I never drink. It disagreed violently with me; so I said, " Very well; you shall have it every day for a week, and I got perfectly used to it. Then I left it off.' There is a rough truth in this; an immense number of people can eat, with apparent immunity, for many years the food they are used to. But this is a question of training the digestion to assimilate and not refuse unaccustomed food, and is quite a different matter to well-digested food introducing poison into the system, if that food contains poison.

A friend staying with me received the other day the following account of a large dinner to mixed dietists, after a meeting on Theosophy, the description being by a semi-convert to strict diet who was present: ' I was much amused to see one odd-looking person after another, with various degrees of dyspeptic appearance, helping themselves, with a pious air of exclusiveness, to one tomato, or a dish of beans, while some took fish, and others refused it with horror, and all agreed that meat was anathema. Certain stolid, wholesome-looking folk ploughed steadily through the whole menu, from soup to fruit; and another entertaining point to the naughty scoffer was the amount of food-talk among the " dietists," while the brutal carnivora had leisure to devote themselves to other subjects! One of the elect, who was daintily regaling himself on an apple and a glass of milk, explained to me, as I ate my salmon and cucumber, the brutalities of the slaughter-house and fisheries, and when I demurred to the statement that animals reared for food have a bad time of it, my opposite neighbour leaned forward and solemnly informed me that " some day I should have to be eaten "! ' Does not this comic account of a modern effort to imitate the old barbaric method of bringing people together by providing them with food and stimulants, rather suggest that, in the future, the one animal function we still perform in public—eating and drinking—may cease to be the pivot of sociability, and that when we wish to see a friend, old or new, we shall some day write: ' Come and have a chat with me ' instead of ' Come and dine, or lunch, or have a cup of tea ' ?

The accusation that dietists talk too much about their food is perfectly true, and must be guarded against; at the same time, I am all for people talking about the subjects that particularly interest them at the moment, for 'out of the fulness of the heart the mouth speaketh,' and the tendency of conversation at many dinner-tables, even when stimulated by meat and wine, often does not rise above the level of sport, scandal, or games. A few years ago it was bicycles, now it is motors that often absorb a whole evening.

The breaking up of uniformity in food may be the initial stage of transition to a higher civilisation, luxury in social entertainment, both in food and drink, having governed the world too long. Just think what it means as showing the change that has come over the world, that a conquering general on his return home should think it desirable to write such an appeal to his countrymen as Lord Roberts' letters to the press on 'treating' soldiers. He wrote two letters, the second emphasising the first; and so much did I honour him for such action, and so important did I feel it to be to spread the knowledge of his wishes in every possible way, that, together with some of my neighbours, I had large posters printed and circulated in the village, giving the full text of the letter, and headed in red ink: 'Return of our soldiers from South Africa.' I take the liberty of including the letter here, 'lest we forget':

'Sir,—Will you kindly allow me, through the medium of your paper, to make an appeal to my countrymen and women upon a subject I have very much at heart, and which has been occupying my thoughts for some time?

'All classes in the United Kingdom have shown such a keen interest in the army serving in South Africa, and have been so munificent in their efforts to supply every need of that army, that I feel sure they must be eagerly looking forward to its return, and to giving our brave soldiers and sailors the hearty welcome they so well deserve when they get back to their native land.

'It is about the character of this welcome, and the effects it may have on the reputations of the troops whom I have been so proud to command, that I am anxious, and that I venture to express an opinion. My sincere hope is that the welcome may not take the form of "treating" the men to stimulants in public-houses or in the streets, and thus lead them into excesses which must tend to degrade those whom the nation delights to honour, and to lower the "soldiers of the Queen" in the eyes of the world—that world which has watched with undisguised admiration the grand work they have performed for their Sovereign and their country. From the very kindness of their hearts, their innate politeness, and their gratitude for the welcome accorded them, it will be difficult for the men to refuse what is offered to them by their too generous friends.

'I therefore beg earnestly that the British public will refrain from tempting my gallant comrades, but will rather aid them to uphold the splendid reputation they have won for the imperial army. I am very proud that I am able to record, with the most absolute truth, that the conduct of this army from first to last has been exemplary. Not one single case of serious crime has been brought to my notice—indeed, nothing that deserves the name of crime. There has been no necessity for appeals or orders to the men to behave properly. I have trusted implicitly to their own soldierly feelings and good sense, and I have not trusted in vain. They bore themselves like heroes on the battlefield, and like gentlemen on all other occasions. Most malicious falsehoods were spread abroad by the authorities in the Orange Free State and the Transvaal as to the brutality of Great Britain's soldiers, and as to the manner in which the women and children might expect to be treated. We found on first entering towns and villages, doors closed and shops shut up, while only English-born people were to be seen in the streets. But very shortly all this was changed. Doors were left open, shutters were taken down, and people of all nationalities moved freely about in the full assurance that they had nothing to fear from " the man in khaki," no matter how battered and war-stained his appearance. This testimony will, I feel sure, be very gratifying to the people of Great Britain, and of that Greater Britain whose sons have shared to the fullest extent in the suffering as well as the glory of the war, and who have helped so materially to bring it to a successful close.

' I know how keen my fellow-subjects will be to show their appreciation of the upright and honourable bearing, as well as the gallantry of our sailors and soldiers, and I would entreat them, in return for all these grand men have done for them, to abstain from any action that might bring the smallest discredit upon those who have so worthily upheld the credit of their country.

'I am induced to make this appeal from having read, with great regret, that when our troops were leaving England, and passing through the streets of London, their injudicious friends pressed liquor upon them, and shoved bottles of spirits into their hands and pockets—a mode of " speeding the parting " friend which resulted in some very distressing and discreditable scenes. I fervently hope there may be no such scenes to mar the brightness of the welcome home.

'I remain, Sir, yours faithfully,

'ROBERTS, F.M.'

HEALTH BOOKS

' Uric Acid as a Factor in the Causation of Disease,' a contribution to the pathology of high blood-pressure, headache, epilepsy, mental depression, paroxysmal hæmoglobinuria, and anæmia, Bright's disease, diabetes, gout, rheumatism, and other disorders, by Alexander Haig, M.A., M.D. Oxon., F.R.C.P., Physician to the Metropolitan Hospital, and the Royal Hospital for Children and Women; late Casualty Physician to St. Bartholomew's Hospital. Fifth edition, with seventy-five illustrations. (Churchill, London.)

In spite of having mentioned this book before, I must re-name it here at the head of these health books, as the more I understand the subject, the more I am convinced that it is one of the most important books on medical science that have ever been given by a doctor to the public.

I am told that it is much disputed here in England, though how it is scientifically disproved I am not able to judge. It has been translated into German, and, with their usual broadmindedness in matters of science, German doctors and scientists are taking a deep and practical interest in Dr. Haig's researches, and are even coming over here to consult him; whereas in England I was lately told by a medical man of high standing and large practice, that 'Dr. Haig stood alone in his opinions, which were not shared by a single member of his profession.' Anyone who knows even as much of the matter as I do, knows this to be quite untrue. How far Dr. Haig's book has been seriously considered and answered in his own country I am not in a position to decide, but I do know that I have the names of over a score of doctors who are working on his lines.[1]

The book is a big one and costs 14s., a sum which may be deterrent to many in these non-bookbuying days. It is hard reading, highly technical, and almost impossible for a lay-mind to read straight through; but as a book of

---

[1] Leaflets of instruction in the best ways of beginning the diet are given by Dr. Haig to his patients, and are to be got by writing for them to A. Haig, Esq., M.D., 7 Brook Street, W.

reference it is singularly clear and instructive, the index being so good that any subject can be readily looked up.

'Diet and Food in Relation to Strength and Power of Endurance.' Haig. Fourth edition. Seven illustrations. This book, costing only 2s., has been prepared by Dr. Haig for the general public in order to make plain and easy the A B C of his theory, so that the many sufferers from the diseases he has specially investigated may begin their studies and have an opportunity of being able in time to judge for themselves.

I must again repeat that all hope of real improvement in the standard of health must come through the intelligent experience of the public teaching the doctors.

Even those who already possess the earlier editions of this book I strongly recommend to get this latest one, as it contains valuable additional pages on bread foods—*i.e.*, not only good home-made unadulterated loaves and biscuits, but all the many kinds and combinations of cereal products, such as wheat, barley, oats, maize, rice. There is also an interesting page of dental illustration, which ought to end the old controversy as to man being carnivorous or omnivorous.

One of the most interesting converts to the diet is Mr. Eustace Miles, M.A., the well-known tennis champion and Cambridge coach, who has frankly given to the public his personal experience in 'Muscle, Brain, and Diet,' a book which contains most useful general instruction about food and health, together with some excellent cookery receipts for the simpler diet—one of which I include in my list of cooking receipts, it being in my opinion much too good to be missed. Two other useful volumes for the student of food as the basis of health, are Mr. Miles' 'Failures of Vegetarianism,' which is far more helpful than all the successes I have ever read of,

because it warns beginners of the errors they are almost
sure to fall into if unguided : and 'Better Food for Boys,'
a little 1s. number of Messrs. Bell & Sons' 'Life and Light'
Series, which brings the subject poignantly home to
parents, guardians, and schoolmasters, and by providing
a table of food values enables everyone, girls and adults,
as well as boys, to calculate the right quantity and kind
of food necessary for health—though nothing but personal
trial and careful observation and patience will teach each
individual the best food for his or her particular constitution and way of life.

'Avenues to Health.'—This last of Mr. Miles' books on
this subject is, perhaps, the most important of all, for it is
brim full of generally interesting information culled from
all kinds of sources, old and new, Western and Oriental,
as to the best means to attain and maintain the highest
physical and psychical health.

My friend, Miss Adela Curtis, writes me the following
notice of three useful pamphlets :

'When first we opened our bookshop, we got a
quantity of little penny books on diet from Mr. Albert
Broadbent, of Manchester, and experimented by putting
one or two in the window, and about the shop casually
among the other books. It may interest you, as much as
it amused us, to learn that these little things were picked
up, looked into, and bought by all classes of people,
whenever we happened to bring them forward. One of
these, " Forty Vegetarian Dinners," of which 20,000 have
been sold, contains an interesting introduction by the
Hon. Mrs. F. J. Bruce, whose testimony as the mother of
a large and healthy family, and the head of a household
numbering some twenty-two members, is exceptionally
valuable, seeing that she has nothing but praise for the
system after sixteen years' practice.

'Another of these dainty pennyworths, now in its

seventeenth thousand, is called " Science in the Daily
Meal," and gives carefully proportioned dietaries for sixteen days, as well as helpful advice for curing common
ailments, and many remarkable extracts on the evils of
modern diet from the works of Professor Atwater, the
great American chemist, Dr. R. Williams of the Middlesex
Cancer Hospital, Drs. Haig, Bouchard, Boix, Virchow,
and Kellogg. This little book has lately been enlarged
to a 3d. edition in paper and 6d. in cloth.

' " Fruits, Nuts, and Vegetables : their Uses as Food
and Medicine," a new 3d. edition of a booklet which has
gone to its forty-fourth thousand, is an admirable compilation, adapting the quaint and half-forgotten lore of
many herbals to the needs of modern life.'

A friend sent me the other day Dr. E. H. Dewey's
book, ' A New Era for Women : Health without Drugs.'
This book seems to have considerable attraction for many
people, but I should say it was rather addressed to those
who take drugs, or who over-eat, as his great panacea for
health seems to be leaving off breakfast—a remedy much
more necessary in America, where they have ten dishes at
that meal. He preaches a great deal what we are all beginning to know and even to believe—that Nature is a
wonder worker. His system of no-breakfast is based on the
fact that sleep never makes anyone hungry, the gnawing
sensation in the morning which people mistake for hunger
being caused by the indigestion of the previous night's
dinner or supper. Food will certainly stop this pain, but
only at the expense of further digestive trouble, and hot
water would be the best treatment of this spurious hunger,
as it would cleanse the stomach of its unwholesome condition and help it to regain a normal and healthy desire
for food. Most people relieve it by biscuits and milk in
the night. Of course, there is such a thing as healthy
hunger in the morning for those who have well digested

their dinner the night before, and who have done several hours' work before breakfast.

'The Perfect Way in Diet.'—A book which for its size contains a greater wealth of scientific information than almost any I know on the subject, is 'The Perfect Way in Diet,' by Mrs. Anna Kingsford, M.D. It is a translation of her 'Thèse pour le Doctorat' presented at the Faculté de Médicine of Paris on taking her degree in 1880, and is just one of those good things which the public is apt to lose sight of in the rush of new books. The treatise opens with a clear and able condensation of anatomical and physiological evidence for man's frugivorous habit, culminating in the following passage : ' If we have consecrated to this sketch of comparative anatomy and physiology a paragraph which may seem a little wearisome in detail, it is because it appears necessary to combat certain erroneous impressions affecting the structure of man which obtain credence, not only in the vulgar world, but even among otherwise instructed persons.' How many times, for instance, have we not heard people speak with all the authority of conviction about the ' canine teeth ' and ' simple stomach ' of man, as certain evidence of his natural adaptation for a flesh diet ! At least we have demonstrated one fact: that if such arguments are valid, they apply with even greater force to the anthropoid apes—whose ' canine ' teeth are much longer and more powerful than those of man—and the scientists must make haste therefore to announce a rectification of their present division of the animal kingdom in order to class with the carnivora and their proximate species all those animals which now make up the order of primates. And yet with the solitary exception of man, there is not one of these last which does not in a natural condition absolutely refuse to feed on flesh ! M. Pouchet, in his ' Pluralité de la Race Humaine,' observes that all the

details of the digestive apparatus in man, as well as his
dentition, constitute 'so many proofs of his frugivorous
origin'—an opinion shared by Professor Owen, who remarks that the anthropoids and all the quadrumana derive
their alimentation from fruits, grains, and other succulent
and nutritive vegetable substances, and that the strict
analogy which exists between the structure of these
animals and that of man clearly demonstrates his frugivorous nature. This is also the view taken by Cuvier in
'Le Règne Animal,' Professor Lawrence in 'Lectures on
Physiology,' Charles Bell in 'Diseases of the Teeth,'
Linnæus, Gassendi, Flourens, and a great number of
other eminent writers. The last-named scientist gives
expression to his views after the following manner : 'Man
is neither carnivorous nor herbivorous. He has neither
the teeth of the cud-chewers, nor their four stomachs, nor
their intestines. If we consider these organs in man, we
must conclude him to be by nature and origin frugivorous,
as is the ape. It may possibly be objected that since,
according to natural structure and propensities, man is a
fruit and seed eater, he ought not to partake of those
leguminous plants and roots which belong rather to the
dietary of the herb-eaters, whose organisation we have
shown to differ in so many details from that of man. It
may be urged that trouble is wasted in proving to what
order man belongs by nature, since with him, alone of all
animals, Art has superseded Nature, and has enabled him
by means of fire, condiments, and disguise, to eat and
digest without disgust, and even with relish, the food of
the tiger, the wolf, and the hyena. Such objections are
not without an air of reason ; and I shall meet them
first by the frank statement that the most excellent and
proper aliments of which our race can make use consist
of tree-fruits and seeds, and not of the plants themselves,
whether foliage or roots.'

Just lately a book has come to my hand which I think will be of the greatest use to those who are staggered by Dr. Haig's scientific language. It is called 'Medical Essays,' by T. R. Allinson, physician and surgeon, but note, as he does on his title-page, that he is 'Ex' L.R.C.P., &c. This book, in 1901, had reached its twenty-ninth thousand, a sale as cheering as it is enormous, showing as it does how its teaching must have permeated all classes. It is in fact a popular mixture of very short articles on health, food, ailments, management of children, and general instructions for right living, and makes the whole subject of physical well-being so plain and straightforward, that I think no head of a young family should be without it. So many are deterred from trying moderation and abstemiousness by the severity of Dr. Haig's measures, that it might help them greatly to read Dr. Allinson's explanations, and begin by trying his less strict form of diet. I should certainly have bought his largest book (price 10s. 6d) if I had known of it. As it was, I got Book F. at 6s. 6d., which is naturally less complete.

Dr. Allinson's main principle seems to be that the greater proportion of disease is brought about by our own ignorance. I know well that in my youth when I ate a large breakfast, which always made me feel uncomfortable, I did it with the sincere belief that I was adding to my strength and working power during the day, whereas it often took me an hour and a half to get over the first effects of the meal, especially in winter. Dr. Allinson states in his preface that what he writes is also 'the beginning, I hope, of a school of healing that will take the place of allopathy and homœopathy.' I feel sure that while medicines are still taken, strict dieting is much impeded in its benefits. Till all drugs are left off, few can judge of what they want in the way of food.

'A Treatise on the Tonic System of treating Affections of

the Stomach and Brain.' By Henry Searle, Surgeon, Kennington Common. Published by Richard & John Taylor. London. 1843.

Amidst the rush of modern medical books, it is not otherwise than sobering and extremely interesting to take up an old book, and see what doctors believed and what they taught sixty years ago. It is one of the most beautiful articles of the creed of the profession, that no knowledge they obtain, no invention they make must ever be kept secret, but always openly given both to fellow-members and to the public. This is a noble idea, and often constitutes the great difference between a doctor and what is called a ' quack,' for the latter is apt to keep his discoveries to himself in order to make money out of them.

There are many notable things in this old book, and the stuffing and tonic system we were all brought up on is well argued; but I mention it chiefly for its strong condemnation of tea and coffee so early as 1843. The increase of tea-drinking in all classes is astonishing. Mr. Searle says, ' Previously to the introduction of tea into this Kingdom, disorders of the stomach were by no means so prevalent as they have been since. Tea in the evening is found particularly refreshing, and is therefore considered an indispensable article of diet; but the refreshing effects of tea are not always unalloyed, most unpleasant symptoms of indigestion being sometimes experienced immediately after taking it. As tea-drinking is a universally established practice, it would be vain to recommend its discontinuance; but it may be strongly urged that tea should be taken in small quantities, and of moderate strength, and that those who are troubled with indigestion should combine with it a large proportion of milk.' Only so did this timid doctor in the early Victorian period venture to advise the public. In my lifetime tea has been entirely discontinued after dinner. Twenty years ago I can remember how the tray used to come in after dinner, and be almost universally refused; and is it not quite possible that the same fate may be in store for the 5 o'clock custom when once the young realise how bad it is for themselves and their children after them? Alas, the same economic danger stares one in the face as with meat, beer, or spirits: hundreds would be ruined in India and Ceylon if the drinking of tea were appreciably to decrease, but I confess I have no sympathy with those who make money by the adulteration of food, or by the sale of beer, spirits, tea, opium, quack drugs, or anything else that brings ruin and misery through bad health to millions of human beings.

Other books that I should like to recommend to students of health are Hutchinson's 'Dietetics,' Parke's 'Practical Hygiene,' Dr. Fernie's 'Herbal Simples' (5s.), Dr. Kellogg's 'Science in the Kitchen' (12s. 6d.), and 'The Stomach,' Mr. A. W. Duncan's 'Chemistry of Food,' and 'Foods and their Comparative Values,' Dr. Lehmann's 'Rational Hygiene,' Dr. Poore's 'Rural Hygiene,' Dr. Dewey's 'True Science of Living,' Smith's 'Fruits and Farinacea,' Huxley's 'Elements of Physiology.'

These and many other books, English and American, on health, education, &c., are kept in stock, as a rule, by Curtis & Davison, 4 High Street, Kensington.

Having given the public this list of books, which, I believe, will help the introduction of a new and a better era, I cannot resist the pleasure of speaking of a French classic, 'La Physiologie du Goût, de Brillat-Savarin,' re-published in 1879 by the Librairie des Bibliophiles with Eaux-fortes par Ad. Lalauze.

My own old edition was of 1841. Most people know his famous aphorism, 'Dis-moi ce que tu manges, et je te dirai ce que tu es.' This book by the friend of the Récamiers has all the charm of the best eighteenth-century French literature.

It is perhaps unkind, but I should be more than human if I withstood the temptation to quote here what he says about the doctors in his day, in his chapter called ' Des Gourmands ' : ' Des causes d'une autre nature, quoique non moins puissantes, agissent sur les médecins : ils sont gourmands par séduction, et il faudrait qu'ils fussent de bronze pour résister à la force des choses.

' Les chers docteurs sont d'autant mieux accueillis que la santé, qui est sous leur patronage, est le plus précieux de tous les biens ; aussi sont-ils enfants gâtés dans toute la force du terme.

' Toujours impatiemment attendus, ils sont accueillis avec empressement. C'est une jolie malade qui les engage ; c'est une jeune personne qui les caresse ; c'est un père, c'est un mari, qui leur recommandent ce qu'ils ont de plus cher. L'espérance les tourne par la droite, la reconnaissance par la gauche ; on les embecque comme des pigeons ; ils se laissent faire, et en six mois l'habitude est prise, ils sont gourmands sans retour.'

In the editor's note to the 1879 edition he has a sentence which I think may form a motto very applicable to the eaters of the simpler foods compared with those who, up to now, have met to consume fish, flesh, and fowl. I claim that in the future the simpler foods may be as dainty and attractive as those recommended by Savarin, and that the contrast between the meals of the future and of the present will be as great as between the French cooking and the Roman : ' Et d'ailleurs, il ne faut pas s'y tromper, la cuisine est véritablement un art, art tout moderne, art tout français, et qui trouve en France son chantre le plus autorisé. Les grands repas des Romains n'étaient qu'un brutal amas de plats gigantesques, où les mets et les ingrédients de toutes sortes se trouvaient confondus dans des sauces dont la seule analyse nous soulève aujourd'hui le cœur. La description du plus beau festin de Lucullus ne peut rien inspirer d'analogue à la douce émotion qu'on ressent en lisant le récit du simple et succulent déjeuner du curé, si onctueusement raconté dans le paragraphe des " Variétés " qui a pour titre " L'Omelette au thon." '

I would recommend all those interested in the instruction of the young to get ' A Short Account of the Human Body,' by Owen Lancaster, lecturer to the Natural Health Society, published by Allman & Son, 67 New Oxford Street, W.C., price 2s. 6d. It contains a coloured picture of a manikin with the skin removed, and folds that lift up and show the position of the internal organs. This diagram is exciting and interesting, and would inspire the young with a kind of reverential feeling for their internal machinery, of which they know so little, and which they treat so often with great unfairness. It has long been said a man is a fool or a doctor at forty. I fear, then, I know many fools, especially among women, who more frequently starve themselves and overwork than men do. What I want is that all, even the quite young, should respect their bodies, and believe that health also means beauty and strength; that all should know the difference between health, or what we are intended to be, and that continual, uncertain, ailing condition so common with young, middle-aged, and old, and to which people actually think it a virtue to submit. How few, even among nurses, know that if anyone is struck down by any kind of accident, the first thing to do, as with a man wounded in full health, is immensely to reduce nourishment and quantity of food! I have several times known more injury to come from overfeeding when all exercise is stopped than from an accident itself. Wherever there is a weak or injured part, there the mischief will settle, especially where the constitution is gouty or rheumatic ;

and this, as should always be remembered, is the case with seventy-five out of every hundred persons.

'Water: How it Kills its Thousands.' This is a pamphlet issued by the Salutaris Company, which is well worth reading, if only for the caution against hard water as a drink for the gouty and rheumatic. Those who cannot afford habitually to buy Salutaris Water may like to know that it is easy enough to have distilled water at home by getting one of the stills supplied by the Gem Supply Co., 6 Bishop's Court, Chancery Lane, E.C.

Would it not be better if, instead of expressing the heartiness of our rejoicings by giving enormous teas and dinners to the poor, which often make them ill or uncomfortable for days after, we were to give them wholesome food to carry away to their homes, which they would often spin out to last a week, and which would help the sick and healthy alike and cost the givers no more?

After the 'Health' chapter, the subject in both my former books that roused most opposition and yet most interest was what I said about girls—their training, relation to their mother, &c. In fact, I have several times been asked to give more of my advice. I never have any 'views,' except as the outcome of my own experiences, and all my dear nieces, real and adopted, have now grown into women and are bringing up their own children.

Opinions on education grow quickly, and as the subject is of perennial interest to me in a general way, I can only mention a book or two which may be found useful and have come to my knowledge, almost by chance, in the last few years. All that is newest, most enlightening, and most stirring seems now to come from America, and I can recommend strongly to all a book, full of concentrated instruction and valuable suggestions, called 'An Ideal School, or Looking Forward,' by Preston Search, United States Superintendent of Schools, one of the 5s. 'Inter-

national' Series, published by Arnold.   An English book,
lately published with an idea of awakening more general
interest in the subject, is 'Education and Empire,' by
R. B. Haldane.

But to mothers, when education is over, I can only
repeat what I have said before—that when all has been
said and done with reference to education, human nature
remains the same, and the really important thing for
mothers to do is to try to know their own children.
They may exhaust themselves in efforts to make their
daughters sweet, attractive, graceful, and marriageable,
but unless they realise that no young people, any more
than the rest of us, can live up to an ideal, the end is
often artificiality and deceit in the children, and bitter
disappointment to the parents.

It seems so easy, in theory, for parents to know their
children; but, in fact, nothing is more difficult.   There
are two courses open to every mother of young children,
say, to the age of fifteen.   One is, to hold up in practice
and precept a high ideal, and persistently encourage the
children always to act up to this standard, so impressing
its importance on them that they come to feel it a proof of
their personal affection for her to try to come up to it, at
least, in her presence, even to the extent of acting a part
to please her.   This way, though often satisfactory in
childhood, generally ends, so far as my experience goes,
in disaster to character, for in insensibly forcing them to
appear to be something they are not, it actually helps to
train them in habits of deception and lying.

The other course is to make up one's mind to the
probability of naughtiness, selfishness, want of good
manners, or any other undesirable but natural expression
of themselves—in fact, to live in what is called the ' Palace
of Truth,' trying to discover the plan Nature has outlined
for them, and helping them to fulfil it by natural growth

rather than forcing them, however gently and skilfully, into some mould of our own choosing. This way
brings a good deal of mortification to the mother and
condemnation of her training from her friends; but I
firmly believe that this method turns out in the end the
truest and best human beings, and if the mother's own
example, supported by the father's, has meanwhile
declared and upheld her ideal of what her children should
be, they will probably realise it in later life.

I am always being told that everything is now very
different from my day, but I still maintain, as I said in
my former book, that all those differences are superficial.
It is but the outward fashions that change. Only three
years ago was published one of the most charming books
of imagination that has appeared for a very long time—
'A Digit of the Moon : A Hindoo Love Story.' A great
many people were disappointed at finding out that it was
original and not a translation; but to my mind it only
adds greatly to its charm and interest that an Englishman
living in India should have been able in these days to
write such a book. It is perhaps too well known to justify
a long quotation from it here. All the same, life is so
full that many miss what they would like to see, and this
description of man's notion of how woman was made
seems to me a literary gem :

'One day, as they rested at noon beneath the thick
shade of a Kadamba tree, the King gazed for a long time
at the portrait of his mistress. And suddenly he broke
silence and said, "Rasakósha, this is a woman. Now, a
woman is the one thing about which I know nothing.
Tell me, what is the nature of woman ? " Then Rasakósha
smiled, and said, "King, you should certainly keep this
question to ask the Princess, for it is a hard question. A
very terrible creature indeed is a woman, and one formed of
strange elements. Apropos, I will tell you a story. Listen!

' "In the beginning, when Twashtri came to the creation of woman he found that he had exhausted his materials in the making of man, and that no solid elements were left. In this dilemma, after profound meditation, he did as follows. He took the rotundity of the moon, and the curves of creepers, and the clinging of tendrils, and the trembling of grass, and the slenderness of the reed, and the bloom of flowers, and the lightness of leaves, and the tapering of the elephant's trunk, and the glances of deer, and the clustering of rows of bees, and the joyous gaiety of sunbeams, and the weeping of clouds, and the fickleness of the winds, and the timidity of the hare, and the vanity of the peacock, and the softness of the parrot's bosom, and the hardness of adamant, and the sweetness of honey, and the cruelty of the tiger, and the warm glow of fire, and the coldness of snow, and the chattering of jays, and the cooing of the Kólila (the Indian cuckoo), and the hypocrisy of the crane, and the fidelity of the chakráwaka; and compounding all these together he made woman and gave her to man. But after one week man came to him and said, ' Lord, this creature that you have given me makes my life miserable. She chatters incessantly, and teases me beyond endurance, never leaving me alone; and she requires incessant attention, and takes all my time up, and cries about nothing, and is always idle; and so I have come to give her back again, as I cannot live with her.' So Twashtri said, ' Very well,' and he took her back. Then after another week man came again to him and said, ' Lord, I find that my life is very lonely since I gave you back that creature. I remember how she used to dance and sing to me, and look at me out of the corner of her eye, and play with me, and cling to me; and her laughter was music, and she was beautiful to look at and soft to touch; so give her back to me again.' So

Twashtri said, 'Very well,' and gave her back again. Then, after only three days, man came back to him again and said, 'Lord, I know not how it is, but after all I have come to the conclusion that she is more of a trouble than a pleasure to me, so please take her back again.' But Twashtri said, 'Out on you! Be off! I will have no more of this. You must manage how you can.' Then man said, 'But I cannot live with her.' And Twashtri replied, 'Neither could you live without her.' And he turned his back on man and went on with his work. Then man said, 'What is to be done, for I cannot live either with or without her?'"

'And Rasakósha ceased and looked at the King. But the King remained silent, gazing intently at the portrait of the Princess.'

HEALTH OF OTHERS

Imaginary conversation between two doctors—Advertisements *versus*
Dr. Haig—Remedies for depression on beginning diet—The old
need not fear change of diet—Tea-drinking and the Chinese—
The Bible and meat-eating—Pythagoreans and the bean—' Better
Food for Boys ' and the schoolmaster—Difficulties of the diet—
Fat and thin women—Mediæval and Greek idea—Individual
cases—Maeterlinck's testimony.

HAVING done my best in the last chapter to tell about
myself, I think it may be of some interest to those who
are trying to help themselves, and who have, probably,
those they love best and all the world against them, that
I should now give some account of my personal observation of others, and the details of a few cases that have
come more particularly under my notice, these having
been kind enough to send me their individual experience.
Some are patients of Dr. Haig, others have been induced
to try the diet from reading Mr. Miles' book, or from
seeing the conspicuous benefit I have derived from the
diet.

For a long time I thought Dr. Haig was the only
medical practitioner of the kind, but the last six months
have convinced me of my error, and now it seems to me
that the idea of wrong food being a great cause of disease,
is quite in the air and widespread. But it cannot be too
often repeated, that the work of Dr. Haig and his col-
leagues is not primarily a specific cure of disease, but a
scientific research into food, their aim being to establish a

high standard of health on a simple and sound foundation
of right diet. Being a doctor, however, Dr. Haig's
experience must necessarily lie among the diseased, or, at
least, the non-healthy, and this gives his system of diet a
most severe trial. I must honestly say that, in all the
cases that have come to me, there has never been one
healthy person ; all have been suffering more or less from
headaches, dyspepsia, anæmia, gout, rheumatism, weakness, &c., in spite of belonging to a class that has every
chance of high feeding in the ordinary way, though many
of them, from non-assimilation, have been as underfed as
the unscientific vegetarians. I confess I am still puzzled
by the apparently good results of the cramming system,
which has been so generally adopted for nervous diseases
by the medical profession, and of which Drs. Playfair and
Weir-Mitchell are the renowned exponents. The benefit
is, I believe, only temporary, and the over-stimulation of
cramming on meat and wine is so dangerous for nervous
temperaments that many cases of death and madness
might be written down as 'victims of the medical profession.' The high-feeding cure for tuberculosis has given
this system such an immense fashion, that in Germany
it is recommended as a *preventive* measure as well as
a cure, it being well known that the weakly form the
best soil for the germs of the disease. I know a German
mother who sent her only and very delicate son of sixteen,
reduced by overwork and examinations, to Nordrach, the
consumptive cure place, to face the possible infection,
that he might be fed on a system which she could not
induce him to follow at home. The result was so satisfactory that he was willing to go back there for his
summer holiday in the following year. The lesson to me
from this, is not that the Nordrach meat diet is best for
the purpose, this being attained with greater ultimate advantage upon the Haig diet (see ' Diet and Food,' p. 108),

but that a person can be made to eat food enough by the help of a mental effort, or by the moral influence of some other human being, such as a doctor, friend, or nurse. The immediate results of the stuffing system tend immensely to support Dr. Haig's theories, as represented in the following imaginary conversation, written for me by a doctor :

'*K.*—If you take only small quantities of food, you may continue some meat, and even tea and coffee, with but little harm.

'*H.*—I grant this to a certain extent, and I know that a little uric acid is better than a great deal; but I can show that it always does harm to take any uric acid, and, personally, I never swallow a grain of it, if I have power to leave it out.

'*K.*—Well, it does not seem to harm me ; I am much better than I used to be, and I am old and yet feel well.

'*H.*—Let us get to facts ; will you tell me, for instance, what your old diet used to be and what your present diet is ?

'*K.*—My ordinary diet might have been 1,200 grains of albumen per day. I then formed 12 grains of uric acid per day, and I introduced some 8 grains more in flesh and tea. I thus had to deal with 20 grains of uric acid a day, and with this I suffered badly from headaches. My present diet is about 800 grains of albumen, and I form 8 grains of uric acid, and introduce in the small quantities of flesh and tea I take some 3 or 4 grains more. I thus have to deal with some 11 to 12 grains of uric acid a day, and have no headaches.

'*H.*—No doubt this is a gain ; but what about your nutrition, and circulation ?

'*K.*—I lead a quiet life, and get up late ; but I am able to walk for several miles every day, and do a good deal of reading and writing.

' *H.*—But your blood colour is nowhere near the proper standard, and your circulation in skin and all organs is slow, so that their nutrition is feeble. Now I know a man who is at present in his eightieth year, and who would do in one hour all the exercise you take in a day (Mr. C. J. Harris, often mentioned in the " Vegetarian ") ; his ordinary day's work is thirty to forty miles on his tricycle, and he not infrequently rides fifty to sixty miles, and on several occasions has done a hundred in the twenty-four hours. His blood colour is good, his circulation is good, he is in every way strong and well nourished ; and the reason is simple, for he takes, each day, his physiological allowance of albumens, quite free from uric acid, while you only take two-thirds of your physiological allowance (800 in place of 1,200), and you further hinder your circulation, and injure your blood colour, by swallowing uric acid, which he does not.

' *K.*—I do not wish to do as much as Mr. Harris, I am content with my life as it is.

' *H.*—But you only live one-half the life that Mr. Harris does, and produce only one-half of the effect in the world ; if you are content with half a life, when a whole one is open to you, well and good ; that is your affair ; but you will not persuade me that the half is equal to the whole.'

The following is a doctor's note on the Mr. Harris referred to in the conversation : ' I may say that I have been watching Mr. Harris carefully for some years, that he always takes a physiological allowance of albumens, and produces a physiological amount of *urea* (which I have repeatedly estimated). There is no doubt where his strength and power come from ; and he gets his albumens chiefly from bread, nuts, and fruit, with very little milk or cheese, and at times none of these last. On K.'s diet such work as Mr. Harris does is impossible.'

Time after time I have heard of various people, and

even doctors, trying the diet and giving it up after a few weeks or months, finding they could not do their work on the non-stimulating food. My family are then much surprised that this does not bring home to my mind a sense of failure. All it does do, is to convince me that very little is yet known of the right mixtures of food ; that there is, in most cases, a more or less severe reaction to be reckoned with from the accumulated uric acid rushing into the blood on the withdrawal of the meat-tea-and-wine diet, and that good results can rarely be expected, except with the young and healthy, under eighteen months to two years time. The change is too radical and permanent to be effected hurriedly.

The hoarding of uric acid in the system seems to me to be Dr. Haig's great scientific discovery, and the one most disputed by other medical men. His work proves conclusively to me that all benefit from German waters, alkaline treatment, &c., is only a temporary palliative, necessitating constant return, with danger of severe relapse in the end. I have known cases of extreme illness, and even death, from missing a course of Carlsbad waters for one season. The advertisements that explain health questions so clearly to the public every morning and evening in the newspapers are all based upon the old theory, that it is possible to clear the whole system of uric acid by alkaline drugs and mineral waters. If the advertisements are right, Dr. Haig's theories are all wrong. The first thing that put him on the scent of error in the old accepted idea, was that none of the usual remedies bring away any uric acid, though, of course, they clear it from the blood, and, therefore, give the well-described sense of relief to the patient. This made him ask, ' Where does it go ? ' He now declares, after years of examination, that it is deposited on the soft tissues, the internal organs, the brain, the muscles, nerves, bone-joints, &c., waiting either to set up disease in them under the action of any favourable external circumstance, or again to enter the blood stream and cause distress as soon as rising alkalinity will dissolve it. He maintains, in his books, that the sole real cure is to stop the introduction of foods containing uric acid, or its equivalent, xanthin, and replace them by a diet which he considers suitable and natural for man.

It is undeniable that, with most people, a few months of the diet causes great depression and weakness, which, for those who cannot take rest from work, is sometimes so serious a matter, that it means giving up the diet or the work. Great relief would be found in these cases if they took, in summer, 20 grains of bicarbonate of soda dissolved in a teaspoonful of cold water and hot water added till the mixture is of the warmth of a cup of tea, or, in winter, a 5-grain dose of salicylate of soda. Neither of these remedies being a tonic, but rather a depressant, it is obvious that the relief experienced must be due to the effect they have in clearing the blood of the uric acid which has been brought into it from the tissues on its way out of the body. They must be kept in reserve for times when rest is impossible, and so enable the patient to continue the diet which is to bring about ultimate cure. The same effect can be produced, by a temporary lapse back to the old uric-acid diet, which is an immense triumph to the enemies of the system, though perfectly intelligible to any student of Dr. Haig's theory. He himself has taken the trouble to publish leaflets expressly recommending a very *gradual* change of diet in order to avoid any severe degree of this depression and apparent exhaustion.

Dr. Haig, Dr. Allinson, and Mr. Miles all seem to think that uric-acid diseases, such as gout and rheumatism, headache, neuralgia, anæmia, epilepsy, would cease to exist if their diet were adopted. For as much as it is worth, my own personal experience and my observation of others under various doctors and treatments absolutely corroborate their views. Doctors are still very much divided in their opinion as to meat being bad for gout and rheumatism, and very many of them tell gouty and rheumatic patients to eat plenty of meat and not much sugar, farinaceous, or carbohydrate substances; so it is not their opinion that uric acid is increased in the system by flesh foods. These very doctors, however, always prescribe the medicines which, as Dr. Haig shows, clear the blood of uric acid but throw it back upon the tissues. Of late years, in severe cases of eczema and shingles, their great salvation seems to be the much-discarded remedy of mercury (calomel). Only the future can prove how far the few are right and the many wrong in this matter. In all probability diet will take centuries to have a fair trial, for the changes it would involve in every department of our modern life are almost unthinkable. I am most anxious that anyone who is at all interested in anything I have to say, not only as regards the improvement of his own health but the benefit to the whole human race, should remember that the subject is one which demands at least a year of close and careful study, that it is in no sense like the German waters, or Salisbury cure, or any other diet prescribed for illness, a temporary affair with the hope of returning cheerfully to luxurious living; it is a road on which may be written what Dante wrote over the entrance to Hell: 'Lasciate ogni speranza voi che entrate.' There is absolutely no return without greater injury from the increased sensitiveness of the body and the quicker realisation of what is injurious food, but I do deny that the road leads to Hell; I should like to add that, even when adopted late in life, it leads to the Paradise of Health. I am also the last to deny that, however strong this belief may be, it does include self-denial even to the old; for, when people are no longer young, habit has become second nature, and with people all round you eating what you have always been accustomed to, it does require a certain amount of strength of mind to refuse all the dishes that a few years ago you would have gladly eaten as the most wholesome food. It has taken me nine years of experience to find out how injurious the most simple food of the flesh-eater, such as the invalid diet of boiled fish and chicken, really is.

A great many elderly people say to me, ' I am too old to change '; only civility prevents me from saying to their face, ' I think you are quite wrong : it is never too late to mend. Many of the symptoms you suffer from would be immensely lessened if you *did* change.' As regards benefit, a great deal, of course, must depend on how much actual disease is already set up in the system, and in the case of old people it is even more important than with the middle-aged to avoid the crisis described above as depression and weakness. On the other hand, it should be remembered that at no other time of life is it so dangerous to overfeed and so little injurious to underfeed as in old age. I believe the time is not far distant when it may be discovered that the great cause of cancer is meat and salt, as leprosy is supposed to originate in eating salt fish in large quantities.

There are no doubt a few people whose healthy faces shine and beam when they tell you they can eat everything. They are seldom really ill, they have beautiful teeth and complexions, strong bright healthy hair, and their nerves are not overstrung. But I think everyone will admit that these people, in all classes, are distinctly rare in civilised Europe in this year of our Lord 1902. Also, if you question them closely, you will generally find that they take mineral waters, aperients, pain-killers, and various other drugs, and are frequently affected by changes of temperature, fatigue, missing a meal, east wind, &c.

Many people quote the Bible in favour of flesh foods, saying, 'If meat had been very injurious, one would expect to find some warning words advising people not to eat it.' According to this line of argument, vegetarians have been hardly used at finding no warning against the deadly nightshade of our damp meadows, or the poisonous berries of the potato which cost Sir W. Raleigh some pangs when he ate them by mistake on first discovering the plant.

In a letter I have received this morning from Dr. Haig in answer to things I had told him, I having been depressed at the little progress made by what I consider the new enlightenment, he says: 'I have now no doubt that the cause will succeed, and that success cannot be too great or come too soon, for there are to-day millions and millions of people who are suffering, dying, or becoming insane simply from ignorance.' I believe this opinion of Dr. Haig's will gain ground, and that twenty years will make an enormous difference in the understanding of everyone on the subject of food. Nowadays, if we see a man drunk, we are half angry and half sorry, but we do not require telling that he has taken too much alcohol to bring him to the state we see him in. I believe that, in course of time, the same knowledge will be ours with regard to right and wrong feeding and its obvious results.

I am fond of distributing Dr. Haig's leaflet, which explains most clearly that the *xanthin* of certain *vegetable* substances—peas, beans, lentils, mushrooms, asparagus, tea, coffee, cocoa—is the equivalent poison to uric acid in fish and flesh. *Apropos* of this latest discovery in the chemistry of food, I was amused to come across the following the other day in that very popular book of Mr.

Marion Crawford, 'The Rulers of the South': 'Hallam
says somewhere that mankind has generally required
some ceremonial follies to keep alive the wholesome spirit
of association. It is hard to say now how many of the
curious rules of life adopted by the Pythagorean brotherhood should be traced to this motive, and many of these
contain more wisdom than appears in them at first sight.
The brethren abstained from eating flesh, as most mystics
have done, but they were as careful never to eat beans.'
So many of our most modern inventions were known to
the ancients that I feel the reason for the Pythagorean
rule 'abstain from the bean' is quite as likely to have
been due to a knowledge of its harmful properties in
dietetics as to a dread of that life of political intrigue
which was symbolised by the bean used by the Greeks in
balloting.

An amusing instance came to me lately of how little
these leaflets are understood even by intelligent and
educated people, when a father wrote to me as follows :
' A cowardly thought arose in my mind of submitting the
leaflet to a friend of mine who is obliged by circumstances
to devote much attention to his inside, but unhappily in
an unguarded moment I remarked to Mrs. G., in the
presence of the most intelligent of our cats, "I am
surprised that you should provide so much fish for the
household, when you know Mrs. Earle's leaflet says that
fish positively *reeks* with xanthin." I don't know how
the leaflet which riots in that particular subject of organic
chemistry came to be torn into little bits, but for me it
disposes of the subject altogether, which is too abstruse
for the man in the street, and needs an expert to confirm
or deny. "Xanthin" may be as real as "microbes," but
is equally invisible, and I have my doubts about the whole
lot of scientific conclusions based upon these insects or
fish or whatever they are.' So carelessly had the leaflet

E

been read that my correspondent can talk of the *xanthin* in fish!

A curious case of the effects of the harm the Salisbury cure has brought about in unintelligent hands was told me the other day by a friend. His mother-in-law lost her teeth, and her gums were too tender to admit of artificial aid. She therefore took pulped meat, often three pounds a day, and strong soups. Between the age of fifty and sixty she became very ill, and a great London specialist was sent for. After seeing her he took the son-in-law aside and said, 'All her symptoms point to a secret store of alcohol. Are you quite certain this is not the case?' Both the son-in-law and the servants were quite certain that she had nothing of the kind. She did not even take wine. But it is interesting to note that the stimulating meat-juices had produced the same symptoms as drink, accompanied by great exhilaration and depression. At sixty she died of starvation, with no organic disease at all.

The question often asked me, and which moves me most, is what I recommend for the feeding of children. There is no doubt that the strongest argument which can be used against drunkenness and debauchery is the hereditary effect it has on children. Doctors now say that the children of confirmed smokers are far sooner injured by smoking themselves than the children of very moderate or non-smokers. It seems to me that this may be an explanation of the younger generation of the present day being so much more quickly and harmfully affected by *tea* than we were. Both coffee and tea have got much cheaper, and their use has been immensely promoted by the preaching of the teetotallers. Clergymen's wives, nurses, and all classes of over-worked women and men of the conscientious type, who would repudiate the help given to them by a glass of beer, habitually stimulate and

injure their nerves by excessive tea-drinking. I think
there is no doubt that Indian and Ceylon tea are more
injurious than even the poor quality Chinese which is
sold in England to-day. This may be an additional
explanation of the increased nervous disease among the
poor, who love the rank flavour of the cheap Indian teas,
and stew them long to extract all the goodness. Let people
who doubt this try leaving off tea for a few months, and
then see how it affects them on taking to it again. One
of the strongest arguments for tea-drinking is that the
Chinese have drunk it for centuries and remain what they
are. I sat at dinner the other evening, next an Englishman who had lived many years in China. I soon began
to question him as to how far he thought the tea-drinking
in China had been injurious to the natives. He said that,
so far as he had ever been able to ascertain, he thought
that it was absolutely harmless, but this probably from
the way they made it. A small pinch of sun-dried leaves
—just as much as they could take up between the finger
and thumb—was put into a small cup, boiling water
poured on it, the cup covered up with its saucer for a
minute or two, and the infusion drunk when it had cooled.
It was then scarcely more than hot water with a slight
flavour of tea. Besides the harmlessness of this mixture,
no doubt the Chinese are better able to take tea through
never having reduced their strength and created complicated diseases by centuries of excessive meat-eating and
alcohol. For poison from the European way of tea-
drinking, as from excessive smoking or drinking of alcohol,
is supposed to be hereditary and is consequently more
quickly injurious in the second or third generation. If
this is so, surely the consideration and study of the effects
that food has on the body is in no sense a ridiculous or
degrading subject. So many people say, 'It is so un-
worthy always to be talking or thinking about food or

E 2

health.' Their one wish is laughingly to put it aside and
shift the responsibility of their own and their children's
health on to the shoulders of the doctors, not even taking
the trouble to learn what the modern advertisements try
to teach them, viz. which foods are digested in the upper
stomach and which in the lower, or that constipation is
the commonest proof of indigestion.

This question of the diet for children seems to me to
be full of difficulty, and I should say cannot be under-
taken without the consent of the father. This he is not
the least likely to give unless he has studied the subject
enough to wish to adopt it for himself as well as for his
children. The first object of ambition to an intelligent
boy or girl is to try to imitate father, and if he constantly throws brickbats at dieting the case is hopeless.
I am sure in a few years the great school difficulty will be
much lessened, and that if a boy is anxious to continue his
regular home food, there will in time be a house for non-
stimulating diet at Eton. One of the keenest vegetarians
I know succeeded in converting her daughter who was
out of health, but had no effect on her son-in-law or her
grandchildren. Where father and mother are agreed, I
think the experiment of bringing up children as vegetarians
may fairly be tried, though probably even those parents
hardly realise the difficulties before them. A great deal
will depend on how they can train their children mentally.
In any case, in the present state of things, the parents
can hardly ever go away from home together for any
length of time, or to a distance which prevents their quick
return, as the children ought not to be left to the sole
care of even the best of nurses. She cannot be expected
not to send for a doctor the moment anything goes wrong,
and he in turn would, from want of practice among
vegetarians, fail to understand the sensitiveness of the
children. The nurse would have no weight in explaining

to him the wishes of the parents; whereas if the parents were present probably the doctor would not interfere with the diet. The plain truth is that it all resolves itself into a question of character and knowledge in the parents. When a doctor finds that he has to deal with men and women who have taken the pains to acquaint themselves with the elements of physiology and the chemistry of food, his tone is very different from the one he uses, naturally enough, to the young and ignorant mother who appeals to him to make her sick child well, as if he worked by miracle, and who submits to be told by him that if she brings up her children on a non-meat diet she is 'a very *wicked* woman.'

My object in repeating these things so often is that radical changes in diet can only be brought about by the public, and children are not likely to take to what their parents and elders do not practise; though if the parents are non-meat-eaters the consequences of home example may act powerfully in this, as in other matters, when they grow to man's and woman's estate. One case, at least, I know where the diet has been successfully carried through school and college and joining a regiment, where the young man was looked upon as rather a curiosity, but was left unmolested.

I am always supposed to be proselytising, and last year I sent to an old friend, a middle-aged schoolmaster, Mr. Miles' 'Better Food for Boys.' He says in reply: 'I have read and thought over Miles' book. It seems to me that the question is mainly a medical one.' This sentence epitomises what so many think and feel, and when I say, 'But apparently the medical profession refuse to give the matter serious thought and experiment,' this closes all discussion, and I am looked upon as a pre-judiced idiot. This I try to bear with the philosophic calm of a non-meat-eater. He writes further: 'To a layman there is much in what Miles says, and for a large number of adults—for any adult, in fact, who leads a sedentary life, I believe his system is beneficial. I am following it myself in moderation—that is to say, I have knocked off meat meals and substituted Plasmon for them, and already dread the idea of sitting down to a big dinner. But if the practice becomes general, what a blow it will be to the social side of life ! You can't seriously sit down to rice and stewed fruit with half a dozen friends.'

My friend's despair about the social side being reduced to stewed prunes and rice is very much as if he invited his friends to cold shoulder of mutton and potatoes. For festive occasions endless combinations and much luxury, attractive to all the senses, could be made. Even for the severest stage of diet—the fruitarian—if a housekeeper spends half what she does on fish, game, poultry, &c., for a dinner-party, she might have the most gorgeous and attractive display of the best fruits to be had in any market, and a cook once trained to the food would produce endless dainty varieties of cakes, biscuits, nut-creams, and rolls. These people who think that the social side of life must suffer with the disappearance of meat from our tables seem to forget that cereals, fruits, and wine are the foods which the highest imagination of man has always thought the most fit for the feasts of the Gods, a detail eagerly seized upon by Burne-Jones for representation in his famous picture on that subject.

To go back to my letter : my friend says, ' I cannot persuade myself that the health of boys suffers from a meat diet. They do not, as a rule, overload the stomach and take strenuous exercise. Of course, it may fairly be said that if you bring boys up on meat they will be less likely to take to the simpler foods later on. But there are great difficulties in the way. Miles' diet cannot be made compulsory, and at a big school two systems cannot well be worked, but I think schoolmasters might do much good by example and by preaching. That is the line I follow. The question which the author raises in his opening chapter is of the utmost importance, but there, again, only a doctor is qualified to give an opinion. Is sensuality less developed in gramnivorous animals than in the carnivorous? I doubt it. Plasmon is becoming popular with doctors. I am convinced that any baddish rheumatic or gouty tendencies would be the better for it.' I should like to say to this last sentence that this attitude towards Plasmon of thinking it a ' cure ' for rheumatism and gout is as dangerous as it is prevalent. In itself this valuable preparation is no *cure*; it is a highly-concentrated proteid food, which enables people to leave off the alcohol, and the uric-acid-forming foods they have been used to. The allusion to animals seems to me rather wide of the mark. All Nature, even in flowers as described in the botanical language which I have known to shock prudish mothers, is apparently wasteful; but, as I understand it, the word *sensuality* implies a *moral* conception—i.e. the antithesis of self-control, and this is a province with which Nature in the mineral, vegetable, and animal kingdoms has nothing whatever to do. What I suppose my friend really means to ask is, whether sex vigour is less in gramnivorous animals than in carnivorous? Had he thought for a moment, it would probably have come to his mind that the bull and the goat are almost proverbially used in art and literature to express this vigour rather than the lion and tiger. In Mr. Miles' books there are frequent allusions to the point, which is always puzzling schoolmasters, as to how to increase physical strength with moral self-control. Schoolmasters solve the difficulty by incessant games, to the detriment of learning. Mr. Miles' solution is non-stimulating food and drink.

I was pleased to see that my friend sympathised with

me about the facility of carrying out the diet in old age, and the benefit of teaching by example.

One advantage of taking to a diet of simpler food late in life is that it lessens to a great degree the family and social opposition, which from the kindest of reasons is almost universal. Abstinence in old age is looked upon as rather a virtue by the young, but when practised by their own contemporaries they consider it as a sign of eccentricity, affectation, love of notoriety, or extreme bad health. This being so, I find that diet looms as an insuperable obstacle with the young. Their experience being *nil*, their knowledge is weak; the diet is often badly managed, the worry is great, and unless the improvement in health is rapid, as in several cases I have known it to be, they lose heart and prefer bad health under a doctor to trying to cure themselves against the approval of relations and friends. No one but an actual invalid, for instance, can stand having to change the hours of meals instituted by our modern civilisation. Unless we are actually ill enough to give in, we must live according to the divisions of the day which suit the occupations of the various members of our family. These change with passing fashions, and, for aught we know, our grandchildren may dine at Queen Anne's hour, which was 5 P.M.

To a bachelor athlete like Mr. Miles, or a curate in the wilds of the country, it seems no hardship, and there is no pulling of the heartstrings in living on Plasmon and fruit; but think of the young wife whose tired husband comes home at 7 P.M. after making, or trying to make, money for her, and finds that she has had her meal of dry bread and nuts and fruit at 6 P.M., and, however affectionately disposed, can only sit and watch him drink his hot soup and eat his daintily-cooked chicken, winding up with iced fruit and cream. The situation is really harrowing, and yet Dr. Haig says one of the best divisions of meals for dyspeptics is 11 and 6. These used to be the French hours for everybody, but that custom is now so entirely changed that when we come across it in hotels our home habits cause us to think the division tiresome and inconvenient. Many husbands might submit to this in their own homes; but a still greater difficulty and irritation arises when the couple go from Saturday to Monday to friends in the country, where the tables groan with luxuries of every kind, and the wife finds it difficult to eat anything, for she has not the courage to ask for milk and cheese, and, the bread being uneatable from newness and baking-powder, she has to wait till the end of dinner for fruit, and suddenly realises, when she is at last eating her grapes, that the mistress of the house is dying to leave the table. I remember well years ago a vegetarian of some distinction who used to dine at our house, and invariably emptied the dessert-dish that happened to be in front of him, whatever it contained. It used, then, to cause me considerable suppressed irritation, but we change so that he has now all my sympathy, and I often long to do it myself. Even with the help of summer vegetables, such as peas and beans, which are dangerous food to those with weak digestions, the young wife has been absolutely starved, and feels it all the more as, being an amateur vegetarian, she will probably, in nine cases out of ten, be underfed and have no reserves to fall back upon in emergency. Even if she has read Dr. Haig's leaflets, or Mr. Miles' lists of food values, she will most likely have done so without the real comprehension which would enable her to apply the facts to her individual case. She might be able to get more *bulk* of food at an ordinary table, but is perhaps prevented from taking it by the haunting fear of getting fat, which I find among nearly all young people of the present day. Obesity is, of course, disease, and I believe is more easily cured by the diet than in any other way, for when I began it I lost two stone in six months ; but plumpness, especially for women, is natural and healthy, and the vanity of wishing to be thin, which has been so encouraged by the Pre-Raphaelite School of Art with its unhealthy ideal of beauty, still holds its sway and accounts for much nervous disease among women.   It is undeniable that the serpentine clothes of the present day look best on a thin peg, but on the whole there is no doubt in my mind that health comes before everything, and for a girl or woman whom Nature means to be fat, to diet herself thin, no matter by what means, is an exceedingly dangerous process which may cause many sufferings and complications, and even end in disease, and is quite as foolish as the efforts of the frog to swell itself to the size of an ox.   At the same time, when I am told—as I often am—'I should so like to take to your delightful diet of vegetables, only it would make me grow so fat,' by people who take butter and cream two or three times a day, in addition to a rich ordinary diet, I cannot help being amused, knowing how easily Dr. Haig's diet can be regulated for fat people.

The old mediæval theory of disregard of the body is still so strongly ingrained in some of the noblest natures that they would far rather be ill now and then than have to think out habitually what is good for them and act up to it, or to believe when they *are* ill that it is their own fault rather than a visitation of Providence which they are to bear with resignation.   They send for a doctor, pay him cheerfully, and hope for better days.   This attitude of mind will change when people are trained from childhood to understand that all sickness not caught from a germ is preventible, and that doing without food for a day, or even two, does far less harm to the healthy than eating wrong food.   To be able to miss a meal without feeling the worse, is one of the surest tests of being in good health. There was a time in my life when I nearly cried with misery if I did not get my tea at 5 o'clock, and felt quite ill all the evening. We must have patience and some courage. Revolutions are never effected in the bosom of peace and perfect concordance.

When I was young I remember saying to a friend, ' I wonder what the new religion will be when it comes? ' She answered, ' Something set to Wagner's music.' This did not seem to me to strike the true note. She had been long in Germany and was devoted to Wagner and his genius. I was not musical, and did not at that time know much of Wagner, but from the little I did know I felt he did not strike the key-note of self-denial which in some form or other is the basis of all religion. It will be curious if the religion of the future should avoid the dangerous asceticism of the Middle Ages which *denied* the body, by going back to the earlier Greek idea which *recognised* it as deserving of all respect, and treated it as a precious instrument to be kept in the highest degree of efficiency and perfection : practising abstinence for the sake of health rather than as a direct means to moral excellence.

In advocating the diet, one of the hardest things to combat, after the opposition of friends and inconvenience to social life, is a certain fondness of food and drink among those who consider that the one legitimate indulgence of life is to eat, if not drink, what one likes. I know several people who never deny that they are much the better for dieting, but who nearly burst with indignation when sitting at a table and seeing others eating all round them, because they cannot do the same. This naïve and childlike side of the question was rather a revelation to me, as all my life I had avoided the things I thought not good for me, however tempting they were. I can't understand eating that which you know does you harm. Modern vegetarianism has enormously grown out of sentiment, but I think it is exceedingly advisable that people who take to the simpler foods for health's sake rather than for sentimental reasons, should at least adopt an attitude of mind which will give the system a fair trial unprejudiced by likes and dislikes, mental irritation being bad for digestion.

A terrible idea has got about amongst my friends, and I might almost say patients, that if they take to the diet more or less strictly, they are not only to be quickly made quite well, but to remain so perpetually. Over-fatigue, change of weather, indigestion, &c., are supposed to have no effect at all on the non-meat-eater. Now I want everybody who undertakes the diet in consequence of ill health to understand that it cannot suddenly repair a debilitated state of the digestive organs. Most of our modern ailments are acknowledged to come from indigestion and non-assimilation. This goes on in many cases long after the excess of uric acid has ceased, and then the patients turn round on me and say, ' Why, after years of this self-denial and all the trouble that this peculiar feeding brings, am I to have this and that ailment?' The probability is that they have frequently not digested their simpler food any better than they did their mixed foods; they also do not get the enforced rest which attacks of illness, often lasting a week or more, imposed upon them. One can only assure them—and this sounds a feeble comfort—that they would have been much worse on the old diet, for I have never met a case which did not admit that any attacks of illness under the diet were very much milder than they formerly were.

It is obviously unfair to judge of a diet which is the best for man in health by its effects on invalids. People with weak digestions have a difficult hill to get over in leaving off meat, as doubtless it is lighter of digestion than any other food containing the same amount of nourishment. But people seldom realise that what may be a gain to the digestion may be an injury to the blood, and will bring round in time the very troubles from which they temporarily escape. For those who are digesting so badly that they are underfed, which can easily be told by colour of gums, loss of weight, excitable nerves, &c., it is most important that they should test themselves by weighing from time to time—say every three months. The nightmare of growing fat, alluded to before, makes many women believe they cannot take either milk or cheese. In these cases there must be a diet within the diet and a careful study of the proteid foods, which give nourishment without bulk.

In cases of severe dyspepsia, Mellin's Food is invaluable. Most people at the first difficulty say, 'The diet is not agreeing with me,' and go back to meat and champagne, which does them immense good for a short time till the increased uric acid in their system brings back their original troubles. All the relations are delighted, the doctor in attendance is jubilant, and the vicious circle again begins. Their winter or spring illness is attributed to cold and east wind.

I, from my experience, am quite convinced that in all cases of dyspepsia, rheumatism, gout, it is impossible to live up to anything like the full physiological allowance of proteid, especially in the case of women who do not work off the evil effect of non-assimilation by violent exercise as most men occasionally do. And here comes the great difficulty, viz. that each person has to find out for himself what amount of proteid he can assimilate, and in what form. Nobody else can know whether Nature is properly performing her functions, and digesting the food taken. Any necessity for a return to drugs, or violently irritating foods like cracked wheat, is a sign that even the simpler foods are not being properly digested.

I have never come across anyone who has given the simpler foods a fair trial of several years, who found any permanent benefit in returning to a meat diet. Doctors, who judge by the *immediate* results, would not agree with this. Most doctors naturally take a pride and pleasure in seeing their patients return to them after a tentative course of the diet, and it is an equal humiliation when good health results and enables them to keep away. Perhaps the health of the community would be improved if we adopted the Chinese and pauper system of only paying the doctors during health and not illness. There is a considerable difficulty in the law of the land being entirely on the side of the doctors. The conversation at my breakfast table this morning would have made a more nervous woman's hair stand on end. Having an invalid patient of Dr. Haig's in the house under a strict diet of bread, almonds, and fruit, a member of the family said, ' It is quite clear we shall have an inquest very shortly in this house. I see the coroner saying with great emphasis, " Are you *sure* that Mrs. Earle is in her right mind ? " ' This arose from a discussion as to whether the patient was being starved to death, or poisoned by prussic acid, the real point in my mind being whether she was taking more proteid food than she was able to digest, although within her physiological allowance, or whether she was less well from being out too much in the damp. Damp certainly has a bad effect on a great many people's digestion. Driving and walking are to my mind much less harmful than the sitting about out of doors which is the invalid rule.

The following statement of his own case was written out for me by a man-patient of Dr. Haig's : ' Present age 42. Occupation sedentary. Lives in London for ten

months of the year. Medical history bad, both on paternal and maternal side. Paternal: gout for generations.
Father suffered from it on and off all his life, and acutely
during ten years in the eyes, necessitating surgical treatment by operation. Maternal: chronic dyspepsia and
liver complaint. Mother's father died of latter at 34.
The man himself up to the age of 34 lived very well and
thought much about eating and drinking, but did not
indulge in either excessively. Was a heavy cigarette
smoker. Suffered during every epidemic of influenza.
In 1894 had an attack of influenza, followed by acute
intercostal neuralgia, which continued for one month, and
ended in an attack of phlebitis in the left leg, which
lasted seven months (of which six were passed in a completely recumbent position, the patient being forbidden
to move without professional assistance), involved the
two soffena veins and the femoral vein, and culminated
in thrombosis, which left the femoral vein blocked permanently for the length of five inches close to the groin.
Attack very severe, the leg remaining considerably larger
than normal size with much varicosity. This attack
was followed at intervals during six years by four other
attacks of intercostal neuralgia (so severe that the dread
of recurrence completely shook naturally strong nerves),
the treatment of which, by means of salicylate of soda
(in 20 gr. powders often taken every hour for eight hours
at a stretch), reduced the patient to a condition of abject
weakness, requiring weeks of the strongest tonics before
strength at all was recovered. Two of these four attacks
were followed by phlebitis in the leg originally attacked,
which, though less severe than on the first occasion, necessitated spells of months' duration in bed or on a sofa. In
September 1899 the patient underwent treatment at Bagnoles de l'Orne, in France, for the after effects of phlebitis.
In November 1899 occurred an unusually severe attack of

shingles on the right jaw and side of the neck and upon the back of the head under the hair. In January 1900 phlebitis attacked the right leg, and the patient was warned by the leading French specialist on the subject, who is one of the doctors at Bagnoles, and came from Paris to see him, that he was in danger of constantly recurring attacks, which might well result in a chronic condition, and that elastic stockings must be worn on both legs from the toes to the knees, and never taken off except in bed. He pointed out that the treatment at Bagnoles was in many ways wonderfully efficacious in restoring elasticity to the tissue of veins, thickened and hardened through phlebitis, but procured no immunity from fresh attacks of that disease. Such immunity could only be obtained (if at all) by improving the condition of the blood, for which purpose he recommended a yearly visit to Carlsbad for the next twelve or fifteen years. The patient adopted Dr. Haig's system of diet on April 21, 1900, and has continued it ever since. He has never had another attack of phlebitis or intercostal neuralgia, and was, in August 1901, authorised by the specialist in question to discontinue the use of elastic stockings altogether, even when taking strong exercise, such as walking or bicycling for long distances.'

The following letter may appeal to those who have to practise the diet under the difficulties of travelling. I myself have not taken voyages in steamers of late years, but in travelling on the Continent I find that if you conciliate the hotel-keeper and make it clear to him that economy is not your object, by paying cheerfully for all you *don't* eat and nearly double for what you *do*, you will find him most accommodating. Italian food adapts itself much more easily to the taste of the vegetarian than the modern French, at any rate in the expensive Paris restaurants ; but there is no doubt that the long table-d'hôtes

where you don't eat are even more trying than the long
tables-d'hôte where you do eat, and this must be just
accepted as a trial which can only be avoided by taking
lodgings.

'July 28, 1902.

' As we have just come back from travelling, you may
care to hear how we got on as to diet. We had foolishly
taken only a small amount of concentrated food with us,
and this very soon ran out. However, we were thus able
to experiment on how far it is possible to live on the
gleanings from 'board-ship *table-d'hôte*. Practically everything is artificial—adulterated milk and butter, frozen
meat, frozen vegetables, frozen fruit. Formerly a goat or
a cow used to travel on the ship, and I believe that is still
the case on some of the very small obscure Continental
lines. The bread is extremely poor, little white rolls that
taste very acid, and are probably made of gypsum, boron,
soda, in fact anything except wheat. The artificial milk
is exceedingly nasty and unwholesome and could only
possibly be taken in very small quantities with tea or
coffee. The only drink that commends itself is hot water.
The unboiled water is never supposed to be safe in the
Mediterranean and mineral waters certainly have a very
bad effect on people unaccustomed to them. It seems to
me, however, that those who have a peculiar diet ought
to expect all these drawbacks, and starvation is not very
terrible for a short space of time. The thoroughly ob-
jectionable thing is the immense length of the meals,
they nearly always consist of nine or ten courses and
take nearly two hours. Imagine munching dry bad bread
for nearly two hours while carrying on a sort of pretence
at eating some of the meat courses, all for the sake of a
little fruit and cheese at the end, these last being the
only courses that are hurried. The only time I got on
well, as regards food, was when the weather was bad.

F

Then the waiter brought me the few items of the *menu*
that I could eat at once and without "reasoning why."
I wish that providing necessaries and not "luxuries"
was the ideal of 'board-ship caterers, as the effort to provide "luxuries" only produces a lot of highly sauced
dishes one more uneatable and unwholesome than the
other. There is yet another unpleasantness; on some
steamers the stores of food are all kept on ice, and when
they are removed from the ice, decomposition sets in very
rapidly, the result being that the dining saloon, besides
its normal, fragrant smells, has a peculiar, drain-like atmosphere. The Continental train services with restaurant
cars have the same principles; they never have fresh
food, even though they may be travelling through the
richest countries in the world, and constantly stopping.
No doubt the reason for this kind of food being supplied
is, that there is a demand for it. The class that lives
most artificially is the class that travels most, and therefore it becomes a matter of business necessity to supply
the kind of food which is most constantly asked for. If
we may be bold enough to give advice to other people
who have the same diet as ourselves, it is that they should
not travel at all, or that they should have a private
yacht, or, if they have to travel in the ordinary way, that
they should take a large supply of nourishing biscuits and
a pouch like Jack the Giant-killer, into which they may
slip everything, and thus keep up the appearance of
having the eating capacity of the average traveller, and
avoid the accusation made by hotel-keepers of undue
economising.'

The following case has been written out for me by
the patient—a young woman whom I have known for
many years, and who has most bravely fought against long
chronic bad health :

' Five years ago, in the winter of 1897, I began a diet

of cereals, milk, cheese, nuts, fruits and vegetables, to
the total exclusion of flesh foods, eggs, pulses, tea, coffee
and cocoa. I was at that time suffering from anæmia,
indigestion, neuralgia, and general weakness. My work
was sedentary, and lasted from 9 A.M. till 7 P.M. For
eleven months I stuck to the diet, appetite gradually
getting so bad that, for the last few months, I used to live
on the less solid foods, as milk, savoury vegetables, and
the more tempting kinds of fruit, with very little bread,
cheese, or nuts. At the end of the time I was so much
worse that I went to Dr. Haig. He looked at the diary
I had kept of my food and shook his head. He then
examined me generally, and said I was "starved." After
a few questions about the food I liked, he said, " Go out
and get a mutton chop and a glass of Burgundy. Then
go into the country for a rest, be out of doors all you can,
take this tonic, and eat as much fish and meat as you
like. The first thing to do is to get a healthy appetite—
then we will consider the diet. But eat anything rather
than be underfed." I carried out these instructions and
went back to him much improved a few weeks later. He
then advised me to change very gradually, substituting
cheese for meat at one meal only—and after keeping to
that for some weeks, to change another meal, and so by
degrees to get on to the diet. This I tried, but failed
again, and then gave up altogether for about a year.
Then I began again, and for two years had a series
of hopeful beginnings and dismal endings. My friends
used to say, " But why change at all when the ordinary
diet suits you ? " It certainly did seem to suit me for
the first three or four weeks when I went back to it
from a non-meat diet, but I had always to be taking
medicines, and sooner orlater came the dyspepsia, with
the consequent exhaustion that made life a burden.
As I woke each morning I used to feel " Here's

another day to drag myself through "—and as I lay down at night I used to be so dead beat that I felt it would be a relief to sleep for good. The change to Haig's diet always brought, at first, a delightful sense of lightness and well-being—a difference as between a well-oiled machine and one clogged with grit and dust— but after a few weeks I used to get too weak to go on with my work, and had to give it up. At last, after rather a bad collapse, the friend with whom I worked bound me over to a meat-fish-wine, &c., diet for twelve consecutive months. When they were over, I started once more very cautiously on the uric-acid-free diet, making a careful study, meanwhile, of food values, mixtures and quantities, and for six months I succeeded beyond my hopes. Every-one noticed how much better I was in every way, and I found for the first time since childhood, that I could dis-pense with all kinds of medicine. Suddenly I caught a bad chill after weeks of excessive overwork, which re-sulted in *mucus colitis* with total collapse, and for two months I was in bed in charge of a nurse. The treatment was injections of carbolic alternately with boracic acid, and this played such havoc with an already enfeebled digestion, that I could take no sort of invalid food, except minute quantities of dissolved Plasmon, without hours of distress. At the end of three months' most careful nursing, my strength had so far improved that I was able to go into the country with a nurse, but the colitis being none the better for the treatment I had undergone, and the doctor who had attended me telling me that it might go on for three years, as it was a most difficult thing to cure, there being no drug that would touch it, I wrote to Dr. Haig and asked him to prescribe. He instantly stopped all injections and put me on salicylate of soda, and in three days the mucus had diminished. At the end of ten days it had entirely ceased, and I then went to stay with a friend of his who asked him to come down from London to see me. This he most kindly did, and the result of the interview was that he stopped the salicylate, which had done its work, and for the dyspepsia, which was still very bad, he prescribed a perfectly *dry* diet, of bread, pounded almonds, fruit, vegetable and cheese, telling me to replace the two latter by more fruit (fresh and dried) and almonds, as I found I could take them. This answered so well that digestion rapidly improved and, with it, strength. At the end of a month I could walk up steep hills on hot days, with pleasure, a thing I had never, at my best, been able to do before. The almonds, even when ground in a nut-mill and pounded by a careful cook, were, I found, liable to aggravate the intestinal inflammation, and as I did not digest cheese well enough to nourish satisfactorily upon it, I experimented upon the almonds till I got a substance as smooth as Devonshire cream, and of this I could soon take 4 oz. a day with comfort. So many people find almonds a difficulty that it may be as well to say how this almond cream is prepared, for almonds are equal to their own weight of meat in nourishment, and for those who dislike cheese, they are far too valuable an item in the list of uric-acid-free foods to be cast aside.

' Almond Cream.—Blanch the almonds in boiling water and grind in a nut-mill. Put about a tablespoonful at a time into a mortar with a teaspoonful of cold water, and work with the pestle till as smooth as butter, no particle the size of a pin's head being left. Mix in a little more cold water till of the consistency of clotted cream, and scrape out on to a dish. It is better to have only a small quantity at once in the mortar, or the smoothness is not so uniform. The nut-mill should be taken to pieces (one easy screw undoes the whole) every day and well scalded, for almonds turn as sour as milk, especially after they have been wet. For this reason, the almond cream should be freshly made every day. For travellers who prefer almonds to cheese, but cannot venture to eat them whole, this form of cream is excellent for sandwiches. Children and invalids, too, can digest them in this way. When I had been three months on this diet, I caught another bad chill, which brought on a severe relapse of the gastro-intestinal inflammation, and again for two weeks I had to have a nurse and be kept in bed. For four days various invalid foods—barley water, beaten white of egg, champagne and soda-water, weak solution of Plasmon, &c. —were tried in teaspoonfuls at the time, and all with equally distressing results. Finally a nutrient suppository was tried, which also proved a failure, and still further reduced my strength. Then, having convinced my nurse that the very things administered with the time-honoured idea of "keeping up my strength" not only failed to do this, but actually drained the little I had by causing sickness and pain, I insisted on a fast of twenty-four clear hours, to give the stomach a chance of righting itself. The thirst, which was intense, I relieved partly by holding a piece of ice in a bit of linen on the lips, and rinsing the mouth from time to time with bicarbonate of soda. Even a spoonful of hot or cold water instantly produced sickness, and there was nothing to be done but leave things alone and wait. At the end of twenty-four hours I took 20 grains bicarb. soda in half-a-pint of hot-water—a teaspoonful every two minutes. After thirty hours I took, with comfort, 2 oz. Mellin's Food of the strength given to infants of four months, and this I repeated three times a day for two or three days, after which the quantity, though not the quality, was gradually increased—the bicarb. being taken an hour before each meal of Mellin's Food. Dr. Haig again most kindly came down from London, and said the starvation treatment was

the only one for the case—there being no risk whatever
in fasting for five days, so long as I had 10 lbs. of flesh
on my bones. His advice for future diet was to drop
almonds entirely till all sign of weakness had disappeared
in stomach and bowels, and to "feel my way" slowly on
to a diet of bread, potatoes, milk, fresh curd cheese, fruit,
with oil, cream, and butter *ad lib.* The bread to be stale,
home-made, white (*i.e.* the wholemeal minus the bran),
the potatoes to be steamed or baked in the skins to preserve the alkaline salts so valuable in all forms of gout
and rheumatism (colitis being a kind of intestinal gout);
the fruit to be ripe and uncooked, and always eaten with
potatoes to prevent acid dyspepsia, and plenty of oil or
butter to aid digestion. After being on this diet for a
month, I dropped everything but bread and fruit, and
recovery became so much more rapid that I have kept to
this permanently with Dr. Haig's full approval, and have
already built up a reserve of strength such as at one time
seemed impossible to all the doctors I consulted.'

A young man I know has had eczema from his birth
and, with every chance of consulting many physicians,
no one had been able to cure it, eczema being one of the
illnesses which, like leprosy, seem, on their own confession, to baffle doctors, though many remedies make it
apparently better for a time. After two and a half years of
a form of the simpler foods which suited him, he seems, for
the first time in his life, to be quite free from it. Having
just heard of this cure, the following story in the papers
(July 1902), of the tragedy in the home of an artisan
(a marble-polisher) seemed to me peculiarly distressing.
The mother had been very despondent because her two
little boys were severely afflicted with eczema. She had
taken them to several hospitals in vain, and at last, in
despair, drowned them in a tub. 'It was God's curse!'
she said to the police; 'the neighbours used to call

their children away from playing with them, and it broke my heart. I put them into a tub of water and put them into the bed. I did not know what I was doing.' A verdict of 'wilful murder' was brought in against her. This probably means a lifetime in Broadmoor. I quote this humble tragedy because I feel that, of all disorders, skin disease is most obviously cured by diet.

After writing these long, and perhaps boresome, chapters on the subject of health, it was with no small comfort that I found in Mr. Maeterlinck's latest book of essays, 'The Buried Temple,' in a chapter headed 'The Kingdom of Matter,' the following passage, which brought me that best of all consolations on a road that is unpopular and strange, namely, to feel 'Verily, I am not travelling alone' :

'We have said that man, in his relation to matter, is still in the experimental, groping, stage of his earliest days. He lacks even definite knowledge as to the kind of food best adapted for him, or the quantity of nourishment he requires ; he is still uncertain as to whether he be carnivorous or frugivorous. His intellect misleads his instinct. It was only yesterday that he learned that he had probably erred hitherto in the choice of his nourishment ; that he must reduce by two-thirds the quantity of nitrogen he absorbs, and largely increase the volume of hydrocarbons ; that a little fruit, or milk, a few vegetables, farinaceous substances—now the mere accessory of the too plentiful repasts which he works so hard to provide, which are his chief object in life, the goal of his efforts, of his strenuous, incessant labour—are amply sufficient to maintain the ardour of the finest and mightiest life. It is not my purpose here to discuss the question of vegetarianism, or to meet the objections that may be urged against it ; though it must be admitted that, of these objections, not one can withstand a loyal and scrupulous inquiry.

I, for my part, can affirm that those whom I have known
to submit themselves to this regimen have found its result
to be improved or restored health, marked addition of
strength, and the acquisition by the mind of a clearness,
brightness, well-being, such as might follow the release
from some secular, loathsome, detestable dungeon. But
we must not conclude these pages with an essay on
alimentation, reasonable as such a proceeding might be.
For, in truth, all our justice, morality, all our thoughts
and feelings, derive from three or four primordial necessities, whereof the principal one is food. The least modification of one of these necessities would entail a marked
change in our moral existence. Were the belief one day
to become general that man could dispense with animal
food, there would ensue not only a great economic revolution—for a bullock, to produce one pound of meat, con-
sumes more than a hundred of provender—but a moral
improvement as well. . . . For we find that the man who
abandons the regimen of meat abandons alcohol also ;
and to do this is to renounce most of the coarser and more
degraded pleasures of life. And it is in the passionate
craving for these pleasures, in their glamour, and the
prejudice they create, that the most formidable obstacle is
found to the harmonious development of the race. Detachment therefrom creates noble leisure, a new order of
desires, a wish for enjoyment that must of necessity be
loftier than the gross satisfactions which have their origin
in alcohol. But are days such as these in store for us—
these happier, purer hours ? The crime of alcohol is not
alone that it destroys its faithful and poisons one-half of
the race, but also that it exercises a profound, though
indirect, influence upon those who recoil from it in dread.
The idea of pleasure which it maintains in the crowd forces
its way, by means of the crowd's irresistible action, into
the life even of the elect, and lessens and perverts all

that concerns man's peace and repose, his expansiveness, gladness, and joy ; retarding, too, it may safely be said, the birth of the truer, profounder, ideal of happiness ; one that shall be simpler, more peaceful and grave, more spiritual and human. This ideal is evidently still very imaginary, and may seem of but little importance ; and infinite time must elapse, as in all other cases, before the certitude of those who are convinced that the race, so far, has erred in the choice of its aliment (assuming the truth of this statement to be borne out by experience), shall reach the confused masses, and bring them enlightenment and comfort. But may this not be the expedient Nature holds in reserve for the time when the struggle for life shall have become too hopelessly unbearable—the struggle for life that to-day means the fight for meat and for alcohol, double source of injustice and waste whence all the others are fed, double symbol of a happiness and necessity whereof neither is human ?'

I owe most of my scientific instruction in the non-meat diet to Dr. Haig, but I was favoured by a leading vegetarian with the following list of medical men in different parts of England who agree with him, to the extent, at any rate, of recommending a non-flesh diet. I thought the list might be of some service to those of my readers interested in the health question:

Dr. R. Edmund Deane, Wilmslow, Manchester.

Dr. Pullar, 111 Denmark Hill, London, S.E.

Dr. A. B. Mercer, Kinclune, Carlisle Road, Eastbourne.

Dr. Charles Reinholdt, 16 Blagrave Street, Reading.

Dr. Charles White, Studley, Upper Parkstone, Dorset.

Dr. W. H. Ackland (J.P.), Fairhaven, Compton Gifford, Plymouth.

Dr. Mitchell, 21 Park Lane, Little Horton, Bradford, Yorks.

Dr. D. Geo. Carmichael, 68 Kenmure Street, Pollockshields, Glasgow.

Dr. Blacker, The Boltons, Farnborough, Hants.

Dr. J. Shaw Lyttle, Dundela, Cilfynydd, Pontypridd.

Dr. Thos. Mowat, Barns Place, Clydebank, Glasgow.

Dr. Benham, Earl's Court Square, London, S.W.

Dr. E. Howard, Curfew Cottage, Datchet.

Dr. W. J. Fernie, 4 Pembroke Villas, The Green, Richmond, S.W. (Author of 'Herbal Simples,' 'Kitchen Physic,' &c., &c.)

Dr. Flint, Scarborough.

Dr. Oldfield, Carn Brae, London Road, Bromley.

Mrs. Fleetwood Taylor, M.D., c/o H. Phillips, Esq., 16 Farringdon Street, E.C.

Dr. George Black, Greta Bank, Chelston, Torquay.

Dr. Hadwen, Brunswick Square, Gloucester.

Dr. *Helen* Wilson, 381 Glossop Road, Sheffield.

Dr. Arnold, Moss Lane East, Manchester.

Dr. Genny, Lincoln.

Dr. Walters, Stonehouse, Glos.

Dr. Johnston, Ambleside.

Dr. Crespi, Wimborne.

Dr. Haig, 7 Brook Street, Hanover Square, London, W

SUPPLEMENT

Home started in Buckinghamshire, under Dr. Haig, for instruction
and practice in uric-acid-free foods—A second home in Hampshire for fleshless diet, recommended by Mr. Eustace Miles, M.A.
—Critical paper, by a patient of Dr. Haig—Dr. Haig's answer to
same.

A FEW fresh facts having come to my knowledge since the
former chapters were written, I am adding them here as
a postscript because I think they will be useful to those
who have read what I have said before.

The news has just reached me that a grateful patient
of Dr. Haig's has started a self-supporting home where
invalids, and those interested in Dr. Haig's theories of
diet, may reside under his personal supervision and teaching.   Miss Florence Jessop, the secretary of the home—
Apsley House, Slough—will give all the necessary information to those who write for it.   The pamphlet, describing
the objects of the home, opens with this sentence · ' Life is
not mere existence, but health and happiness.   The object
here is life in its fullest sense.'   With Dr. Haig's
characteristic eagerness to share his knowledge and experience with all who care to learn, he has provided that
doctors shall be taken in, free of charge, for Saturday to
Monday visits, that they may get to understand about the
place.   This is only one more instance of what I have
noticed ever since I have known him, viz. that he tries to
teach everyone all he knows in the belief that, if only they
understand his theory, they will be able to keep their own

health in order and greatly control the chances of disease. The terms of the home are from three to five guineas per week, which include all comforts, and Dr. Haig's attendance on his bi-weekly visits. Poor patients can place themselves in his care at the Metropolitan Hospital, or the Royal Hospital for Children and Women, at both of which he is a visiting physician.

I was told that, after the publication of my second book, some people said, ' I wonder what Dr. Haig pays Mrs. Earle for puffing him in her books ? ' This deliciously characteristic question of the age amused me immensely. I wonder what these same people will say now? It seems a pity that there should be anyone who does not understand the difference between personal interest of any sort, and the impersonal interest in scientific work which promises to benefit the whole world, if those who believe in it help to make it known.

Mr. Eustace Miles, M.A., is thoroughly recommending another home called Broadlands, at Medstead, Hants, and says that he may be used as a reference, as it is just the kind of establishment to which he alluded in a recent letter to the ' Daily Mail.' Particulars can be obtained from Miss Houston, the secretary. The terms are two guineas a-week, which at first sight appears cheaper than Apsley House, but as this charge does not include medical attendance and instruction, there is no great difference.

It being my fixed idea that the food question rests with the intelligent public and not with the doctors, I asked a friend who for over two years has taken immense interest in the question both for himself and others, to be kind enough to put on paper his present attitude with regard to diet. My friend sent his critical paper to Dr. Haig, who has most kindly taken the trouble to write for me his explanation of the difficulties raised.

'*23rd October*, 1902.—You ask me to give you a *résumé* of my present views on diet. I am still as firm a believer as ever in Dr. Haig's conclusions as to the part played by uric acid in causing disease, and in that authority's contention that it must, from any point of view, be desirable to avoid as far as possible all food which in itself contains uric acid, or one of the other acids producing analogous effects. On some points, however, I am in a critical frame of mind towards Dr. Haig's opinions, or rather what I believe to be his opinions, which conceivably may not be quite the same thing, seeing that I am only a humble amateur follower of the expert to whom I owe so much. And, firstly, as to cheese and milk. Whether by reason of the casein or of the lactic acid, which both contain, it seems to me to be clear that a large number of people cannot digest these substances, at any rate not in the large quantities originally recommended by Dr. Haig; and many of the cases treated by Dr. Cantani, the Neapolitan specialist, seem to prove that milk and cheese are the direct cause of gout in some constitutions. It may well be that these substances must, for such people, be placed in the same category as eggs, which, on Dr. Haig's own showing, contain no uric acid, and yet, according to him, produce the same effects as if they did. I am told that Dr. Haig has ceased ordering milk and cheese so freely as at one time he did; if that be the fact, I am glad to hear it, for I incline strongly to the opinion that their habitual use by many people has been responsible for many mishaps. Then, as to the amount of food which each person should eat in the twenty-four hours. I doubt the reliability of Dr. Haig's method of computation—I mean, the method of multiplying the weight in pounds by nine, and treating the product as the number of grains of proteid which should be eaten per diem. *Prima facie*, I should have expected that people

would differ from each other in digestive idiosyncrasies
quite as much as in other respects, and my experience up
to the present time seems to me to show that the assumption is well founded. If this be true, a rule whereby
the amount of food is ascertained for each person in an
identical manner cannot be sound, because it takes no
account of his or her individual peculiarities. Again,
the circumstances and conditions of life frequently vary
very much in the course of a few years ; at one time a
man may work more physically than mentally, at another
the reverse may be the case ; at one time he may be living
in a hot climate or in a stuffy town, at another he may be
living in a cold climate or in the country ; at one time he
may be leading a life of constant worry and strain, at
another he may be enjoying a tranquil existence. I do
not believe that these changes are represented by sufficient,
if by any, changes in the weight, so as to make it the
proper, or even a possible, criterion ; while I do believe
that those changes materially affect the digestive capacity.
At any rate, in my own case and in that of my wife, we
were not in a satisfactory state, when eating at the rate
prescribed by Dr. Haig's method, during the first two
years, whereas we began to improve from the moment
that we began to eat considerably less. My proper
amount, according to Dr. Haig, is about 1,500 grains of
proteid, whereas I am eating at this moment, and for the
last two months have been eating, under 1,200 grains. It
may be that two months do not afford a sufficiently long
test ; but against that objection I set the fact that the
experience of everyone whom I know (including yourself) tallies with mine of the last two months. The
matter is all important if the true rule is that every
particle of proteid taken during the twenty-four hours,
which is in excess of what is actually required just to
make up the loss occasioned by the energy, mental or

physical, expended during the twenty-four hours, or which, for any reason, the individual is unable to metabolise, acts as a direct poison. This is the rule laid down by Dr. Cantani, as I understand the matter, and I am convinced that it is the true rule, and that the disregard of it explains many of the failures of the diet, and still more the quasi-failures. A friend of yours and mine suggested to me on Sunday last a way of maintaining the reliability of the method of calculation laid down by Dr. Haig, and yet of admitting that it often gives unsatisfactory results in individual cases. This was to regard Dr. Haig's amounts as the proper ones for the normal person, but to hold that very few people who take to the diet are in a normal condition at the start, and take a longer or shorter time, and generally longer time, to become so, and that beginners should regard Dr. Haig's amounts as the goal towards which they should work, but should not expect to be able at first to eat in accordance with them. My reply is that the assumption that people are not in a normal condition at the beginning is almost certainly true, and that the suggestion may be as sound as it is ingenious, but that it does not represent Dr. Haig's view, as I understand it, and that, further, a method of calculating the proper amount of proteid which is not a reliable guide at the precise time when people most want such a guide—namely, at the beginning, when they have no data of their own to go on—can hardly be considered satisfactory.

'I would further point out that the weight, to which Dr. Haig's method has to be applied, is the normal or proper weight, not the weight at any given moment, as indeed must be the case, for otherwise the fatter a person got the more he would be bound to eat, irrespective of all other consideration, and *vice versa*. Now everyone does not know his normal or proper weight, seeing that we have not all been in the habit, from early youth, of being weighed at frequent intervals, and of taking a careful note of the result : we are therefore not in a position to strike an average, and so ascertain the normal weight. In such circumstances the only thing to be done is to take the weight according to the height, but Dr. Haig himself regards that method as by no means satisfactory. Again, all methods of computation necessarily depend for their success on the accuracy of the tables of analyses, which purport to show the proportions of proteid and other things contained in the different foods. Now it is obvious to anyone who has looked into the question at all that hardly any of these tables agree, and sometimes they differ materially ; and even when they do agree, there remains the question whether the person consulting them can extract the whole of the proteid shown by the tables to exist in a particular food ; in other words, whether he can fully digest it.

'In short, from whatever point of view you consider the matter, nothing can take the place of personal experience ; and therefore, while it is desirable, and in truth unavoidable, to accept Dr. Haig's method of calculation, and to assume the accuracy of some of the food tables to begin with, as one must have some basis to start on, it must be clearly understood that, if the result is not what it should be, the blame is not to be attached to the simple foods, or to the principles inculcating their adoption, but to the manner in which the patient is applying them in his or her case.

'One of the surest signs in my opinion that the individual application is wrong, is constipation, whether the patient has previously been a sufferer from that scourge or not. Of course, if the patient, on mixed foods, has never been troubled with it, and it began only with the adoption of the simple foods, the proof of mistake in in-

G

dividual application is the more emphatic, but in any case, as I maintain, it is only a question of degree. The usual experience is, I fancy, that the adoption of the simple foods is followed at once by a marked improvement in the behaviour of the bowels, which causes the heart of the neophyte to rejoice, and makes him think that the problem is solved for good and all. Such may indeed be the happy fate of some of us, but more usually, I think, the initial improvement gradually slackens, and finally dies away altogether, and the *status quo ante* the change of diet returns more or less. To all persons in that position I say that they should not be discouraged, that they should not think that the diet is in principle at fault, but that they should think that their individual method of application is wrong, and should never rest until, by experiment, they have discovered exactly what is the cause of the trouble. I do not think that Dr. Haig looks at this complaint, as it ought to be looked at, namely, as one of the most immediate and natural proofs that the diet is not being properly applied, except, of course, in cases where there is reason to diagnose structural or other organic defect in the bowels themselves. The result of not doing so is that a person, who adopts the simple foods, is very liable to bring over from his former condition and training the orthodox attitude, both among doctors and laymen, that constipation is one of the mysteries of Nature, which must be tolerated with resignation and fortitude, or treated with violence. Yet it seems to me much more likely that it is merely due to the fact that the food is presented to the bowels in a condition not suitable to their task, and that, either because something, which ought to have been removed by the process of mastication and stomachic digestion, has been left in, or *vice versa*, and is a kindly warning of Nature, which is meant to show us either that we are taking the wrong foods, or that we are

taking them in the wrong quantities, or that we are com-
bining them wrongly.

'I expect you will say that, if the whole matter is
really a question of individual effort and experiment, there
is very little chance of right principles of diet ever spreading generally, seeing that most people expect their doctors
to do all the thinking for them on the subject of health,
and would decline even to entertain a proposition to
change their diet, unless at least one could assure them
that they had only got to follow a cut-and-dried programme. I quite agree ; but then I do not think, and never
have thought, that most people are at all likely to be convinced in any circumstances at the present stage. What
I hope will happen is that the experience of the teetotallers
will be repeated, and that a sufficient number of people
will become convinced of the evils of the mixed foods, to
bring the necessary pressure of public opinion to bear
upon the medical profession to force them to review their
pronouncements in the light of the new data, which are
slowly accumulating. When this has been done, and the
majority of doctors advocate a non-flesh diet, at first for
invalids and then generally, the lay public will follow
their lead, as a matter of course, in time. Considering
that the profession is now contemplating the establishment of sanatoria for the treatment of consumption at
the North Pole, and remembering what they would have
thought of such a proposition, say, to be quite safe, forty
years ago, we need not despair. Meanwhile we should
none of us lose an opportunity of drawing the attention
of our medical friends to their monstrous impertinence in
claiming our respect and implicit belief professionally, so
long as they, as a class, are at least as unhealthy as the
rest of us. What should we say to a seaman who obviously could not sail his own boat, if he wanted to come
and take charge of ours ?

' P.S.—By the way, the Cantani I have referred to above was a doctor of that name who practised at Naples and died in 1898. He was immensely thought of by the Italian medical world, and devoted himself specially to gout, rheumatism and kindred disorders, in the treatment of which he obtained some remarkable successes, principally, as it seems to me, by his skill in applying the rule which I have quoted in this letter. He admitted meat in small quantities into his dietary, and absolutely excluded milk and cheese; but as I have not yet been able to get hold of any of his books, which have not, I believe, been translated, and have only seen an English work, which purports to give a *résumé* of his theories and practice, I do not feel myself in a position to speak more definitely on the subject.'

Dr. Haig answers these criticisms as follows :

' In making a few comments upon your friend's letter, I should like first to mention some points which cannot be lost sight of without getting into hopeless confusion.

' (1) Certain diseases are due to excess of uric acid in the body and blood.

' (2) I propose to get free from those diseases by as far as possible ceasing to swallow uric acid.

' (3) As a practical result of this a daily excretion of uric acid amounting on meat and tea diet to, say, 20 grains, can in eighteen months to two years be reduced to 10 or 11 grains a day; and with this reduction the diseases due to uric acid gradually diminish and cease to trouble.

' (4) This result has nothing whatever to do with individual constitution, age, digestion, or metabolism; it is simply a matter of swallowing or not swallowing some 8, 10, or 12 grains of uric acid each day.

' (5) Quantity of food on a diet on which one has lived

from childhood is a matter of habit : one unconsciously
learns as a habit to eat enough ; but even here one may
learn to eat too much. On a new diet one must have
some rule to prevent any serious mistakes ; but no rule is
absolute, it is a rough guide to be adjusted by each individual. The rule of 9 or 10 grains of albumen per
pound of body weight per day is the one that was taught
in all text-books of physiology when I was a student
twenty-five years ago. These twenty-five years of use
have not enabled me to convict it of serious error ; on the
contrary, I believe that for a rough guide it is very close
to the truth, and that investigation would show that when
the weight to be nourished has been properly calculated
(and this, of course, requires skill and experience), there
are not many people who do good work in the world and
maintain their weight, strength, activity and blood colour
at the normal standard, on less than 9 grains of albumen
per pound of body weight per day.

'(6) No one has, I believe, found any xanthin or uric
acid in milk or cheese, and the taking of one and a half
pint of milk and one ounce of cheese every day does not
interfere with the reduction of uric acid from 21 grains
down to 10 per day (see No. 3). Some German chemists,
on the other hand, have found xanthin in the yolk of eggs,
and the taking of eggs does at once raise the uric acid above
10 grains per day in proportion to the quantity of egg
taken. Milk and cheese are therefore uric-acid-free foods :
yolk of egg is not. Thus far we have nothing to do with
individuals, all are absolutely alike; those who swallow
uric acid suffer from its effects, those who do not swallow
it do not suffer.

'(7) But when we come to deal with the available uric-
acid-free foods, it is quite a different matter : here every
individuality in health, disease, deformity, age, and condition has to be considered in deciding which foods offer

special advantages for each individual case; and here there is scope for endless experience and discussion, which I shall not attempt even to outline in a letter.

'I can now, however, indicate shortly what I should say in reply to your friend's remarks. As to milk and cheese, I know some who do best on large quantities of milk, and I know others who do best on none ; such fluid diet is very bad for some diseases, very good for others ; there can be no rule for all. The harm that Cantani saw in gout and rheumatism from taking milk and cheese was possibly due to taking them in excessive quantity, or to the acids and salts in cheese, or to the dyspepsia which is in some people produced by milk ; but gout and rheumatism are but a small part of the diseases produced by uric acid ; and the harm that milk and cheese did in these was not due to introduction of uric acid. I have known cheese cause a relapse of rheumatism ; but this was because it contained much salt. I often purposely put patients on too large an allowance of albumens at first as in some diseases it is safer to be overfed rather than underfed ; and later, when the patient has got over the change of diet, the quantity can be reduced.

'As I say in "Diet and Food," but few people want more than 1,400 grains per day ; but it is easy to over-estimate the exercise taken, and also to over-estimate the correct weight, or proper quantities of tissues to be nourished, as distinguished from fat which does not want extra nourishment. No doubt I often make mistakes in these matters in spite of my best endeavours. No rule is or can be absolute which has to take such data into account. If strength and blood-colour keep normal, I am content no matter how little food is being taken ; but for those who maintain strength and colour over years my figures will not often be found to be very far out.

'I do not know anything about Cantani's metabolism.

I do know that, if I do not swallow uric acid or take too much food—i.e. above the 9 grains limit (an error which may be committed on all diets)—I do not suffer. I quite agree that each person is to some extent a law to himself, especially as regards digestion ; and to a lesser extent as to quantity ; but the great physiological law of quantity is much less often broken than might be imagined. As to kinds of food each must try for himself, only general rules can be stated, and your friend is quite right in saying that the mistakes of individuals in details do not affect the great principle of the exclusion of uric acid. I am also inclined to agree with him that constipation is often a question of finding the particular uric-acid-free foods that suit the individual in question. Still even here it is a rule that milk and cheese are constipating (and this may account for part of their evil influence in gout and rheumatism as noticed by Cantani), while some breadstuffs and most vegetables and fruits have the opposite effect. Constipation may precipitate an attack of gout or rheumatism, but it does not cause it; and it can only precipitate it when an excess of uric acid is present in the blood; and Cantani's patients would often have some excess of uric acid in the blood as they took some meat, so that milk and cheese might precipitate gout in them, but not in one of my patients; and the case mentioned above was a beginner who had not got free from uric acid. Many meat-eaters suffer from constipation, but not all, and one must, of course, give a different diet to those who suffer and to those who do not. Those who can take breadstuffs freely generally do well to live on them ; but many, myself among the number, cannot take enough bread, and must add milk and cheese to a small extent or starve. In my case this is due to an error in education, for if I had been brought up on breadstuffs I could easily have lived on them now. No one ever swallows uric acid with impunity; but each, as he learns this and comes to change his diet, must decide which foods suit his disease his stomach, his teeth, his age, and his habits, which last have grown to be a part of him.

(Signed)   ' A. HAIG.'

I read this autumn, according to my custom, the epitome in the ' Times,' of the speeches which a number of our greatest doctors addressed to their pupils on the reopening of the schools, and it is a remarkable thing that in those speeches there was not one word of instruction, or, indeed, the smallest allusion to the fact that it was either desirable or possible to keep people from being ill at all.

Just as I reached the end of my notes comes the thrilling account of the Berlin conference on Tuberculosis. It is a disappointment to me to find that there has not been sufficient investigation made during the last eighteen months to confirm or refute Dr. Koch's theory that the bovine tubercle is not transmissible to man. It is, however, cheering to find that Dr. Koch is himself as firm as ever in his belief that overcrowding, damp, want of fresh air and sunlight are the predisposing causes to consumption, and that we must abolish these unhealthy conditions if we would successfully fight the monster which preys on so large a proportion of our population. He also refers to a point I have often been surprised to see so much neglected, viz. that while people will take care not to drink unboiled cow's milk, they cheerfully eat butter and cheese without any sterilising precautions.

GOATS

Goats at Naples—Possible solution for milk difficulty in rural districts—A toothless generation—Ignorance as to nourishing value of
separated milk—Mr. Hook on goat-keeping—Personal experiment
—Roast kid and *agneau-de-lait*—Reasons for prejudice against
goats—Suggestions for the philanthropic—Immunity of goats
from tubercular disease—Day at Guildford—Almonds—The
Astolat Press—Mr. Gates' herd of Toggenburg goats—Feeding of
goats—Chemistry of food to be taught in elementary schools.

WHEN walking in April in the streets of Naples, I came
across a large herd of goats being milked from door to
door, and it suddenly flashed upon me with great force
that, if the English people could be persuaded greatly to
increase the keeping of goats, especially in rural districts, it
might be possible to arrive at some solution of the problem
which faces everyone who gives thought to matters of
health, viz. the serious deterioration in the physique of
the people of this kingdom. There is no more convincing
sign of this than the fact, universally acknowledged I
believe, that a whole generation is growing up which have
hardly any teeth left by the early age of twenty. I am
told that not by any means the least of the sufferings of
the soldiers in South Africa was toothache, and I remember it was suggested in some newspaper that the War
Office should provide false teeth for the recruits ! This
state of health is by no means surprising when it is
remembered how many children and young people are
now brought up on baker's bread and stewed tea without
any milk at all ; and, strange as it may seem, I believe a

child might have a better chance of health if it were
brought up on moderate quantities of pure beer, instead
of what the village mothers of to-day give their children.
Milk is looked upon in our country districts as an extravagant luxury sometimes ordered by the doctor instead of
cod-liver oil, and even when milk is ordered, there is
great difficulty in getting it, for it is an easily verifiable
fact that farmers prefer to keep their skim milk for young
stock to selling it to the poor—and the new milk is all
sent to the towns, or sold to well-to-do customers. Of
course, skim milk would in no sense take the place
of cod-liver oil, neither does skim milk do for babies
unless enriched by some fattening food ; but most people
are hopelessly ignorant of the value of separated milk;
they think the nourishment of milk is removed with the
cream, whereas all the proteid is in the casein of the
separated milk, there being none whatever in the cream
or butter. For this reason it is obviously useless to
depend for nourishment on cheeses made from cream,
such as Camembert. I know one village in Suffolk
where the proprietress offered the poor the separated milk
of her dairy as a gift. This they refused, as they thought
it quite worthless, and only 'pig's food,' a very different
thing in their minds from food that was good for pigs.

In a village not far from here a friend told me that
she had been helping a man seriously ill of consumption.
After he was removed to a home, the clergyman said the
wife and five young children were very badly off. She,
not sending food from her own house except in cases of
grave illness, offered to give the mother two quarts of new
milk a day for the children. After a few days the mother
came up, and said she did not want the milk at all, as she
had no use for it. It is almost inconceivable that a
mother could so ignorantly refuse what was good for her
children ; it only helped to confirm me in my opinion

that the use of milk and the knowledge of its value are absolutely dying out in English villages.

On my return home after my visit to Italy, I made inquiries as to the most recent book that had been published in England on goat-keeping. This brought to my hand the excellent little book by Mr. Bryan Hook, son of the artist whose beautiful sea-scapes delighted the eyes of my generation on their visits to the Academy. The book is called 'Milch Goats and their Management,' and is published by Vinton & Co., 9 New Bridge Street, E.C., 1896. For anyone interested in the subject of goats, and especially for those meaning to keep them, there could hardly be a better guide than this book, prompted as it was (see preface) by 'a firm conviction of the advantage that might be derived from a wider cultivation of the milch goat in this country, and an affection for the most intelligent, engaging, and picturesque of our domestic cattle.' Mr. Bryan Hook has had so many years' experience in goat-keeping, that his testimony is exceptionally valuable, and intending goat-keepers will do wisely to follow his advice as to choice of breed, feeding, housing, and general management, the only point in which I would beg to differ from him being in the construction of the goat-house. Mr. Hook recommends a raised path down the middle of this, and a gutter on each side for drainage, which is to flow into a pail or tank sunk outside the buildings, and be used for garden manuring. Mr. Gates, the well-known owner of the pure Toggenburg herd of goats near Guildford, suggests that a better way, as an economy of labour and an improvement in hygiene (see results of experiments by Dr. Poore in dry treatment *versus* water-borne sewage, in his invaluable book 'Rural Hygiene'), is that the raised paths and gutters be omitted, and the floor simply strewn with sawdust or dry garden earth. This absorbs all liquids, prevents smells, and is easily swept up each morning when the house is brushed out, and if buried superficially is one of the finest fertilisers. This system has been followed by Mr. Gates, and he recommends it in preference to the water-swilled gutters. Certainly nothing could have been sweeter than his goat-house, which I visited one summer afternoon when it held ten goats. The sparred floor of the stalls is better raised eighteen inches from the floor, and the spars should be placed half an inch apart instead of one inch, as young kids are apt to get their very slender hoofs in between the wood when it begins to soften, and a broken leg may be the result of spars too wide apart, these young creatures being astonishingly nimble even from birth.

Mr. Hook says : 'Goats are probably more subject to rheumatism than most other animals, and I have known them to be so acutely affected when heavy in kid that they were unable to rise, and almost unable to walk when on their legs.' I quote this sentence because it strikes me that goat-keepers may find that this tendency to rheumatism is caused by too high feeding, especially with oats, peas, and beans. Horses in the confinement of the stable often suffer in the same way, and are unable to stand the damp when turned out into the fields, not because that gives them the rheumatism any more than bad weather gives it to us. They have it in them from wrong feeding : the damp merely develops it.

In consequence of this book I bought one goat, a hornless black and white crossbred, so that I and my gardener might gain a little personal experience—I always dispute that 'a little knowledge is a dangerous thing,' a little to my mind being better than none—and find out how much trouble and benefit there is in keeping goats on a very small scale.

My first experiment after the birth of the two kids the

next spring—a boy and a girl—was to give to the children about the place a goat's-milk tea. They all, with one exception, said they liked it very much. The exception was the youngest, almost a baby, who was probably not hungry. After this the goat's milk was divided between the children of the gardener and coachman, and such honoured guests who were fond of milk and yet totally disbelieved that goat's milk was without taste or smell. Without a single exception every one of these pronounced it excellent, and some preferred it to cow's milk as a constant drink. This may be due to the fact that, as Mr. Hook says, the fat globules in goat's milk are so much more minute than in cow's milk that it is lighter on the palate and easier of digestion. Servants—who in my experience are the most conservative-minded of all classes—have now found out the good qualities of goat's milk so markedly that they are glad not to allow it to go out of the house.

I can see the cautious reader—probably a male— saying, 'I wonder why she didn't take it herself?' It was because I did not wish to mix goat's with cow's milk in the separator, and I always drink separated milk; for though not a large milk consumer, I prefer it without its cream, taking that portion of the milk by preference in the form of butter. Those who drink coffee or tea will find that goat's milk gives both these a much better colour in the cup than even the best cow's milk.

My little kids I did not want to keep. I tried to sell them both in the neighbourhood, but no one would give me anything for them, dead or alive, so I had the Billy killed and served for a Saturday to Monday party which I knew had no interest in the simpler foods. Having seen them immensely enjoy it, I called to their notice the *menu*, on which was plainly written 'roast kid,' whereupon a young man at my side told me the following anecdote.

He had lately dined at a very expensive restaurant where
the *menu* contained for the roast, *agneau-de-lait*. He,
having a frugal mind, and thinking of his own lambs in
the country which sold for very little, asked to interview
the chef after the feast. He said, 'Well, sir, I have no
objection to telling you that *agneau-de-lait* is not to be
procured in England, or at any rate is enormously expensive. This dish of which you have been eating is
only a kid.'

I could not bear to kill the useful Nannie, so I gave
her away to the man who had had charge of her mother.
He used her as a pet for his children to drag a tiny cart,
and is now so delighted with the milk our goat gives that
he is going to keep his own for the same purpose. This
year the same mother only had one kid, a very pretty
little female, which I am rearing and keeping. The one
kid instead of the usual two may be due to the mother
having been principally fed on grass. She does not give
nearly as much milk as the better-bred goats, but this may
be partly because she is not so highly fed. She is pastured
in a fresh spot every day and is given leaves, &c. In
winter I find chopped mangold is a favourite food.

Several cases have come to my knowledge of poor
people refusing to keep a goat even when it had been
given them, because they say it is so troublesome and
destructive a creature. My interpretation of this would
be that it is no good giving a goat to poor people, unless
you give with it a strong collar, chain, and tethering pin,
for unless this is done it is always breaking loose and
doing damage. These things are expensive for poor
people to buy—the collar and chain costing 7s. 6d.—they
try to manage with bits of rope, or any other makeshift,
and the result is unsatisfactory.

Mr. Hook says in his book which supplements the
larger work on goats by Mr. Pegler, that this prejudice

against goats is largely due to the fact that English cottage people get their small experience of goats from buying them out of the Irish herds which are brought over now and then. 'Unfortunately, it is from such inferior animals that an estimate of the whole species is generally formed, the over-sanguine goat-owner becoming disappointed by the wild nature, mischievous habits and scanty produce of the animal that he has purchased on the assurance that it will give two quarts of milk a day, and live anywhere and anyhow.'

I was walking one evening on a Surrey common, and saw a little boy just taking a goat into a cottage. My interest being very keen, I began asking him why he kept it. He said, 'For milk.' My next question was, 'What do you do with the kids?' He answered with a grin, 'Our goat don't 'ave no kids: she ain't 'ad none fur seven years.' This surprising answer so floored me that I meekly said, 'I suppose she doesn't give *much* milk?' He said, 'No, not much.' I afterwards heard that the little cottage contained a family of fifteen children and mother and father.

There is no doubt a difficulty in the poor keeping a goat for each family, unless they are near a good common with grazing rights, which exceptional circumstance is practically not worth reckoning. My idea is that as philanthropic people abound everywhere, some of these might be the goat-keepers of the district, and either send the goats, in charge of a boy, round to the cottage doors to be milked there in a small measure, to be charged for at a remunerative price, or have the goats milked at home at an early hour in the morning and evening, so that the children might fetch the milk before and after school-hours. The charitable, on the other hand, who do not fear to give—such as the clergyman, the nurse, the district visitor, &c.—might keep goats for their own use and give away milk for sick children and other urgent cases.

Only time and patience will get rid of the prejudice against goats which exists in all classes, and may be partly due to experiences at Gibraltar, Malta, Corsica, and parts of Switzerland where the goat's milk is almost undrinkable. This unpleasant smell and flavour is caused by wrong feeding, by dirty hands in milking, and by letting the milk stand too long. The delicious *bruccio*, of Corsica, is a fresh curd made from goat's milk. Julius Cæsar openly avowed he preferred the plebeian goat's cheese to the greatest delicacies of the table, and the newspapers say President Loubet does the same.

In the dim distance it does not seem to me impossible that useful foods may be made from sheep's milk, especially while lambs are still killed for meat.

One of the chief recommendations of goat's milk in these days of nervousness about the danger of cow's milk, is that goats are among the few animals entirely exempt from tubercular disease. Sir William Broadbent, writing on the prevention of consumption, says, 'It is interesting to note that asses and goats do not suffer from tuberculosis.' It is a continual surprise to me that goats are not kept to supply the consumptive sanatoriums, and I hope this most important measure may be adopted at the King's new Hospital, for the prejudice of the patients might be met in the same way as the French chef met the demand for *agneau-de-lait* as stated before.

The other day a friend came to see me who had last year been interested in my goat-talk. She told me she had bought a goat for her baby and was going to buy another, as both she and her children liked it so much. She said with pride, on my showing her my comparatively ugly black and white hornless mongrel, that *her* goats were beautiful fawn and white Toggenburgs. I felt humbled, and she said, ' Don't you know the goat-farm at Guildford, kept by Mr. Gates, the head of the West Surrey Dairy Co. ? ' A few days afterwards I and a friend, who was herself anxious to keep goats, started for Guildford. Arrived at the station I suddenly remembered that I wanted to order a book, and trying the penny in the slot post-card box, found to my disgust that the cards were unstamped. I think the sale of these post-cards would greatly increase if, instead of two comparatively useless cards at a penny, one, stamped ready for use, were sold at the same price. The diminution of profit would probably be covered by the additional sale. We walked up to the dairy and felt a little flat when told that Mr. Gates was not there and lived three miles away. The clerk suggested that we might talk to his nephew. This we accordingly did, and heard that Mr. Gates was going to sell the whole herd, as it had been a hobby and he no longer had time to devote to it. It being part of the vegetarian creed to be cheerful under disappointment, we resolved to spend our two hours before the return train in loafing about the picturesque old town.

Our first excitement was finding the Market Place filled with a detachment of Engineers, whose carts looked rather like Chinese junks and whose Boer hats and rough costume, to our imagination, conveyed the impression that they were just back from the front with all the South African dust and sunburn still thick upon them. On inquiry we found they had only returned a short time ago, but they had been on manœuvres in the neighbourhood, and the junks turned out to be pontoons, the dust being good Surrey sand.

Passing beautiful fruit shops, so rare in the villages near me, we bought two very cheap market baskets and proceeded to load them with fruits of all kinds which had suddenly flooded the market from the lateness of the

season, and were plentiful and cheap. Our next search
was for almonds, but walking up the High Street I
was suddenly glued to the window of a curiosity shop
by the sight of a gorgeous blue and green fish, different
from any I had in the china aquarium of fish which swim
over my ugly hall stove on the whitewashed wall at
home. Finding the price moderate, I yielded to the
temptation to make the fish share the basket with my fruit.
We then went to the best looking grocer's shop we could
find, and my friend, who is an almond-fancier, asked if
they had any Jordans. Fortunately for us they had
bought in an extra large supply in the season, and had
plenty of this kind at 2s. per lb., a great improvement upon
the Valencias at 1s. 8d. The disappointment sometimes
caused to housekeepers by receiving bitter almonds, with
the possible result that the nourishing item of a guest's
meal has to be left out, may be guarded against by
ordering ' second quality Jordan almonds,' which are
1s. 10d. per lb. at the Stores. The twopence extra is well
worth while to those who have once appreciated the
difference between them and Valencias, which are the
same in shape and size as the bitter almonds, so that it is
impossible for a cook to know she is preparing uneatable
food if she gets the latter by mistake. It is extravagant
to buy the best Jordans at 3s. per lb., when the second
quality are as good except for size. The shape, a very
long oval, is the same, quite unlike the squat oval of the
Valencia almond.

It is curious to notice that when the derisive enemy
accuses one of living on ' nuts and apples,' he is generally
ignorant that almonds *are* nuts, and far the most nourishing of the whole nut family. Even when people do
realise this, they are filled with dread at being ordered to
consume a vast amount of prussic acid, having vaguely
heard that this poison is extracted from almonds. It

may be worth while here to go back once more to our friend 'Chambers's Encyclopædia,' and quote a few facts on the subject :—

'Bitter almonds contain the same substances (as sweet), and, *in addition*, a substance called amygdalin, from which is obtained a peculiar *volatile* oil. For the preparation of *Fixed* Oil of Almonds either bitter or sweet may be employed. The cake which is left after the expression of the *fixed* oil from bitter almonds, contains among other matters a portion of two substances, called respectively amygdalin and emulsin or synaptase. When the cake is bruised and made into a paste with water, the synaptase acts as a ferment upon the amygdalin, and one atom of the latter resolves itself into two atoms of *volatile oil* of *bitter* almond, one atom hydrocyanic acid (prussic), one atom grape sugar, two atoms formic acid, seven atoms water.'

The volatile oil is not originally present in the bitter almond. The nut does not contain a trace of the oil ready formed, so that the oil is purely the product of the fermentation of amygdalin.

It may be suggested that this change might be brought about by fermentation inside us; but in an interesting paper in the 'Herald of Health' for April 1902, by Dr. E. P. Miller, there is an account of the two so-called ferments, the digestive or inorganic fermentation *versus* organic fermentation, and he says : 'The term ferment is not one that should be applied to the enzymes spoken of as the unorganised ferments that are elaborated within the cells of the glands producing them, for they are not in reality ferments, but simply digestive agents provided to prepare the nutritive constituents of food for absorption and assimilation.' Mrs. Wallace's excellent monthly magazine the 'Herald of Health' is full of information of all kinds on health topics. The last page gives a set

H 2

of useful general rules 'for the physical regeneration of man,' with which I am in great sympathy, a sympathy which I cordially extend to the motto of the paper : 'Life is not mere existence, but the enjoyment of health.'

To go back to my account of the day in Guildford. In the shop where we bought our almonds was a stall presided over by an American girl with a chafing-dish and several varieties of American cereal foods and specimen dishes prepared from them. What attracted me, as I had a vegetarian coming who always asked for farinaceous *bulk* which I avoid when alone, was a bundle marked 7*d*., and called 'Nouilles lactées Suisse,' or Swiss milk vermicelli, which shows its Swiss intelligence by instructing the public as to the percentages of its component parts in an analysis signed by Dr. Bertschinger of Geneva. It is not otherwise than a noticeable sign of the times that in Germany and Switzerland prepared foods have to be analysed and certified by first-rate chemists. I have never come across this with either French or English foods. In the case of England, at any rate, where adulteration is so common, I think all patent medicines and foods should be certified by a Government inspector.

Having filled our market baskets, we found them so heavy that we left them with this young lady till our return, and, with all the joyful feeling of touring in a picturesque foreign town, we walked on to find the shop of the well-known Mr. A. C. Curtis, author of 'A New Trafalgar,' and founder of the Astolat Press. It was a slight shock to our æsthetic sensibilities to find the shop in a chaotic bustle of 'sale,' which we forgave when we learnt that it ensured our finding Mr. Curtis himself on the premises ; and as my friend too was a writer and a bookseller, he welcomed us with all the friendliness of a fellow-craftsman, and fetched out from the back of the shop a case of his dainty little editions. Amongst these was his ' In Memoriam ' on vellum published at 10s. 6d. net, of which he himself lately bought back the few remaining copies he could find among the booksellers at an increased price, as the book is now out of print and scarce.

The present fashion for these miniature libraries—as seen in the success of the ' Temple Classics,' the ' Bibelots,' and the new ' Unit Library,' which brings the great classics of all nations within the reach of English peasants— is very indicative of the stress of the times, which means pocket volumes for the busy workers who would perhaps never read at all but for the snatched intervals between work.

The name ' Astolat Press ' suggested to us on our return to write and ask Mr. Curtis the origin of the assertion that Astolat—the home of Elaine—was the old name of Guildford. This was his reply : ' In the " Morte D'Arthur," book xviii. chap. ix., Sir Thomas Malory says of Sir Lancelot, " And then he rode so much until he came to Astolat, that is Guildford, and there it happed him in the eventide he came to an old baron's place that hight Sir Bernard of Astolat." There is no doubt that Tennyson identified Astolat with Guildford, and used the present ruined keep in his mental pictures. And he might well fancy Elaine watching the ford by St. Catharine's for the flash of the knight's armour as he rode from Winchester up the track we call " The Pilgrim's Way," but which is one of the earliest roadways in England, and existed long before St. Thomas of Canterbury's day. The Astolat Press is quite a small affair and inhabits, in what was once Archbishop Abbott's stable, an Elizabethan red-brick building, with solid walls, oak beams, and square-paned leaded casements. The loft makes a capital compositors' room, and the solid ground floor a good foundation for engines, machines, and hand press.'

After the first day at Guildford we felt rather crest-fallen at having learnt so little that was new about goats. I wrote to Mr. Gates, and he quickly answered that we might come over at once and see his herd. We were received with the greatest kindness and hospitality, and every question was most cordially answered. His goat-house was a picturesque thatched building, the floor strewn with sawdust, the animals being tethered in little stalls on a raised platform of battens half an inch apart, and about two inches wide. Apart from the obvious gain in cleanliness, this raised structure is a great convenience in milking and saves a special milking bench.

The goats were nearly all pure Toggenburgs—beautiful deer-like creatures with fine fawn-and-white coats—and when I asked why they were being sold all together, thinking in my ignorance that it would be better to spread them about, Mr. Gates told me that he was anxious to sell them to someone who would keep the breed pure, as the Swiss Government had now forbidden their exportation. They have been purchased by Mr. R. Sugden, Longden, near Rugeley, Staffs. Mr. Sugden intends to keep the breed pure.

The average yield of these goats in full milk is two to two and a half quarts each daily, the actual quantity given by Mr. Gates' herd of five in one season having been 714 Imperial gallons, or forty-two full-sized railway churns. In addition to his household supply of milk and butter, Mr. Gates has sent out more than 1,000 bottles of milk, sterilised for travelling, besides selling the fresh milk for infants at Guildford. The price he charged—viz., eightpence a quart—is an interesting contrast to the price at the London dairies.

Further details may be gathered from an illustrated

article entitled 'A Dairy Farm in Miniature,' by Mr.
Bryan Hook, in 'Country Life' of April 8, 1899. The
goats were photographed, but the pictures do not do them
justice.

The impression made on one by these goats is that,
compared with other breeds, they are as racehorses to cart-
horses. Probably with increased knowledge and interest
in the subject, the English goat will be improved—it
being a most useful creature for all who cannot afford the
special breed. Mr. Gates' little goat farm had the advantage that it was adjoining a common where the goats
could be turned out in charge of a boy. We also noticed
that those in a field were allowed to run loose, a great
improvement on the tethering system, as these creatures
love change. Mr. Gates told us that this was quite safe
when once they had grown used to a place, for they are
so intelligent and friendly that they attach themselves to
people and places like dogs. It was interesting to us to
notice that, although there were eight milch goats and
two or three kids in the house, the place was as sweet as
a well-kept stable—the he-goats, the only offenders in this
matter, being kept apart in a field.

On the subject of feeding Mr. Gates told us that he
grew lucerne on purpose for them, as they were fond of it
for a change, and it was a most useful fodder, as he cut it
three times over in the summer, and it grows almost as
fast as it is cut.

I have found that other useful foods are comfrey, sunflowers, summer prunings of apple and pear trees, hedgerow cuttings, sweet chestnut leaves, and the leaves of
the globe artichoke. In Italy cows are fed on artichoke
leaves, but I cannot persuade my pampered Jerseys to
eat them.

According to many people, for the last two years I
have had goats on the brain, which is only a variety of

the more usual accusation that for many years, alas! for them, I have had diet on the brain. These accusations at first really distressed me, as no one feels cheerful under the implied supposition that senile decay is coming upon one with rapid strides. The fact is that goat-keeping is merely a variation of my interest in diet and the improvement I hope it is to bring about in the health of the modern world. It has been an immense gratification to me to see that there has been such a very general growth of interest in this subject during the last few years. We see it affecting all classes from the highest to the lowest— our statesmen, our clergy, our men of science, almost the entire Press, and last, but not least, the King himself. The cloud of ignorance about food shows signs of breaking up and dispersing. How complete it can be, even among intelligent well-educated people, was illustrated to me the other day when talking about food values over a tea-table with a trained village nurse and a friend much interested in the subject. Neither of them seemed to understand what I meant, and one of them suggested, ' Surely the most nourishing food is that which digests most easily.' I answered, ' You may easily digest fruit and vegetables, but the actual food value of what you have digested is very small indeed.' In fact, they were entirely unaware that there was such a thing as chemical analysis of food, and a scientific knowledge of the subject which they might have mastered with half the labour and time spent over complicated crochet stitches and lovely drawn-thread needlework. A friend of mine who has lately had a serious illness, told me that neither of the first-rate trained nurses who attended her had ever heard of such a thing as being able to calculate the number of grains of proteid necessary for the day's nourishment, or to decide at a glance, at a well-spread table, what food on it is the best to choose.

Nothing has encouraged me more than finding this summer that the matter has been taken up in the highest quarters, and has been introduced into the Revised Instructions for the Public Elementary Schools (1902). As few people take the trouble to get Blue-books, I venture to quote the following paragraphs from Appendix V. on Cookery:—

'**V.—The dietary value of the food** and cost of the materials should be taught at each lesson, if only one course of Cookery lessons is being given. When the arrangement is that the girls attend Cookery classes for two or more successive years, the dietary value of food should not be taught till the second year.

'**Second year.**—Instructions should be given on the various food stuffs, *i.e.*—cereals, pulse, fruits, vegetables, meats, and fish; beverages. The dietary value of food. Digestion of albumen, starch, fat. More advanced dishes should be demonstrated and practised at each lesson, illustrating over again the Primary Methods taught in the First Year course.

'**Third year.**—Complete dinners should be cooked by groups of children attending the class. The price of the dinner and the number of persons for whom it is intended should be written on the blackboard. Instructions should be given on:—

(*a*) Expenditure of wages on food.
(*b*) The making of preserves.
(*c*) Use and abuse of tin foods.
(*d*) Vegetarian diet.
(*e*) Preparation of food suitable for infants.

'The scholars should have practice in drawing up *menus* of dinners suitable for an artisan family, stating the price and season of the year.'

This will instruct in the Elementary Schools, but I

hope the time is not distant when no educated child will sit down to any table whereon food is displayed without a perfect knowledge of the simple rules as to the nourishing values and right combination of food, and when a young mother will no more dream of asking her nurse whether she shall give her rather delicate offspring fruit or no fruit, than whether it shall go naked or clothed.

WHOLESOME FOOD ON THREE SHILLINGS A WEEK

'Cornhill' budgets—Food reformers and lentils—Taste for savoury foods—Nervous appetites—Cabinet Minister and charwoman— The healthy foods—Maeterlinck's appeal against meat and alcohol—Food values—To feed a family of four on 12s. a week—Nut milk—A week's *menus*, and cost—Ditto, with once-a-week cooking—Advantage of living in country—Goat's milk at a London dairy—Cheapest and healthiest diet at 2s. 4d. a week—To wean servants from the beef-beer-tea faith—Possible purpose of meat-eating phase in evolution—A philanthropist's experiment—Amateur farmers—A pair of Bushey Art Students—Receipts.

In 1901 the Editor of the 'Cornhill Magazine' published a series of five articles called 'Family Budgets,' beginning in April with that of the workman at 30s. a week, by Mr. Arthur Morrison, and ending in August with 10,000l. a year, by Lady Agnew. Between these came, in May, a lower middle-class budget of 150l. to 200l. a year, in June an income of 800l. a year, and in July my own, which I now republish, on 1,800l. a year. The reviewers found great amusement in the idea that there could be the smallest difficulty in living on such an income, but the proper adjustment of medium or large incomes is often a more complicated matter than the management of one which provides only for the necessities of life. A friend of mine whose inheritance of 2,000l. a year was stated in the newspapers, received a letter asking for a considerable sum out of it, on the plea that the loss of it could not be felt out of her 'boundless wealth.'

The ' Workman's Budget ' is, I think, the most interesting one of the series, as it deals with the hardships of town life on an income which is ten shillings a week higher than the usual bare subsistence of a pound a week.

My friend, Miss Curtis, has kindly sent me the following suggestions on wholesome food for the poor and the depopulation of rural districts—huge difficulties, which have been dealt with in a most thoughtful and stimulating way in a book of collected essays by various writers, and published by Fisher Unwin in 1901, under the title of ' The Heart of the Empire.'

' The question of wholesome food for the poor is not in itself a difficult one : the real obstacle lies in the prejudice against a non-meat diet, which is often to be traced to the want of knowledge and sympathetic understanding of the tastes of the poor, in those who champion the economic way of living. Social reformers urge lentils as the article of food which gives the maximum of nourishment at the minimum of cost ; but, apart from the little-known fact that the xanthins of lentils and all pulse foods are now suspected by experts to be as unwholesome as those of flesh foods, and therefore to be ruled out from the dietary of all who wish to control as far as possible the causes of disease, we have to face the fact of the people's dislike to all *porridgy* foods. This dislike cannot be lightly dismissed as a fad : the plain truth is that the appetites of the people are indicative of their constitutions, and these have changed during the last century of meat and tea diet.

' Imagine the under-nourished, over-worked mother of a family, after a hot day's washing or charing, sitting down with appetite to a mess of lentils. One might as well expect a delicate, hard-worked Cabinet Minister to enjoy a summer luncheon of boiled bacon and beans. The Cabinet Minister chooses lobster salad and a whisky-and-soda: the charwoman chooses tinned salmon and tea— if she can get it.

'The poor like fried food—a bit of fish ready cooked from the shop, a rasher of bacon, a pig's fry, anything crisp and savoury—and failing this, they like the tinned stuffs which give an excuse for just the piquant dash of vinegar or pickles which their jaded appetites require.

'Perverted tastes! says the reformer. Yes, but why are they perverted? Surely the whole conditions of life in our big towns are perverted from the way of health, and it is unreasonable to expect unhealthy men and women to have healthy appetites.

'The charwoman's husband, if he has the luck to earn his living by outdoor labour, may have a lentil-hunger; but how can she be ready for stodgy food after a day, or rather a lifetime, spent in wrestling with dirt in a stuffy house set in acres of stuffy streets?

'Before we can expect people to eat lentils and beans we must see that they live under conditions which produce a healthy hunger, and towards this the Garden City Association, 77 Chancery Lane, W.C., is working in a way that deserves the support of all who wish to improve the present state of our towns and cities.

'I have lately been asked to advise the best dietary for a family of four, consisting of father, mother, and two young children, in circumstances that allow of only 3s. per head a week for food, and since this is a case which may easily come into the experience of any who work amongst the poor, I include it here on the chance of being able to offer some useful hint. By "best" I understand "healthiest," so I exclude all such articles of food as are shown by Dr. Haig's researches either to produce uric-acid diseases, or to aggravate a previously existing tendency towards them. This cuts off all flesh foods, together with eggs, the pulses (peas, beans, lentils), tea, coffee, cocoa, and alcohol : and leaves milk, cheese, nuts, cereals, vegetables, and fruit, to which may be added Plasmon, which is milk in a dry, concentrated form, requiring discretion in its use, for if taken indiscriminately it may, like all highly nitrogenous (albuminous) foods, cause indigestion. A propos of the many objections from intelligent people to its being an "artificial food," it may be stated that Plasmon is no more artificial than the strong stock which a cook prepares by allowing the water to evaporate. All cooking is artificial to primitive man, and Plasmon is milk so cooked that the water evaporates and the condensed nutriment remains. Being made from skim milk, which explains its extreme cheapness, it is deficient in the fat and sugar present in new milk, and for this reason Plasmon should not be substituted for new milk in feeding children. All young animals require fat and sugar, and a calf brought up on skim milk may be big and bony, but is never so well-favoured and thriving in condition as one which has been even partly fed on whole milk. Moreover, fat, so necessary to good digestion, is very evenly distributed in new milk, and for this reason alone it is, as a rule, unwise to upset nature's balance when feeding children. To adults, invalids, travellers, and athletes, Plasmon is an immense boon, and it should also do much to solve the problem of right feeding for the many whose incomes are insufficient for their needs.

'Here, for a moment, I would beg leave to refer to Maeterlinck's recent splendid appeal against meat and alcohol. In it he says, "A little fruit, or milk, a few vegetables, farinaceous substances . . . . are amply sufficient to maintain the ardour of the finest and mightiest life."

' If anyone ignorant of food values should try to reform
his diet by this statement, it is probable that he would
come to grief, although a student of dietetics would avoid
disaster by substituting *and* for *or* in the sentence above,
and by taking all four kinds of food to make up his
dietary. Fruit can hardly be made an alternative to
milk, the respective albumen values being too disproportionate. One pint of milk = 262 grains albumen or
proteid; but it would take over 2 lbs. of fruit to get this
amount of nourishment, and so great a bulk of watery
food would be likely to upset digestion somewhat
seriously. If dried fruits such as figs, dates, raisins, and
French plums were used, a much smaller quantity—
$\frac{1}{2}$ lb.—would give the necessary albumens; but most
people, when told to eat fruit, think of the fresh kinds,
which are more tempting to sight, touch, taste, and smell.
Only if " fruit " were read in its widest sense to include
nuts, could it be fairly substituted for milk, and very few
general readers would be likely to remember this. Milk
itself is rather too bulky a food to depend on for the day's
proteids, and should be supplemented by a dry form such
as cheese or Plasmon, or by nuts—the following list
showing roughly the order in which the simpler foods
take rank as compared with meat, and with each
other :—

$1\frac{1}{2}$ oz. meat = 1 oz. cheese = $1\frac{1}{2}$ oz. almonds = $2\frac{1}{2}$ ozs.
                           oatmeal.
   ,,   ,, = 4 ozs. bread = 4 ozs. dried fruit = $\frac{1}{2}$ pint
                           milk.
   ,,   ,, = 17 ozs. vegetables or fresh fruit = 140 grs.
                           proteid.

' To return to our family of four. If it is obliged to
live in London, milk at 4*d.* a quart is too costly an item
to allow of the parents drinking it, and even if an effort

were made to give the children ½ pint each every day, our 12s. would be seriously diminished.

'A cheaper drink, and one which Professor König says contains 268 units per lb. more nourishment than fresh milk, is the Nut Butter, sold in 1 lb. tins at 8d. by the International Health Association of Manchester, which supplies all the nut and cereal foods invented by the Battle Creek Sanatorium group of American food re-formers, headed by Dr. Kellogg. This nut butter is to be had from any vegetarian stores, and can be mixed with water to the consistency of cream or milk as desired. It is not at all greasy, and can be used as a substitute for milk in soups, puddings, &c. Children would prefer it sweetened. It has a roasted pea-nut colour and flavour, and next to Plasmon and Protene is the most highly nourishing drink I know, containing a greater proteid percentage than milk or raisin tea. Containing as it does a good proportion of nut fat in the finest possible emul-sion, this nut milk would obviously be far more economical as a food-drink than cow's milk, and for half the cost—viz., 8d. a week—the children could be given twice the amount, or 1 pint each a day instead of ½ pint.

'For breakfast, then, oatmeal or barley porridge eaten with bread and marmalade or treacle, the children being trained to drink their nut milk in *sips between* the dry food, instead of washing it down half chewed, which is the usual result of the common habit of pouring milk over porridge—a bad plan which effectually prevents the cereal from being mixed with the saliva necessary to its digestion. If the adults cannot at first manage the porridge, fried bread and potatoes, or rice rissoles with a little Plasmon powder mixed in them, will make a savoury and sustaining meal, and if well masticated with bread there would be no craving for drink till an hour or two later, when water, hot or cold, is all that is necessary for health; but as beginners often sadly miss the pick-me-up of the hot tea at breakfast, the following week's *menus* allow for Plasmon, hot bran tea, barley water, or any similar drink. To those who can afford it, Mellin's Food is an excellent substitute for tea, coffee, and cocoa, if made with a good deal of water, and not much milk. It is a malted food, and a great help to weak digestions.

'For dinner, cheese eaten plain with bread, vegetable, and salad, or grated and mixed with some cereal like macaroni, rice, hominy, or ground maize.

'For supper, bread with dried and fresh fruits, either plain or cooked into one of the many forms of pudding.

'If the fruit supper be not liked, as is possible enough in the early stage of the diet (except by children who, if healthy, always enjoy it), some variety of vegetable with cheese or Plasmon may take its place. The week's *menus* here given have been found successful, and the cost is not above 12s. a week for quantities enough for two adults and two children, the albumen values being calculated at 1,400 grains a day for the man at hard work ratio, 1,300 grains a day for each child at growing ratio, and 1,200 grains a day for the woman, supposing her to be thirty-five years old.

'The receipts for these and other dishes will be found at the end for convenient reference in cooking. They admit of endless modification and enrichment with cream, butter, &c., but are given in a cheap form as the most useful.

SUNDAY.

Breakfast, 8 A.M. Fried polenta. Bread. Jam. Plasmon drink. Milk.

Dinner, 1 P.M. Baked potatoes. Sage and onions. Apple sauce. Bread. Date pudding.

Supper, 6 P.M. Cheese. Bread. Lettuce salad.

MONDAY.

Breakfast, 8 A.M. Fried bread. Oatcake. Treacle. Plasmon drink or barley water. Milk.

Dinner, 1 P.M. Savoury polenta and grated cheese. Cabbage. Bread.

Supper, 6 P.M. Barley and raisin pudding. Grated nuts. Bread. Bananas.

TUESDAY.

Breakfast, 8 A.M. Oatmeal porridge. Milk. Bread. Jam. Plasmon drink or bran tea.

Dinner, 1 P.M. Rice croquettes. Bread. Potatoes.

Supper, 6 P.M. Cheese. Bread. Radishes. Apple dumplings.

WEDNESDAY.

Breakfast, 8 A.M. Fried hominy. Milk. Bread. Treacle. Plasmon drink. .

Dinner, 1 P.M. Vegetable stew with barley. Bread.

Supper, 6 P.M. Cheese. Oatcake. Celery. Dates. Bread.

THURSDAY.

Breakfast, 8 A.M. Oatmeal porridge. Bread. Jam. Plasmon drink. Milk.

Dinner, 1 P.M. Macaroni à la tripe. Cheese. Stewed tomatoes. Bread.

Supper, 6 P.M. Grated nuts. Bananas. Bread.

FRIDAY.

Breakfast, 8 A.M. Fried potatoes and onions. Milk. Bread. Treacle. Oatmeal drink.

Dinner, 1 P.M. Savoury nut cutlets (or stewed chestnuts). Bread. Potatoes. Celery.

Supper, 6 P.M. Cheese. Beetroot. Bread. Baked apple pudding (or jam roll).

SATURDAY.

Breakfast, 8 A.M. Barley porridge. Bread. Oatcake. Jam. Milk. Plasmon drink.

Dinner, 1 P.M. Macaroni and cheese patties. Braised onions. Potatoes or carrots. Bread.

Supper, 6 P.M. Grated nuts. Dates. Bread. Currant dumplings.

Cereals	.	1 lb. daily per head, 11*d*. a week.			
Dried fruit	2 ozs.	,,	,,	3*d*.	,,
Cheese	. 2 ozs.	,,	,,	4*d*.	,,
Nuts	. 2 ozs.	,,	,,	4*d*.	,,
Vegetables and fruit	.	.	.	6*d*.	,,
Jam or treacle	.	.	.	2*d*.	,,

$$2s.\ 6d.$$
$$4$$

$$10s.\ 0d.$$

½-packet Plasmon . . . 8*d*.
Nut milk and nut butter . 1*s*. 4*d*.

$$12s.\ 0d.$$

' The prices given are not such as the West-end house-keeper is accustomed to, but happily, as stated in Mr. Morrison's article in the "Cornhill" for April 1901, the truth of which can be proved by a walk through any of the Saturday night slum-markets, the poor have facilities for buying cheaper than is dreamed of by those of comfortable income, and where the housewife who prides herself on catering for her family at 10*s*. a head, pays 1½*d*. each for bananas, the slum-sister gets them at three and sometimes even six a penny.

' For the town-dweller in winter, chestnuts are not dear if bought by the stone or half stone at Co-operative

Stores, for although by the pound they are a somewhat expensive food, the good ones rarely being under 4*d.* and often 6*d.* per lb., in quantities of 14 lbs., the finest dark-skinned Italian chestnuts are sold at the rate of 2*d.* a lb., and there is no more delicious and nourishing dish than stewed chestnuts. Children like it as kittens like cream.

'The amount of cooking may be urged as a drawback to such a diet, for it presupposes a wife at home all day, and anxious enough to make the meals attractive not to mind the extra work, though this might be reduced by making two meals a day instead of three ; for instance, at 7 A.M. and 2 P.M., or at 8 and 3, 9 and 4, 10 and 5, 11 and 6, or 12 and 7, according as is most convenient. This scheme is being tried by all classes in America with singular success. But there are many women who have to be out all day, and cannot therefore attend to cooking. For such, another dietary of wholesome and nourishing food is given for the same cost, and which will only involve a once-a-week cooking—*i.e.,* of bread-stuffs, including cakes, biscuits, and tarts to make a pleasant variety. Potatoes, too, might be boiled or baked in skins and used for salads as required, if kept in a cool place—outside on a window-sill, for instance, provided it did not overhang a drain !

'Home-made bread a week old is such sweet eating that those who have to study their digestion take care to have no other. The success of the diet at this cost depends upon the house-mother making the bread, for it is half the cost and four times the nourishment of baker's bread, and the cheaper the flour she buys the better it is, the expensive "pastry whites" and "Vienna" flours containing more starch and less proteid than the yellowish house-hold "seconds." For frying purposes the cotton-seed oil used by fried fish shops is the cheapest, and a quart once bought will last for months, being used over and over again. This has not been estimated for in the list, because its cost being spread over so long a time is fractional per week, and the original outlay may be regarded by the housewife as part of her cooking " plant," like frying-pan or kettle. The nut butter in the week's expenses refers to cocoanut butter, which is a pure and excellent substitute for butter, lard, or dripping for pastry, biscuits, cakes, &c. The nut butter from which the nut milk is made will not do for these purposes or for frying, it being pulse-like rather than fatty.

'The Rippingille oil stoves are admirable for bread-making, as for all other cooking operations, and are cheaper than coal.

SUNDAY.

Breakfast. Bread. Barley flour biscuits. Jam. Plasmon drink. Milk.

Dinner. Bread. Cheese. Celery. Ginger cakes.

Supper. Bread. Dates. Grated nuts. Oranges.

MONDAY.

Breakfast. Bread. Oatcake. Treacle. Plasmon drink. Milk.

Dinner. Bread. Cheese. Watercress. Jam tarts.

Supper. Bread. Bananas. Nuts.

TUESDAY.

Breakfast. Bread. Maize biscuits. Marmalade. Plasmon drink. Milk.

Dinner. Bread. Cheese. Lettuce. Parkyn.

Supper. Bread. Apples. Nuts. Figs.

WEDNESDAY.

Breakfast. Bread. Wheatcake. Jam. Plasmon drink. Milk.

Dinner. Bread. Cheese. Potato salad or watercress. Currant short cakes.

Supper. Bread. Nuts. Oranges. Dates.

THURSDAY.

Breakfast. Bread. Barley wafers. Treacle. Plasmon drink. Milk.

Dinner. Bread. Cheese. Onions. Marmalade tarts.

Supper. Bread. Nuts. Figs. Apples.

FRIDAY.

Breakfast. Bread. Oatcake. Marmalade. Plasmon drink. Milk.

Dinner. Bread. Cheese. Radishes or beetroot. Ginger biscuits.

Supper. Bread. Nuts. Bananas. Raisins.

SATURDAY.

Breakfast. Bread. Maize biscuits. Jam. Plasmon drink. Milk.

Dinner. Bread. Cheese. Potato salad. Currant loaf.

Supper. Bread. Nuts. Dates. Apples.

' The following is an excellent diet for dyspeptics who wish to reduce cooking.

Breakfast, 10 A.M. 4 ozs. bread.

8 ozs. potato.

4 ozs. curd cheese with green salad.

½-pint milk.

Dinner, 5 P.M. 4 ozs. bread.

8 ozs. potato.

16 ozs. fresh fruit.

½-pint of milk.

Oil, butter, cream, ad lib.

' This gives a total of 1,208 grains proteid, which is the physiological allowance for a man or woman of 9 stone 9 lbs. leading an active working life.

' The details are as follows :—

4 ozs. curd cheese	= 400	grains proteid or albumen.
8 ozs. bread or cereals	= 272	,,	,,
16 ozs. potatoes . .	= 137	,,	,,
16 ozs. fruit . . .	= 137	,,	,,
1 pint of milk . .	= 262	,,	,,

$$1,208$$

' The cost of this diet to anyone living in the country with a couple of goats, a patch of vegetable garden, some fruit trees, and an acre of arable land, would be literally nothing but the labour of working the land and caring for the goats. Honey, too, could be added, and with due exercise of forethought in bottling, drying and storing surplus fruits and vegetables, the supply during the winter months could be secured.

' To those fortunate enough to be able to get it, goat's milk is highly to be recommended, for the minuteness of the fat globules makes it easy of digestion, and being thinner to the palate, it is appreciated by those who dislike the *fulness* of unseparated cow's milk. If the demand in London were at all in proportion to the worth of the article, the present ridiculous price charged for it would fall to something within reason. Last summer I went into one of the principal branch offices of a well-known London dairy and asked the price of goat's milk. I was told it was 4s. a quart, and in reply to my involuntary exclamation at such an exorbitant charge, the amiable young woman in charge said, " But think how many lives it has saved, Madam! " It did not seem to occur to her, or to her employers, that this was the very

reason why it should be within the reach of all who
need it.

'Cheapest and healthiest of all the forms of diet I have
experimented upon is that of bread and fruit. With
bread, dates, and apples, it is easy to live on 4*d*. a day,
and get the full proteid ration for active working life for
a body weight of 9 stone 4 lbs. A big man might have
to spend 6*d*. Life on this diet is easier and happier than
on the régime of meat, fish, game, wine, tea, &c., which
costs eight times as much, for there are no digestive
troubles on bread and fruit, unless one makes the mistake
of taking them in a wrong proportion, or of eating stupid
kinds of bread and unripe or overripe fruit; and when
one thinks how digestion controls such things as depression, headache, irritability, nervousness, exhaustion,
and other minor ills which go to make some lives almost
unbearable, one cannot help wishing that all suffering
people could be persuaded to give fair trial to a diet of
the simplest foods.

'Unhappily, for the physical well-being of the next
generation, the countryman who now possesses a large
well-stocked garden, and, as in many of the villages in
the West and South of England, often an orchard and bit
of pasture as well, has so little knowledge of how to make
the best use of his belongings that he sells his produce in
order to buy unwholesome food such as tea, bacon, beer,
butcher's meat, and tinned abominations like lobster,
salmon, sardines, and potted meats. There is little
reason to suppose that the parents of to-day will change
their habits of diet, but much may be hoped for in the
way of a more intelligent and less wasteful order of things
in the next generation, now that the beginnings of the
chemistry and economics of food have been introduced
into the School Board curriculum. And if, meanwhile,
the educated and leisured classes would take the trouble

to look carefully into these matters, and begin to practise a more enlightened system of dietetics, such as can be begun without any household upset of drastic change, the servant class would at least see that masters and mistresses no longer believe that life depends on eating meat three times a day.

' A great opportunity in this direction might be seized with advantage by the wives of country clergy. The wife of the present Bishop of Japan, when living at Andover, used to have a group of young village women to " high tea," or early supper at the Vicarage every Sunday evening, the household servants being given a holiday, and the meal being prepared and shared by the family and guests.

' A similar plan could be made the occasion for introducing the simpler foods in many an attractive guise, and if the supper were given on a week day, it could be arranged to follow an informal cookery demonstration. Servants, like children, are very imitative, and will take to a new idea much more quickly if it is not forced upon them. One mother of a young family lately lit upon a happy plan in this connection. Her husband took to the simpler diet for health's sake, and she tried it from curiosity. The children soon began to beg for the food they saw their parents eating, and at last, as a great treat, they were allowed a non-meat dinner twice a week, with the result that before long they chose no meat at all.

' This surely is a good way to treat the subject with regard to servants. Let them see that you are well and able to work without meat and stimulants, and they will gradually lose their faith in beef, beer, and tea—a faith for which the example of the upper classes is entirely responsible, and which the bulk of the medical profession still supports.

' It has taken hundreds of years of bitter experience of

disease to awaken even a small percentage of the race to seek more wholesome food, and it will take centuries yet to convince the majority. Meanwhile there are interesting signs that the course of evolution for humanity leads through a meat-eating stage, and although one may gladly endeavour to save suffering to those who are ready for the change, one can possess one's soul in patience with regard to the world at large. Our very blunders towards the truth have their place and purpose in fitting us to appreciate the truth when we find it. The age of excessive meat-eating is helping to produce a highly nervous race, which is apparently the material required for the next stage of evolution—the stage in which the psychic force is to dominate the physical.

'Only by the apparent perishing of one order can another and higher order be born. Perhaps when physical suffering and disease have reached their limit, we shall be ready to receive and to obey the laws of a saner and loftier way of life. When a man like Virchow says, "The future is with the vegetarians," one realises that with all their errors, and they are legion, they are the pioneers whose blundering efforts may be compared to the old shoes, and empty tins, and broken shards which go to make the foundation of the new road along which the whole race will some day travel.

'To try forcibly to evade the intermediate phases, heart-rending as some of these must be, is as foolish as any other premature interference with the natural laws of social growth. For instance, I heard lately of a benevolent landowner who, distressed by the overcrowded condition of the slums, and the dearth of people in the country, built a model village with delightful cottages and gardens, recreation hall and library, church, club-room, inn, baths, and everything else that could be desired for health and happiness. He then transplanted

slum-dwellers who were known to be in actual desti-
tution — literally starving — and provided them with
varieties of wage-earning occupation under the most
healthy and liberal regulations. At the end of a year his
village was deserted; all the starving slummers had gone
back, of their deliberate choice, to the misery whence he
had taken them. Such facts would at first sight appear
to be baffling. They do but illustrate the old adage,
" You may take a horse to the water, but you can't make
him drink." Those slum-people had not yet had their
fill of slum experience, and were by no means thirsty for
the refreshing life of the country. Wretched as they
were, they had not reached the limit of their capacity for
enduring squalor and getting some mysterious good out
of it. Their turning-point had not yet come. Their
evolution had been forcibly interrupted, and their blind
instinct in returning to take it up where it had been
broken off was a sound one, strange as it may look on
the surface.

‘ A friend, who for years had had slum-dwellers driven
down in brake-loads of fifty for summer days in her fields
and gardens, told me once that her heart had never ached
for them with quite such intolerable bitterness since she
had heard from their spiritual pastor, that as they drew
up one evening at the entrance to their alley after one of
these outings, one of them had said, as if voicing the
general sentiment, " The country's fine for a 'oliday, mates,
but, arter all, this smells like 'ome ! "

‘ There are many athirst for a country life, but they are
not necessarily dwellers in the slums, though they may
be found there, as in any other quarter of our crowded
towns. Perhaps we should waste less time in futile
regret over the depopulation of our villages if we could
look upon our cities as the great mills of evolution into
which the slow, massive, yokel strength has inevitably to

be drawn and ground "exceeding small," even to the dust of physical wreckage, that the nervous matter of the brain may be developed and exercised at all costs. It is those who have gone *through* the mill who are ready for the country: whose nervous systems have been developed to the point of exhaustion by the strain and grind of city life for, it may be, two or three generations, and whose brains are as restless and alert as the ploughboy's are dull and apathetic, whose nerves need repose as much as his need tension. The weary governess, the neurotic dressmaker, the dyspeptic bank-clerk, the anæmic student, the worn-out mother, these are they who crave for the country as prisoners crave for air and light, and these are the types which, in my experience, are to some extent counterbalancing the current that sets from village to town, for these are among the applicants for 'small holdings,' in the hope—sometimes forlorn enough—of making a livelihood out of market gardening, chicken, bee, flower, and fruit farming. Many are the mistakes, and grievous the disappointments and even failures which they must suffer; for as a rule they have little capital, less health, and no experience, and yet, in spite of all, so intense and deep-seated is their instinct for country life that they often manage to struggle through the first few years of hardship and make the modest living they desire.

'To those whose experience has brought them into more or less intimate touch with widely different classes of the community, there are not a few indications that the farming industries of England are being recruited from social levels entirely unlike those of old days.

'It used to be almost a joke among the Bushey art-students that if anyone married before his profession could support him he turned cottar-farmer; and certainly some of them succeeded as such in the face of overwhelming difficulties, by the very simple and sensible device of throwing conventional ideas of fitness to the winds, and doing in England, without false shame, the work they would have done as a matter of course if they had gone " out West," or to the Colonies, to "take up land." These men and women, in most cases of delicate constitution and highly nervous temperament, have pitted themselves against conditions which to the labouring and artisan class would have seemed hopeless, and by sheer force of the cultivated intelligence that comes of good birth and breeding have won against long odds. One couple who started with 100*l.* capital and *no* income, on a tumble-down little farm of twenty acres, of which ten were so "foul" that no farmer in his senses would have taken the place—the docks and thistles having to be scythed down before the horse could be coaxed to pull the plough through them!—have done so well that they have now moved on to a sixty-acre farm in the next county which has been under intensive culture for many years.

' The full story of their experiences—some of them as comic as others were tragic—I hope to tell elsewhere another day.'

RECEIPTS

' **Polenta Cutlets.**—Stir one pound Indian maize meal into slightly salted boiling water, adding two Spanish onions chopped very fine, an ounce of butter or oil, three ounces grated cheese, a little pepper, and herbs, if liked. Mix well, and put into a double pan and cook for half an hour, when it should be stiff. Turn out into a dish and press it into convenient shape for cutting up when cold into fingers, which dip in milk and breadcrumbs, and fry brown. Can be served with sauce if desired.

Unfermented Bread.—To every pound of flour allow half-pint milk and water (quarter-pint each). Mix lightly as for pastry, no kneading being required. Form into small rolls or fingers and bake in moderate oven on a pastry tray.

If wheatmeal (*i.e.*, the flour of the entire wheat grain) is used, care should be taken to have it very finely ground, as bran flakes are irritant and relaxing and very unsafe in some forms of digestive trouble, though equally good and useful in others. The indiscriminate recommendation of cracked wheat porridge and whole-meal bread is hardly wise in these days of gastro-intestinal delicacy, as seen in the many cases of appendicitis in adults and colitis in children. House-mothers should know when to give the right food to those in their charge.

It is a good plan to keep in the house several different grains, such as wheat, barley, maize, oats, rice, in two forms—*i.e.*, in the whole grain and as flour—and to make bread or biscuits of both mixed. For instance, pearled wheat, or the grain stripped of its outer bran, should be cooked slowly for three or four hours in a double pan with water enough to swell it without breaking the grain into a mash; this cooked wheat is then mixed with ordinary wheat flour to the consistency of bread or pastry, and is shortened if liked with cocoanut butter, or ground nuts, and left plain, or sweetened and flavoured with dried fruit, spices, &c., and rolled out and cut for baking into biscuits or finger rolls. The same method can be followed with all the other cereals except maize which in the whole grain is too hard in this country. Maize meal mixed with wheat flour is very good in biscuit form, but is best cooked first, as if for polenta without the savoury seasoning. All these whole and ground cereals can be got from Bax & Sons, millers, 35 Bishopsgate Street Without, E.C. The whole groats (the entire oat grain) are particularly good for porridge, but should, like all grains, be cooked several hours in a Gourmet boiler, or duplex boilerette, or other form of double saucepan which prevents burning. The duplex boilerette is made by Mr. Wellbank, Duplex Boilerette Works, Banbury, Oxon.

Date Pudding (cheap).—Wash dates quickly in hot water, dry, stone, and chop them, mix with double their weight of breadcrumbs, and a little sugar. Add skim milk or weak Plasmon solution till of pudding consistency. Steam two hours in buttered basin. Ground nuts and grated lemon peel may be added, if liked, or raw coarse oatmeal, and soaked tapioca, with a little flour, may be used instead of the breadcrumbs.

Macaroni and Cheese Patties.—Take one ounce macaroni well boiled, cut very small, and add one large tablespoonful of grated cheese and the same of cream or nut butter. Season with salt, pepper, and mustard, if liked. Make some short paste, roll out thin and line patty pans. Fill with the mixture and cover with paste. Bake a light brown. Instead of patty pan, the paste can be doubled over the mixture and fried as fritters if preferred. Butter and milk can be substituted for cream, and it is quite good without either, but in that case the macaroni should be rather moist and the cheese fresh and soft.

Macaroni à la Tripe.—Boil some macaroni till soft. Drain and put aside. Cut in fine rings as many Spanish onions as will equal the macaroni in bulk. Fry in butter till quite tender, but not coloured. Remove from pan, and make a sauce by adding flour, milk, pepper, salt, a dash of nutmeg, and a torn bay-leaf. Replace onion rings in this, and simmer gently for twenty minutes ; then remove the bay-leaf and add the macaroni, and heat through. Serve with grated Gruyère or other cheese. Chopped parsley may be substituted for the bay-leaf.

Parkyn.—One pound sifted oatmeal, one pound

treacle, one pound coarse, brown sugar; quarter-pound butter. Ground ginger to flavour. Mix, and bake in very slow oven in flat cakes the size of a saucer.

Barley Water.—Four ounces pearl barley, two quarts water. Thoroughly wash the barley, add the water and boil till reduced to one quart. Strain through hair-sieve (or muslin) previously scalded, and press through some of the barley to thicken. Time to reduce, three to four hours. Can be flavoured with lemon peel and juice, sugar or honey, apple peel and pips, or any fruit juice. Stewed rhubarb juice is good.

Savoury Vegetable Stew with Barley.—Chop up carrots, turnips, onions, potatoes, celery, tomatoes, and fry in butter. Add pearl barley and plenty of cold water, with seasoning of parsley, thyme, bay-leaf, mace, pepper, and salt. Stew in double saucepan till the barley is swelled and thoroughly tender. Serve with fried dice of bread.

Fried Hominy.—Cook hominy in the usual way (soaking overnight in cold water, and boiling like porridge till soft) and put aside to cool. Then cut in any shape preferred, fry brown in oil or cocoanut butter, dust with sugar or salt, and serve.

Stewed Chestnuts.—Slit the skins of the chestnuts, and put them into cold water. Bring to the boil, keeping lid tight, and cook about ten minutes. Lift out a few at a time and remove both outer and inner skins. When all are done, put them into a clean saucepan and cover with milk or milk and water. Stew slowly till the nuts break and are coated in a smooth creamy sauce. Time, one to two hours. An old saucepan used for nothing else should be kept for blanching them, as chestnut skins discolour it badly.

Apple Dumplings without Suet.—Soaked tapioca and butter make a good substitute for suet in boiled pudding crust. Pare and core a large apple for each dumpling required. Fill the centre with a clove, a little sugar and a bit of butter. Cover with paste, tie in a floured cloth, and boil thirty to forty minutes. For the crust use as much soaked tapioca and butter in equal parts as you would have used of suet.

Stewed Tomatoes.—Skin one pound tomatoes by blanching in boiling water for five minutes; then put them into a stewpan (earthenware for choice) with two lumps of sugar, a bit of butter the size of a walnut, and a little salt. No water. Stew them in their own juice gently for twenty minutes.

Baked Apple Pudding.—Fill a pie-dish with alternate layers of sliced apples and bread crumbs, seasoning each layer with bits of butter, a little sugar, and a pinch of mixed cinnamon, cloves and allspice. Pour over the whole enough water and treacle mixed to moisten it, cover with crumbs, stand in a baking tin of hot water, and bake till apples are soft, three-quarters to one hour.

Barley and Raisin Pudding.—Take the barley left from making barley water, put it into a buttered pie-dish with a handful of washed raisins. Cover with milk or dissolved Plasmon, and bake slowly one hour, or till the barley has swelled to top of dish.

Curd Cheese.—Take two pints new milk, curdle it either by slow heat, or by rennet, lemon juice, fig juice, or bruised nettles. Turn the curd into a cheese cloth or butter muslin (coarse canvas will do), previously scalded, tie loosely and hang up to drain. After three or four hours tie again tighter. In twelve hours it is fit to eat, but if preferred it can be pressed and turned every day till as firm as ordinary cheese.

Nut Cutlets.—Boil two ounces of butter in rather less than half-pint of milk. Add three ounces of dried browned

K

breadcrumbs (or brown breadcrumbs). Cook till it does not stick to the pan. When cool add two ounces of ground walnuts, almonds or Barcelona nuts, seasoning to taste, and a little chopped onion or chives. Mix thoroughly and shape into cutlets or balls. Roll in flour, or in egg and breadcrumb, and fry in butter. Serve with tomato sauce, walnut gravy, or for children with bread sauce.

If the same dish is prepared with white breadcrumbs, and a little good melted butter sauce used for mixing, it looks and tastes very like chicken croquettes.

Soufflé Potatoes with Nuts.—Take out the inside of as many large baked potatoes as you require. For each potato add one ounce of ground nuts, a dessert spoonful of cream or a bit of butter the size of a small walnut, pepper, salt, and if liked a seasoning of onion juice and parsley. Beat thoroughly to a smooth creamy consistency, put the mixture back into the potato skins, bake till very hot, and serve.

A very savoury flavour is given by frying some sliced onions in the butter and adding the butter alone to the potatoes, but the colour is a little darkened by doing this.

Brazil Nut Soup.—One pound of ground Brazil nuts stewed for twelve hours in two quarts of water; flavour with celery and a few fried onions. Add one quart of boiling milk. Pass through a strainer, season, and serve with fried bread dice.

A nourishing dish for a child or invalid is good bread sauce to which has been added two ounces of ground almonds, well pounded in a mortar. Serve with fried dice, or spinach.

Baked almonds slightly salted, and ground, make excellent sandwiches.

Rice Croquettes.—Take cold boiled rice and mix it with fried chopped onion, a few breadcrumbs, pepper, salt, chopped parsley, or mixed herbs flavouring, and enough melted cocoanut butter or nut milk to bind all together. When cold, shape into cakes, and fry in deep boiling fat.

For this, as for all sorts of savouries like nut cutlets, fritters, &c., a frying basket and plenty of oil or fat are necessary for good cooking. It also needs experience, or a frying thermometer, to know when fat is at the right heat for deep frying.

Braised Onions.—Peel large or medium sized onions in warm water to prevent the volatile oil from affecting the eyes, place them in a baking dish with butter enough to baste them well, and bake three hours, when they will be brown and tender. Remove them on to a serving dish, pour hot water into the baking tin, and with a wooden spoon rub off all the dark brown caked juice, thickening if liked with a little flour, in which case the tin must be placed over heat enough to boil the gravy and cook the flour. Pour over the onions and serve. This is a delicious dish which puzzles meat-eaters, as they think they are eating a rich meat sauce.

Raisin Tea.—Take half a pound good raisins and wash well in cold water. Cut them up roughly to free the pulp in cooking, and put them into a stewing jar, or Gourmet boiler, with one quart cold water (distilled, for perfection). Cook three to four hours, when the liquid will be reduced to one pint. Press all but insoluble skins and stones through a fine, scalded sieve, and use either hot or cold. If too sweet, a little lemon juice may be added, but it is best without for invalids and children. This drink is of the same proteid value as milk, and is so much more easily digested that it is being used successfully in many cases of gastric disease where both milk and vegetable or meat soups are impossible. It is not recommended to meat-eaters, as the sweetness might cause bilious disturbance ;

but to scientific vegetarians and fruitarians it is invaluable as a nutrient drink.   For those who suffer from cold it is very warming.

Nut Cutlets (another way).—One cupful grated bread, one cupful each of grated almonds and walnuts or Brazils, one teaspoonful powdered mace, one tablespoonful grated onion juice, one teaspoonful powdered mixed herbs, salt and pepper if liked, enough good melted butter sauce to mix and bind together.   Stir all thoroughly well, and allow to go cold.   Shape into cutlets or balls, flour, or egg and breadcrumb, and fry in a basket in deep fat at 350°.   A few seconds will turn them brown, and they are crisp and dry, not the least greasy.   A very nourishing and savoury dish which should be eaten with vegetables in the same proportion as if the cutlets were of meat.'

EIGHTEEN HUNDRED A YEAR

REPRINTED FROM THE 'CORNHILL MAGAZINE'

A YOUNG friend came to see me not long ago, and after a short period of a somewhat shy reserve he looked up with a beaming, happy face, and said, 'I'm going to be married.' It all sounds so simple, these few words, and yet what do they not mean in two young lives! I responded with a smile and the ordinary platitude of, 'I am very glad, and especially so for your mother's sake, for it will give her great pleasure.' As we talked on, I naturally came to the prosaic, elderly question, 'What have you got to live upon?' His answer came short and straight enough. 'With what my father left me and my salary, I shall make up 1,600l. a year, and the lady I am about to marry, I am told, is to have 200l. of her own.'

'That will do well enough,' I said, 'even if you have to live in London. The most pessimistic objector to early marriage can hardly say that love need fly out of the window on such an income as that. But, all the same, wealth is comparative,' for everything depends on position and what there is to keep up. The young man, being of the cautious, prudent type, asked, 'And what do you think I ought to save yearly on such an income?' I answered, 'From 200l. to 300l. a year.' He, not differing, but yet interrogatively, replied, 'I wonder why? I shall have more later on. Why is it necessary to save at all, and not just fit my expenditure to my present income?' This opens so large a question that on my young friend's departure I asked myself why I think as I do about it.

There seems to me a point of resemblance between saving and the very different occupation of gambling. Why is it that gambling has always, in all countries and at all times, been condemned by wise and prudent people, and saving (that is to say, not living up to your income, but leaving a margin more or less wide, which you intend to add to your capital) been approved? It cannot be only that in ten years or so you should be 2,000*l.* or 3,000*l.* richer. The approval of the saving and the condemnation of the gambling are directed, I think, to the mental attitude of the gambler or the careful man, rather than to any practical result to them personally of their conduct. The saving recommended is in no sense the spirit of the miser who piles up wealth for which he has no use, but a cautious guarding of expenditure which provides for future children, against a rainy day, or enables a man later in life to better his house or his furniture, or to increase the enjoyment of his holidays. To adjust income and expenditure exactly is extremely difficult, and anyone who does not pitch his estimate of expenditure below his income is almost sure in practice to exceed it. Of course, it is much less important to save on a more or less assured income (for no income is absolutely assured) than it is to save on an income which is almost entirely derived from salary, and dependent on a man's life or health or the success of the business in which he is engaged. To save ever so little is very much better than keeping elaborate accounts. If, at the end of the year, the savings are there, no doubt remains that the expenditure has been, as regards essentials, well regulated— though getting as much as can be got out of the money spent is quite a different matter from making both ends meet. It is, all the same, interesting and good to remember what can be done at a pinch, and how the upper working classes live in comfort on an income where thousands of impoverished gentry would simply starve on double the sum. The fundamental principle which governs the lives of the working classes is to ignore to-morrow— to live from hand to mouth and day to day. And it is on this point that gentility with a very small income is often perverted by not recognising the merits of the principle when circumstances make it a necessity. This seems to me worth considering, although I recommend the opposite principle as the one generally most admirable to practise with a larger income. The working man does his best for the moment, assumes that his children when reared will do likewise, and the rest he leaves to Providence, or chance, or whatever the unknown quantity may be called. The 'gentle' reared man, I say, cannot be happy unless he has a security against fate, not only for himself but for his family. It is a fine idea in many ways, but it can perhaps grow into an exaggeration. The serious handicap to the 'gentle' man is the education of his children. He must pay through the nose for it, or his children are apt to sink into a class to which they do not rightly belong and for which they are quite unfitted. The working man starts his children as he started himself, with nothing more than the education provided by the nation, and their own power to work.

The expenditure of an income of 1,800*l.* a year will vary a great deal in detail according to whether it is spent in London or the country. I shall therefore consider the two separately, taking London first. Of course, the most important item is house-rent, and requirements and taste differ so widely that it seriously affects the whole income.

The old idea was that house-rent should absorb only a tenth of the income; but this in London is practically impossible—an eighth is nearer the average nowadays.

Even this will vary very much with circumstances—the requirements and wishes of both parties. The wife constantly holds to living within easy reach of her family and friends, and the husband's wishes will be much affected by his kind of work. Saving of time in getting to work may be of great importance, necessitating the use of cabs. The house rent, which, on an income of 1,800*l*., in most cases had better not exceed 200*l*., including rates and taxes, may very easily mount up to 350*l*. When this is the case it is well to commit the extravagance boldly, and so secure a house in a locality which is practically a certain let, if circumstances make this desirable, or if the expenditure of any one year has exceeded the average. There seems to be a very general impression that living in a better locality and a more central part of the West End is an actual economy ; this may be the case if cabs are much used, but if the Underground or 'buses be the usual mode of locomotion, very little is saved except time, which in the case of the woman does not generally affect the income. At one period of my life, influenced no doubt by the growing so-called artistic fashion, I had a great dislike to the old two-roomed back and front house ; but I am now inclined to think that on the whole, especially in small houses, it is the best plan of house building for London. It gives more room, more convenience, and more air than any of the modern houses, arranged on what is considered a superior system—viz., blocking up the middle of the house with staircase and landings, all more or less dark, and which divide all the rooms from one another. Corner houses are, in my opinion, to be avoided, as they are always stuffy, a draught through being not easily obtained. Flats are not suited to young married couples.

The boudoir or morning room so vaunted by agents seems to me very superfluous for young married people.

In early married days and in winter, for reasons of
economy, the husband being out all day, there seems no
reason why the wife should not share the man's sitting-
room. But if the drawing-room *is* used, she must live in
it, or it will have an unbearably stiff appearance. The
great advantage of the two-roomed house, with the
absence of a dividing landing on the bedroom floors, is in
case of illness. No one who has had to experience any
kind of nursing fails to appreciate the great importance
of rooms that communicate, and much suffering is often
spared to the nervous child who feels the presence of its
nurse in the adjoining room, and sees the gleam of the
nursery light through the half-closed door. Besides, in
the busy modern London life, those who have lately
become one will feel it an advantage, by no means to be
despised, that they can talk at all sorts of odd times
through the open door, and discuss life's little difficulties,
which are often created by a non-understanding of the
circumstances. When man and woman are joint masters
in the small details of everyday life and the just ruling of
servants and children, there should be the comprehension
of what Mr. Morley calls 'government by discussion,
which is now counted the secret of liberty.' George Eliot
says somewhere that 'a man with an affectionate disposition who finds a wife to concur with his fundamental
idea of life easily comes to persuade himself that no other
woman would have suited him so well, and does a little
daily snapping and quarrelling without any sense of
alienation.' How true this is; but also, how infinitely
better is it that this should be done upstairs than in the
drawing-room or dining-room, possibly before servants
and guests.

Having now given my opinion on the preferable style
of house, for the sake of argument I will say that the
young couple decide on the more fashionable locality, and

weight their income with a disproportionately high rent. Under these circumstances I think the disposition of their income and general expenditure would work out into something like the following table :—

		£	s.	d.
I.	Rent, rates, and taxes	360	0	0
II.	Housekeeping, including living, washing, lighting	550	0	0
III.	Repairs, insurance, cleaning, painting, &c. . .	100	0	0
IV.	Coal	60	0	0
V.	Dress (man and woman)	200	0	0
VI.	Wages, including beer, for four servants . .	130	0	0
VII.	Wine	60	0	0
VIII.	Stamps, newspapers, stationery, &c. . . .	30	0	0
IX.	Doctors, dentists, accidents, journeys . . .	100	0	0
X.	New house linen	20	0	0
XI.	Charities	40	0	0
		£1,650	0	0

Cabs, amusements, and presents will have to be saved out of clothes or journeys ; with so heavy a rent, putting by money some years will be very difficult. Here I must add a grave word of caution against a practice, only too common I fear, of running into debt over the process of furnishing. A wise man ought to have money in hand before he decides to marry. If he has no savings, it is better to take some portion of his capital and pay his bills, returning it by degrees out of his yearly income. In this way he begins fair on strictly ready-money principles, by which I mean paying everything weekly, an impossibility if stray bills keep coming in. Any bill that cannot be paid weekly should be paid quarterly. One bill I fear often postponed is that of the doctor. I think it would be immensely to the advantage of both doctor and patient if it were a more received custom that the general practitioner should be given his half-guinea, like the M.D. his guinea, at the close of each visit. Few people have any idea of how unjustly doctors are treated as regards their bills, hardly liking to complain when they are neglected

or even forgotten altogether. At the time of the illness
no fee is ever grudged, but doctors' visits carelessly in-
dulged in are apt to run up a very heavy bill, which causes
considerable and unjust irritation at Christmas. Receiving bills, paying bills, and running up new bills poison
the first weeks of the new year to a great many.

I enter into no details with regard to servants' wages,
as on this subject also opinions vary widely as to which
department is to have the experienced and expensive servant. Speaking in a general way, every maid represents
an additional sixty or seventy pounds a year, and every
man another seventy or eighty. These sums cover all
expenses connected with a servant, including wages. It
is generally worth while to increase wages to keep a good
servant, and few things are more extravagant than changing servants ; but no one gets what he wants by offering
wages above the average. If for an exceptional case
wages are raised, always go back on changing the servant
to the sum you originally gave. In the eighth volume of
Mr. Charles Booth's wonderful book, 'The Life and
Labour of the People in London,' there is a chapter on
domestic indoor servants which gives considerable information, and which, I think, all young householders
would do well to read. It is with no small surprise one
realises how very limited in number, as compared with
population, are the people who can afford to keep any
servants at all. Mr. Booth says, 'With three servants—
a cook, parlourmaid, and housemaid—a household is complete in all its functions ; all else is only a development
of this theme.' Most of my young women friends will
be surprised to hear that he gives the lady's-maid no place
at all, and of course she is the easiest servant to suppress
without altering the style of living or inconveniencing
the husband in any way. A large class of people who
keep three servants, even if they increase them to four,

add a kitchenmaid, or an up-and-down girl, rather than a lady's-maid. I am inclined to think that in early years of married life a lady's-maid, besides being a great comfort, partly pays for herself by the saving of dressmakers' bills, and turning old things into new. It is fancy things made at home that really pay, not petticoats and underlinen. The lady's-maid, too, must undertake the mending of house linen, an important duty, as very few housemaids are to be trusted to do any fine needlework at all ; though if one afternoon a week is set apart under the lady's-maid's superintendence the housemaid would probably be quite capable of sewing on buttons and doing necessary repairs both to house linen and the husband's underclothes. A woman who is obliged to have all her things made out will find the allowance of 100*l.* a year insufficient, if she is to be well dressed. It would mean buying ready-made shop clothes or going to inferior dressmakers. The beauty of dress is not so much what it costs, as the individual representation of the wearer's mind and taste. No one, putting aside the very best dressmakers, can carry this out so well as an intelligent maid at home. This also applies to the dressing of children. There is perhaps no time in a woman's life when she can be so well dressed on what is now called 'a small allowance' as in the early years of her married life. She has her trousseau to work from, and if she is sensible even in London she will go little into society beyond dining out: it should be her object to reduce the number of entertainments for which different dresses are required, it being much more difficult and expensive to dress suitably than smartly. London clothes, luckily, do quite well for Saturdays and Sundays in the country; though they are most inappropriate for real country life.

In a small establishment the only servant who is likely to be hard-worked, and therefore deserving of special consideration, is the single-handed cook. I am all in favour of beginning life with a young cook who has been kitchen-maid in a good kitchen, and who is willing to let even her inexperienced mistress be housekeeper, regulating expenditure and diminishing dining-room luxuries, as what can perfectly be done without is what swells the weekly books. There is no economy in stinting the daily food, either for the dining-room or the servants. Servants who come to a certain class of master and mistress look upon good feeding as their due just as much as sheets to sleep in or the wages which are handed to them quarterly. A lower class of modern servant clings to having her wages paid monthly. This request should be yielded to as little as possible, as it tends to make saving more difficult, and saving is specially desirable for domestic servants, who, with unaccustomed luxury, have much in their daily life which undermines their moral sense. One of the best ways of easing the cook in her work is the foreign method of servants having their meals after they come out of the dining-room. Servants' breakfast must either be before or at the same time as the dining-room. Luncheon at 1.30, and sent out as quickly as possible, gives the servants their dinner. If the master's dinner is after 8 o'clock it comes very hard on the servants, as it makes their supper so late. Reducing the number of courses (by which I mean having the food brought up at the same time, more in the style, though mercifully reduced in quantity, of the suppers of our great-grandfathers) is conducive to health, an actual economy, and gives the cook less to dish up and wash up. A foreign fashion introduced of late, and becoming almost universal in England, is the serving coffee after luncheon, which servants now copy by having tea after their dinner. Both really injure digestion, but tea is far the most unwholesome, and chemically turns meat into lumps of iron,

the justification, though they don't realise why, of the
male hatred of ' high tea ' ; consequently, this habit of tea
after dinner will only increase the almost universal
dyspepsia. The teetotalling propaganda has much to
answer for, so difficult is it in life not to fall from Scylla
into Charybdis, and now tea is adding its poison to the
alcohol which has so fatally undermined the health of our
towns and villages. A friend of mine told me the other
day that the doctor said that half his patients on her large
estate in the North of England were due to excessive tea
drinking. This is owing, as in Ireland where the mad-
houses are half filled by tea-drinkers, to the stewing of
the tea leaves. Low diet makes the poison much more
active : this obliges non-meat eaters, to the great surprise
even of themselves, to renounce tea-drinking altogether,
even when the leaves are quickly infused. Keeping servants up at night makes early rising an impossibility.
The young couple must decide which they prefer. The
lady must study books, go to stores to learn quality of
goods and their prices, and not be ashamed to ask the
advice of her contemporaries, which is generally more
valuable from those that are poorer, than from those that
are richer, than herself.

A short experience will teach us that, broadly speaking, our friends are divided into two classes—those who
complain much of their poverty and the expense of everything, and those who apparently live in the lap of luxury,
stinting neither themselves nor their servants, and who
yet maintain that their books are lower than seems
possible from any reasonable calculation. This may be
the result of some different method of keeping accounts,
or that the house is run by some very experienced housekeeper, cook or man-servant, or governess, who gets
the uttermost farthing out of every bargain. And the
economy is further magnified by the lady herself, who,

in giving a sketch of her ordinary expenditure, frequently omits some important item. Such friends are
apt to send us home in a very depressed state of mind,
which is not without use, for it rouses us out of our own
carelessness. In the case of a cook without a kitchen-
maid an ordinary dinner cannot be well served, even for
a small party, without some outside assistance, and I think
it would be better in London to try to find a girl who
lives at home, and who would be glad to make money and
be willing to come in on certain occasions, than the usual
old and experienced charwoman, who takes her own line,
instead of submitting to the training of the young cook.
Many young girls on leaving the Board schools might be
quite willing to do some work which does not oblige them
to leave their homes. All assistance of this kind is a
growing difficulty; the bridging of a gap between what
are called the respectable poor, and the class just below
them, still remains a curiously unsolved problem. We
are always hearing of people so poor that they must be
supported by charity, whose children are now educated
in the national schools in a way that should fit them to
become under-servants and supply an ever-increasing
demand throughout the country, and yet it is more
difficult to get under-servants than ever. The expense of
occasional outside assistance in the kitchen will be very
inconsiderable, and if the young couple frequently dine
out the cook will have many evenings to herself. The
cook and the man-servant should not both be out at the
same time. In the kind of household I am attempting to
describe it is certainly in the food department that with
care the greatest saving can be made without discomfort
or parsimony. The young housekeeper must feel that
this depends on herself, and need not feel ashamed to
take to heart the words of Socrates : ' I am distinguished
from others and superior to others by this character only,

that I am conscious of my own ignorance,' and so naturally to be aware of our own lack of knowledge is the first step towards a better state of things. A constant acquiring of knowledge is the one thing that redeems housekeeping from being intolerably uninteresting ; without this its daily monotony is very trying to many characters. I heard the other day of a lady who said she was getting rid of her cook for no fault, but merely because she was so tired of seeing the same face every day. I do not believe that sort of feeling would come over anyone who tried daily to teach her cook something.

Let every young housekeeper do her best to simplify life. It will only add to her powers of hospitality, which should always be without competition—nothing should be done with the idea of surpassing others. The great use of stores and wholesale shops is the knowledge they give of the market price of goods, and, as a matter of fact, often when things are cheapest they are best, as they are then plentiful and in their prime. Never buy anything out of season, is one of the best rules, and on the whole I think it makes the best living. Season in London does not apply to what grows in our gardens. There is, of course, a season for imported goods, such as cranberries, oranges, or the beans from Madeira, and these things vary considerably in price from week to week. Never forget one of my favourite precepts, that if luxuries are bought they should be of the very best, and come from the very best shops. This applies especially to wine, which can only be got good through the thought and knowledge of the host, not forgetting that bad wine is by no means always cheap wine. Flowers have become a somewhat wasteful luxury in London. I see no reason why a table should not be made to look quite as pretty with plate and china, without any flowers at all, or say

one Japanese arrangement, which means trouble and
taste, not expense. I remember, some years ago, thinking how beautiful a flower decoration was on a dinner-
table, and on expressing my admiration to my young
hostess, she said, ' I am so glad you like them. They
were so cheap. I do not think the whole lot cost more
than 5*l*. ! '

To return to the household. By far the easiest
servant to secure is a housemaid. This by no means
implies that she knows her work. Even trained housemaids are sometimes wrongly trained, and the mistress
of the house is often very incapable of teaching even the
simplest rules of how to keep a house well aired and yet
clean ; the London housemaid's idea is to keep it clean
by shutting the windows. The right way is to keep
windows open day and night, and dust certain pieces of
furniture several times a day. It is both economical and
clean to make an arrangement with the laundress to do
the maids' washing at so much a week a head, instead
of giving ' washing money ' to the servants themselves.
I also think it of great importance that the beer money,
instead of being paid weekly, should be added to the
wages and paid quarterly. If masters and mistresses
only realised the number of young servants who have
been taught to drink by being tempted to help themselves
to the brandy and whisky on the cold grey mornings when
they come to their work, masters and mistresses would
be more careful to lock away these things before they go
to bed ; as, alas ! even those who believe that taking
spirits before going to bed is very injurious to health, are
forced by the laws of hospitality to produce the decanters
before the departure of their guests.

Nothing oils the machinery of the household such as
I have been describing so much as the introduction of a
very young footman or boy, and in case of the man-

L

servant being a soldier servant it becomes almost a necessity, for he has his master's uniforms to attend to, and is often called away by various duties into the country. The boy cleans knives and boots, carries coal, which is injurious heavy work for women ; sleeps in the pantry if the man-servant is married, which often means a better class of servant. The boy's help also enables the housemaid to keep entirely to her own work all day, very desirable in London, instead of being called down constantly to answer the door. Once more referring to the list of expenditure for 1,800*l*. a year, the items marked No. III. will be thought by many to be over-estimated at 100*l*. a year, but this is certainly not too much, taking one year with another, if the house, which is so constantly the case in London, has to be painted outside every three years. Others may think 60*l*. a year too heavy a charge for coal, but this must include wood for lighting fires, no inconsiderable item in a London house. 20*l*. a year for keeping up the stock of linen is rather under than over what I should deem necessary, unless the young *ménage* is very much better set up than is usually the case in England. My estimate for living in London leaves us with the very narrow margin of 150*l*. a year, but my calculations were upset by finding on inquiry that the rent of even small houses in good localities south of the Park are so much higher than I expected. Some years ago excellent houses were to be had on the north side for 125*l*. a year. After all, everything I have said resolves itself into what applies to every income—i.e., to pretend to yourself that you have less than you have got, and then live at that rate, and you will always be rich. This is the honest and comfortable way of living, but it does not perhaps always appear worldly wise, as experience shows us that feckless and extravagant people generally get paid for, somehow, and very often at the expense of those who are careful. It will always be so more or less, and is only the old story of the Prodigal and the son who stays at home.

Living in the country on an income of 1,800*l.* a year changes the expenditure in many ways ; in some more expensive, and in others cheaper. Shall we not throw into the balance all the unbuyable luxuries the country gives us ? Sun and light, air and cleanliness inside and outside the house, winter hoar-frosts and summer's radiant colours, flowers growing at their own sweet will, the song of birds and the ceaseless interest of insect life. ' Life is sweet, brother, there's day and night, brother, both sweet things ; sun, moon and stars, all sweet things ; there's likewise a wind on the heath '—so says old George Borrow in ' Lavengro,' and few men's writings can take us into the country as his do. The ' wind on the heath,' what is it not worth ? Keats, too, speaks of it in his own gentle way :

> To one who has been long in city pent,
>   'Tis very sweet to look into the fair
>   And open face of heaven,—to breathe a prayer
> Full in the smile of the blue firmament.
> Who is more happy, when, with heart's content,
>   Fatigued he sinks into some pleasant lair
>   Of wavy grass, and reads a debonair
> And gentle tale of love and languishment,
> Returning home at evening, with an ear
>   Catching the notes of Philomel—an eye
> Watching the sailing cloudlet's bright career,
>   He mourns that day so soon has glided by ;
> E'en like the passage of an angel's tear
>   That falls through the clear ether silently.

For the young, the energetic, and the ambitious, towns are best at first, and they can gravitate towards the country as they grow older. This surely is one of the greatest inducements for saving, and in no sense is it a mean or lowering object. We will begin now with our changed table of expenditure for life in the country. The most important reduction will be in the house-rent.

TABLE—COUNTRY

		£	s.	d.
I.	Rent, rates and taxes	180	0	0
II.	Housekeeping (living, washing, lighting)	450	0	0
III.	Repairs, insurance, cleaning, painting .	100	0	0
IV.	Coal	80	0	0
V.	Dress (man and woman) . . .	180	0	0
VI.	Wages, including beer (four servants) .	130	0	0
VII.	Wine	50	0	0
VIII.	Stamps, newspapers, stationery . .	30	0	0
IX.	Doctor, dentist, accidents, journeys .	100	0	0
X.	New house linen	20	0	0
XI.	Charities	40	0	0
		£1,360	0	0

This table shows considerable reduction, and, if saving is not very necessary, a pony, carriage, and groom can be added, besides the obligatory garden, which, well done, including wages and all expenses, must be counted at 150*l.* a year : so the table now stands :—

	£	s.	d.
Carried forward , ...	1,360	0	0
Garden	150	0	0
Pony, carriage, groom	130	0	0
	£1,640	0	0

This does not include the initial cost of buying the pony and carriage and setting up the stable.

With these luxuries the margin is as narrow as the London one. Any careful housekeeper will find it easier to make reductions in the country, though it will probably be at the expense of having friends to stay, which is one of the pleasures of living in the country, minimised by the fact that it often interferes with your pursuits, care of poultry, garden, &c., as very naturally the friend who

takes the trouble to come and see you exacts your un-
divided attention. One of the expenses of country
hospitality is not only the laundry bill, but the wear and
tear caused to good linen from always being in the wash-
tub. I confess to often feeling considerable sympathy
with the landlady of olden times who felt it such a pity
to send sheets to the wash, and gave orders to damp them
a little and iron them out.

Furnishing in the country can be done even more
simply and sensibly than in London. If washing house
linen is more, cleaning of curtains and chintzes, &c., is
infinitely less ; three months of London making things
much dirtier than a year in the country. The great
secret of sensible and yet pretty furnishing is observation
and keeping your eyes open. People as a rule notice
nothing, and come into a house and garden almost as if they
were blind, and it is curious to observe how the awakening comes when they are going to furnish for themselves.
This selfish impetus should not be necessary. The want
of training of sight, scent, and hearing are among the great
deficiencies of civilised education, and I fancy this defect
has been keenly felt in South Africa. The newspapers
have commented upon this subject from time to time, and
I noted from one the other day that ' as a matter of
scientific fact there is little difference in the powers of
vision of different races. The difference lies in the
faculty of detection, and this is a matter of training and
constant practice. Two men have equally good sight,
but one, by reason of the necessity of his daily life, will be
able to detect an unusual object, whilst the other will be
entirely unable to recognise anything abnormal. This
being so, it is all the more necessary that the training of
the eye should form a very important part of a soldier's
education in the art of war. Scouting and judging of
distances will have to be reckoned with in any practical

scheme of army reform.' This is no doubt perfectly true, but the training of the eye and the quickening of the powers of observation should begin with both sexes from the very earliest age. And I am convinced that no one can manage a house and garden well unless these faculties are highly developed.

Poultry keeping in the country is a pleasure and an interest, but it hardly pays unless eggs are sold in the winter and chickens in the early spring. If the garden is carefully and knowingly stocked to supply the wants of every month in the year, the saving in the weekly books is considerable, as nothing ought to be bought, except potatoes, and the plentiful supply of vegetables for many months in the year considerably reduces the butcher's book. Everyone who has space in the country should keep pigs; nothing so prevents waste or actually pays better. Buying two young ones twice a year—once in January, selling them in May, and buying again two more in May, giving them nothing till well on in October but green vegetable garden refuse and the wash-tub. This tub must be kept carefully clean—that is to say, no meat or tea-leaves or coffee-grouts. It is quite a common cook's trick to throw into the pig-tub the heads and insides of game or poultry; this is quite wrong, and will easily produce diseased pigs. In October acorns and chestnuts are good for them, but they make the meat hard unless the pigs are fed on barley-meal for quite three weeks before they are killed; part must be sold and part preserved for home consumption. Curing hams from a good receipt, with a careful cook, not only thoroughly pays, but produces far better hams than those usually bought.

A garden makes a very great saving in the weekly books, and enables a family to live well with much less meat. A leg of mutton is a much more economical joint than a shoulder or a neck; but for a small family it generally means so much cold meat. The following method of dealing with a leg of mutton, out of one of my old books, may be suggestive to some :—

First day.—Cut off the knuckle, boil it slowly, cover it with caper sauce ; serve with mashed turnips and carrots.

Second day.—Cut a steak off the large end and broil it ; serve with maître-d'hotel butter and fried potatoes or onions.

Third day.—Cut some cutlets off the side near the knuckle, breadcrumb, and fry ; serve with brown sauce. Purée of greens or purée of chestnuts. Beetroot hot or cold.

Fourth day.—Bone and stuff the fillet, which is to be roasted (put the bone into the liquor that boiled the knuckle) ; serve with the roast-meat, jelly, jam or apple sauce, mashed potatoes, white beans and salad.

Fifth day.—Hash part of the remainder in a good brown sauce made with the reduced liquor from bone, thicken with burnt flour, and add minced olives, gherkins, or mushrooms ; mashed potatoes.

Sixth day.—The remainder minced in shells or small pots, breadcrumb, and brown in the oven; serve with cold potatoes fried up, macaroni and sultanas, or rice.

FEBRUARY

NOTES FROM NINE MONTHS OF A SCRAPPY JOURNAL

1901–1902

Forcing cut branches—*Amygdalus Davidiana*—Early spring flowers under glass—Bulbous irises—Epimediums in pots—Letter to Westminster on railway carriages and tuberculosis—Congress on tuberculosis in 1901.

February 20th.—I left home some days ago. A few snowdrops and a few green blades of daffodils were coming through the ground: that was all. Spring seemed very far away, but it was not really so. For a short time one must help nature, and have faith. During the last three years I have considerably developed the practice of forcing cut blossoming branches in water, in the little hothouse. Even in December if the weather is wet and bad, the *Jasminum nudiflorum* is far more effective and flowers more in a mass, if treated in this way, than left on the plant. I saw a year or two ago at one of the Drill Hall shows the *Amygdalus Davidiana*, which is a Chinese early-flowering shrub and exceedingly lovely. I bought one which has grown satisfactorily. It can be cut in January, and the bright brown stems, covered with buds, flower from one end to the other, after cutting and putting into water in the stove. It is a plant I thoroughly recommend to anybody living in the country in winter or having flowers sent up to London. *Prunus Pissardii*, which here the birds strip of buds, the common pink almond, *Forsythia suspensa*, all do excellently cut and brought on in this way, by which I mean, they do very much better than if left to flower out of doors. All stalks must be peeled. The blackthorn, *Prunus spinosa*, and its double variety also answer if picked in the same way a little later on in the year. The Drill Hall exhibitions, interesting as they are all the year round, always attract me most in the spring, as the plants are better adapted at that season than any other to the kind of miniature pot-cultivation required for shows. Here everybody is able to see what these plants ought to be when grown in perfection, not in heat, but under glass which saves them from wind and weather. All those who try to grow fine Christmas roses in light soils know well how difficult it is, whereas the imaginative writer who speaks of gardens from the poetical point of view rather than the practical, alludes to them in a lordly kind of way as 'flourishing' of their own accord in mixed borders or even in woods and shrubberies. As a matter of fact, though they hate being disturbed, I have found strong nourishment when their leaves are growing and protection from the weather in any but very exceptional winters, absolutely necessary for producing even a moderate success. I have never seen them growing in their natural state on Alpine slopes, as M. Correvon so charmingly describes them in his poems 'Fleurs et Montagnes':

LA ROSE DE NOËL.

Sur les flancs escarpés du riant Salvatore
    Et sous l'âpre frimas
De l'hiver, j'ai trouvé la neigeuse Hellébore
    S'étalant sous mes pas.

Sa fleur cherchait abri sous le sombre feuillage
    Bronzé par les autans,
Et dans son pâle éclat on pressentait le gage
    Des beaux jours du printemps.

Oh ! rose de Noël qui fleuris sous la glace
    Pendant les jours mauvais,
Dis-moi comment tu fais pour demeurer vivace
    Sous les brouillards épais.

Dis-moi d'où tu reçus la douce quiétude
    Que je lis sur ton front,
Dis-moi qui t'a donné la ferme certitude
    Que les beaux jours viendront.

Je voudrais comme toi garder l'âme sereine
    Dans les jours de malheur,
Et quand survient le deuil et la lutte et la peine
    Toujours croire au bonheur.

Pâle soleil d'hiver qui fleuris l'Hellébore
    Au matin de Noël,
Viens chasser mes brouillards et puisse ton aurore
    Toujours luire en mon ciel.

The loveliest things at the February and March Drill Hall shows, are the small early bulbous irises. They are most interesting plants, more so than orchids I think, but I must own they do not last nearly so long, either cut or on the plant. It is the great merit of orchids, even those kinds that are easy to grow, that on the plant and off it their flower branches last so long. An excellent way to support the flower in the pot is a small cane, split open at the top, holding the branch, and stuck in the earth of the pot. I am now going to try growing some of the small bulbous irises. They should be planted in pots or pans in loamy soil, no manure, and treated exactly like hyacinths or narcissi, covered with cocoanut-fibre in a cold frame, matted at night to keep out the frost, and then brought into the greenhouse early in February and carefully watered, to be induced to flower. *I. reticulata* and *I. persica Heldreichii* are the best, I am told, but there are several varieties, and all are pretty. Almost the greatest winter excitement of our gardens is the beautiful *I. stylosa*. People who manage

them well seem very successful with them from December
to April, but, like most irises, they seem to do better if
close to some other plant, which means, I suppose, dryness through the summer. They will grow well on the
sunny side of a yew hedge, but never flower well till the
second or third year. The foliage is thick and untidy, and
one has to look for the flowers or one may miss them,
and for the house they are best picked in the bud. I am
going to try the little Snake's-head of Italy, *I. tuberosa*,
in pans. It seems too dry for it here out of doors.

In these days of horticultural crocuses I do not
think the original type plants are cultivated nearly
enough in our gardens. Early in March this year, and
it was a late bad spring, they were most beautiful at Kew.
C. etruscus, *C. biflorus*, *C. chrysanthus* were the names
on the labels. I see *C. Imperati* and *C. calvigatus*, for a
cold frame, are recommended in 'English Flower Garden.'

Epimedium rubrum and *E. macranthum* make charming greenhouse plants in early spring, potted up in
September (I imagine) and kept in a frame. The little
dwarf *Narcissus Bulbocodium* and *N. cyclamineus* look
charming in pots and pans. All these species seem
much more interesting than a large number of ordinary
hyacinths, double or single.

Mr. James Rhoades sent me this little poem the other
day, saying that he should have dedicated it to me had
he known me sooner. It appeared in the 'Westminster
Gazette,' June 1899 :

> Are flowers the very thoughts of God
>   Made visible to bless ?
> If so it be, O happy ye
>   Who such a faith confess,
> As, led by April, blossom-crowned,
>   Ye roam o'er vale and hill,
> With every here a cowslip found,
>   And there a daffodil !

> Are the birds' songs but jets of joy
>   From the Eternal Bliss ?
> If it be true, O happy few
>   With such a faith as this,
> As, thrilled by many a feathered throat,
>   Ye roam o'er hills and vales,
> With every now the cuckoo's note,
>   And then the nightingale's !

Travelling to London the other day by rail, third class, the extreme dirt of the carriage recalled to my mind a paragraph I had seen in the ' Westminster Gazette,' and made me write the following letter to the Editor. I republish it here because it seems to me that the subject cannot too often be brought before the public, who are still alarmingly ignorant of and careless as to the infectiousness of tuberculosis.

' SIR,—I noticed in your issue of December 23 a short paragraph bringing to our notice how much room there still was for improvement in the methods that could be adopted for the prevention of the spread of the tubercles which cause consumption. The ignorance of the general public on the subject, even the educated public, is the only excuse that can be offered for the constant neglect of the most ordinary sanitary precautions. I live in what might be called a distant suburb, and often travel backwards and forwards to London in a third-class carriage ; so do a number of young people of both sexes, and so do rough workmen who know no better and who expectorate freely on the floor of the carriage. Many of these may be gravely ill from tubercles without being the least aware of it, or of the danger to which their spitting may expose their fellow-travellers in all sorts of accidental ways. Both old and young generally insist on all the windows being closed in winter. Spitting can only gradually be put a stop to,

as it is done by a class who do not even understand how
dirty it is, and who consider they have a right to do it if
they like.

'The proper cleansing of third-class carriages depends
on the railway companies, and as it is done now it would
be considered quite insufficient for a cattle-truck. A dirty
broom is inserted and the surface dirt partly swept into a
dust-pan. This enables what is left to dry the quicker,
increasing the real danger, for the infection only spreads
when the expectoration becomes dust. We have heard
much abuse of the fashionable long skirts in the streets
of London, but seated in a carriage no woman's dress, be
it ever so short, can help touching the floor. A poor
woman will probably dry her skirt at the kitchen fire and
brush it either there or in her bedroom ; the maid of a
rich woman is likely to do exactly the same ; and in this
way thousands of grains of poison may be spread in the
house, and the boy or girl just leaving childhood behind, is
the accidental victim of the mother's journey. Surely it is
an imperative duty to clean and disinfect carriages, and
it seems to me directors should have felt guilty of a very
heavy responsibility, which they did not thoroughly
understand, when they refused to allow the National
Association for the Prevention of Consumption to put up,
free of expense to the companies, an intelligent notice in
the railway carriages about the dangers of expectoration.
A better way of educating the public I cannot imagine,
while giving offence to none, though certainly necessitating a more intelligent way of keeping the carriages
clean. A notice put up in the booking office is absolutely
useless, as no one has time to read it there. Much the
same may be said of the long-delayed notice in our post-
offices, where it is generally hung in an out-of-the-way
corner with postal directions about mail days to all the
ends of the earth. I trust greater knowledge may do

away with this conservative prejudice which dislikes nothing so much on most health subjects as truth and light.'

It is rather curiously typical of the national attitude towards important legislation on health subjects, that, although the King has taken such lively interest in the Society for the Prevention of Tuberculosis, and done all that a Sovereign could do to bring the subject to public notice, assisted by the most influential people in the land, yet that railway directors, heads of firms, and that hopelessly immovable body, the general public, should take so little alarm and pay so little attention to all that has been put before them. Last July (1901) the press for many days teemed with accounts of that very remarkable meeting of medical and scientific men, the Congress on Tuberculosis. I attended one or two of the meetings, and it was my privilege to be there the day Professor Koch made his famous statement on the difference between the tubercle in man and the tubercle in cattle, and stated his belief that the two things were not interchangeable. How the scientific world will decide about this in the future we do not yet know. The great interest to me, and one of a peculiarly dramatic kind, was to be overlooking a large hall crowded with people well versed in the subject, who were quietly told in very broken English a fact which they had not expected to hear. The greatest oratory would, perhaps, hardly have called forth so evident a thrill through the whole audience, though courtesy necessitated the suppression of either surprise or opposition. Apparently, from all statistics, England though, perhaps, the most afflicted with this terrible scourge, is yet the most backward in sanitary precautions. M. Brouardel gave some interesting statistics as to what is being done in France

He delivered his lecture in his own language with all the charm and crispness of his nationality. All the lectures delivered at this Congress are, I believe, still to be got from the Secretary of the Society in Hanover Square.

As with so many other improvements in general health, the future rests not so much with the medical profession as with the general public, but they must understand before they can act. How few people believe that common colds without inflammation ought to be treated like consumption with an open-air cure, not only for the good of the patient but also for the benefit of other people, as fresh air is the great destroyer of infection. In the case of having to lie up for an accident or an operation, the same thing applies; if the patient lies day and night by an open window the difference to general health on recovery is hardly believable compared with having been shut up in the ordinary sick-room with the windows only opened once a day for ten minutes.

MARCH

Valescure—Tree heath and briar-wood pipes—Fragrant herbs and thyme carpet—Aloes and agaves—Cork trees—Fréjus—Ruins of the Tuileries : De l'Orme and Bullant—Meissonier's picture of the burnt Tuileries—Cannes—Eucalyptus trees—La Mortola— Arrival at Florence—Turban ranunculus—Mino da Fiesole— Letter about Florence.

HAVING the chance last year of letting my house to a friend for March and April, I was tempted to do that which I had always declared I would never do—viz., go to the South of France at the time when everybody else goes there. On March 1 we started, a party of four ladies, for the Grand Hôtel, Valescure, arriving at the St.-Raphaël station after the usual day and night journey. We had left great cold behind us, and found the weather grey and mild. But we were told it had been very cold, and it certainly was soon bitterly cold again. Snow was on all the hills, the almond-blossoms were brown, and the shrub we call mimosa, really yellow acacia, looked pinched and unable to bloom. Our hotel being situated on the side of a hill with a beautiful open view of plain, mountain, and sea, surrounded by delightful pine woods, had a great charm of its own, and for those who have no objection to hotel-life I cannot imagine a more satisfactory resting place in fine spring weather. But I dislike hotels of all kinds, and it was singularly unlucky, as I hoped to live outside all day, that the mistral blew hard during the whole ten days I was at

Valescure, and the weather was colourless, though dry.
On the other side of St.-Raphaël, down by the sea, there
is another hotel, the Grand Hôtel Boulouris, built on a
perfect site for those who like being near the sea, and
even more completely solitary than the one at Valescure.
I gathered that it was not so well heated as the Valescure
hotel. This I think an advantage, for though I like fires
in the rooms, I hate hot passages, as they always make
people afraid to open the windows.

All down to the shore and up in the woods by Boulouris were magnificent tall plants of what I thought was
Mediterranean heath : the flowers were only half open
and injured by frost. I suppose most people know that
from the roots of this tree heath are made the favourite
wooden pipes known as briar-wood pipes. The name
briar is a corruption of the French word *bruyère*, meaning
heath. The collecting of these roots seems to be a trade
wherever the plant grows, and the 'Encyclopædia Britannica' says that 'in Italy they are taken to Leghorn where
the roots are shaped into blocks, each suitable for a pipe,
the cutting of the wood so as to avoid waste requiring
considerable skill. These blocks are simmered in a vat
for twelve hours, which gives them the much-appreciated
yellowish-brown hue of a good briar-root. So prepared,
the blocks are exported for boring and finishing to
St.-Claude (Jura), in France, and to Nüremberg, the two
rival centres of the wooden-pipe trade.' In Moggridge's
' Flora of Mentone,' he refers to this and other European
heaths as follows : ' *Erica multiflora* abounds near
Toulon and Hyères, but from the Esterelles mountains
to Genoa I only know of three small patches of this
pretty plant, all of which are near Nice, one round the
tower of St.-Hospice, and the others at Bellet in the
valley of Magnan. . . . *Erica mediterranea* is never
found on the shores of the Mediterranean, but grows in

M

South-Western France, Western Spain and Ireland. . . .
The English often call *Erica arborea* (Linn.) the
"Mediterranean Heath." This is a mistake. . . . It
abounds along the coast from Marseilles to Genoa, and
though usually cut down, it is occasionally allowed to
take its natural course when it becomes a small tree
8 to 10 feet high. At Cannes, the wood is used for
turning, and of it are made the briar-wood pipes which
are imported into England. *Erica arborea* and *Erica
scoparia* are to be found growing together, but the former
flowers in February or March, and is quite out of blow
before *E. scoparia* first expands its greenish flowers in
the end of April. Even when out of blossom, the tree
heath may always be distinguished by its hairy branches.
. . . It is also found in the Canary Islands, Madeira,
Portugal, and the Pyrenees. *E. scoparia* endures a
varied temperature, but is more confined in its distribution, being found only in Northern and Southern France,
Spain, Italy, Corsica, Sardinia, and Dalmatia.'

How curious it is this distribution of plants, especially
when the spreading and flourishing seem to depend not
entirely on temperature ! Temperature being the ordinary cause of the stopping of vegetation towards the
north, the nature of the soil must, I suppose, be the
reason in this case for such fantastic skipping of whole
districts.

At Valescure, the woods and open spaces are covered
with heaths that have been cut down, and many low-
growing sweet-smelling herbs. The dryness, however,
combined with the bitter cold winds we were having, held
back the scents, and one walked on thyme and sage and
other sweet-smelling herbs without even perceiving it,
and brushed through myrtle and bay without their giving
one anything of their fragrance. In a resigned tone we
said to each other, ' March is March everywhere, even

on the Mediterranean shores,' and I secretly wondered why I had left my own little greenhouse full of sweetness and spring flowers for the euphemistically called 'Sunny South.' I know a place in England where nature's sweet-smelling carpet is imitated by planting a wood path with garden thyme instead of turfing it. It is kept low and somewhat even by perpetual clipping, and it is a pretty idea that the scent is given forth in summer by the present fashion of women's trailing skirts. This seems to be a rare exception to the well-known evils of the present long dresses. It is well for all sensible women never to forget the story of the German professor who sent his wife and daughter out for a walk through the town, and on their return let them see through his microscope the vitality of the microbes they had brought back with the dust. Dry sunshine is necessary for the sweet smelling of thyme, and many other herbs, whereas flowers generally smell sweeter for dampness in the air.

In my youth what I called aloes I now know to be agaves. Agaves come from South America, and die when they have flowered, whereas the aloes come from South Africa and do not die after flowering. These agaves used to grow wild and uncultivated on the tops of walls, thickly crowded together—mothers, children, and grandchildren. The offsets were of all sizes, and the largest plants only occasionally flowered as the last supreme effort, probably when the nourishment from some well-cultivated neighbouring field drained down on to the wall top. Now these agaves are, in a way, cultivated. That is to say, they are taken up and divided and planted along the roads as we do with our *Saxifraga pyramidalis* when we want to make them flower better on our rockeries. This causes the much more frequent blooming of the larger plants by the side of the roads, and does away with the tradition that they only bloomed once in a hundred years. But the death of the plant is much more pathetic and apparent than it used to be. Alone it throws up its enormous flower stem, unsupported by all the young growth which, when left to nature, would be around it. It prematurely falls over and is cut off with its seeds scarcely ripened and the spot is empty, no young plants being there to hold it up and to flourish round the decaying stem. Anybody wishing to grow agaves in pots for ornamental use had better take off the young growths every year as this immensely increases the rapid development of the principal plant.

There are fine old specimens about Valescure of the cork-tree, *Quercus suber*. They have a quaint picturesqueness all their own from the removing of the bark once in about six years for the making of cork, which is one of the old industries of the country. This treatment, that would kill all other trees, does these no harm. ' Chambers's Encyclopædia ' gives the following details about cork-trees :—' Spain and Portugal chiefly supply the world with cork, and in these countries the tree is often planted for the sake of the cork. . . . The acorns are eatable, and resemble chestnuts in taste. The bark in trees or branches from three to five years old acquires a fungous appearance, new layers of cellular tissue being formed, and the outer parts cracking from distension, until they are finally thrown off in large flakes, when a new formation of the same kind takes place. Cork intended for the market is generally stripped off a year or two before it would naturally come away, and the process is repeated at intervals of six or eight years. The bark of the young trees and branches is either useless or of very inferior quality ; it is only after the third peeling that good cork is produced. The removal of the cork, being not the removal of the whole bark, but only of external layers of spongy cellular tissue, all or greater part of which has ceased to have any true vitality, and has become an incumbrance to the tree, is so far from being injurious, that when done with proper care, it rather promotes the health of the tree, which continues to yield crops of cork for almost 150 years. . . . Besides the use of cork for stopping bottles, casks, &c., it is much used, on account of its lightness, for floats of nets, swimming-belts, &c. ; and on account of its impermeability to water, and its being a slow conductor of heat, inner soles of shoes are made of it. All these uses are mentioned by Pliny; but the general employment of corks for glass bottles appears to date only from the fifteenth century.' The process of trees casting their bark can be seen every autumn in London from the plane-trees on the Embankment or the Parks. This is the reason why plane-trees flourish in cities where other forest trees die. In these days cork carpet is most useful and made from the refuse cork for protection against cold and damp from stone or brick floors. I think, as with ordinary oil-cloth, it would probably pro-duce dry rot on wooden floors.

My friends knew Valescure well, and were always taking me to spots where, in ordinary years, the wild flowers grew—iris, anemones, and many others. I had to console myself by looking at books, and when I came home I bought Moggridge's ' Contributions to the Flora of Mentone; the Winter Flora of the Riviera from Mar-seilles to Genoa.' The 'winter' I had had, but the ' flora ' was denied me. The book is a charming one for anyone spending some time along that coast. The illus-trations, though not up to the level of those in old books (it was published in 1871), are far superior to many botanical coloured illustrations of a later date. I know two novels which treat of this part of the world : ' The Individualist ' by Mallock, and one of the stories in Bourget's ' Homme d'Affaires.'

Fréjus can be seen from the hotel windows. The Roman remains give a great interest to the countryside, and had I remained longer I should certainly have bought a book called 'The Romans on the Riviera,' by W. H. Bullock Hall (Macmillan, 1898). Roman ruins are always very attractive to me. Those about Fréjus are of a late date, and not unlike the specimens found in England, having evidently been built for utility, and because Fréjus was a seaport in those days. Now the sea has receded for several miles.

One garden I did see which had an interest all its own, not for its flowers, which were not out, but for the fact that Miolan Carvalho, manager of the Paris Opéra-Comique during the Second Empire, had built this villa, and brought to the garden many fragments from the ruins of the Tuileries after the burning in the time of the Commune. They were dotted about the garden without very much purpose, and the tall, white, beautifully-carved stones looked rather sad, I thought. So classical are they that many people who know no better mistake the garden for a villa of the Lower Roman Empire, so pagan are the ruins and so picturesque. I particularly liked the Avenue where a lion by Caïn looked disdainfully down upon the tourists. These remains seemed out of place in this garden, and accentuated to me the everlasting regret I have always felt that the great and rich French Republic did not have the courage, in spite of their depressed condition after their reverses, to rebuild the beautiful fire-scarred ruin of Catherine de Medicis' palace and use it for the chief of their State, with all its historical recollections behind it, including among these the wild despair of the destructive Communards. I believe many people have thought that the Tuileries gardens looked much better without the tall palace. That may be; but the Elysée, which is now the official residence of the President, is

certainly not a sufficiently large and handsome palace for
the head of the State in a country like France. The
fragments of ruin in this Riviera garden were full of
carving and ornamentation, and one wondered to whose
hand they belonged of the two builders of the Tuileries,
De l'Orme or Bullant. Mrs. Mark Pattison, in her
'Renaissance of Art in France,' says that, at the death of
De l'Orme, Bullant was called upon by Catherine de
Medicis to carry on the construction of the Tuileries
(1570). 'But it must be remembered that when De l'Orme
died he had nearly completed the centre pavilion with
its crowning dome, as well as the two wings to right and
left. It is difficult to say precisely what was contributed
by Bullant, but most probably the whole or a part of the
pavilion, which originally terminated these two wings on
the north and south. In their erection he followed, we
must suppose, the plans left by De l'Orme, for in the
main the two pavilions were of a piece with the rest of
the building. The Ionic order, the " ordre féminin,"
specially affected by De l'Orme, ranged below ; and above,
along the first floor, ran the Corinthian. The proportions
and character of the general features he left unaltered,
but in the decoration and adjustment of parts, Bullant
took a certain license. . . . To his eyes the polite elegance
of De l'Orme's work, the refined inspiration of Lescot's
design as expressed in the neighbouring courts of the
Louvre, seemed wanting in fervour and meagre in enrichment. De l'Orme had panelled his surfaces with delicate
pilasters ; Bullant detached along the front innumerable
columns, and niched between them royal coats-of-arms.
De l'Orme had pointed blank spaces here and there by
the application of a highly-outlined frame, a touch of
ornament, a scutcheon faintly profiled. Bullant cut
bravely and unsparingly broad sweeps of decorative line
in high relief. The ornament which encircled every

blazoned niche was burdened to the utmost with elaborate details. Wreaths garlanded the sides of the windows: crowns filled the spaces above them; crowns, surmounted with points of *fleur-de-lis*, were again repeated at the summit of intermediate shields, uplifted by the out-stretched arms of accompanying figures.

'Yet this luxurious superabundance of ornament never degenerated into vulgar profusion. Exquisite delicacy of sight and touch gave truth of accord to all the parts, and traces of brilliant execution not long ago revealed the same rare quality of light and happy touch which still enlivens the gracious courts of Écouen. Until the hour of the Commune brought fire upon their walls, these two pavilions stood in grace scarcely impaired by time or restoration. The alterations which took place during the reign of Louis XIV. resulted in a complete refashioning at the hands of Le Veau and Dorbay of the centre portion, and the two wings completed by De l'Orme. But the two pavilions raised by Bullant suffered only the loss of the mansard roof, which was replaced by a more convenient attic, and the addition of some ill-calculated ornaments carved upon the shafts of columns originally plain. A sun- and moon-dial, one such as those described in the treatise "De l'Horlogiographie," which has been affixed to a blank side-wall facing south, necessarily disappeared in contact with the extensions of the Louvre executed by Henry IV., but the rest remained even as Bullant had left it, the effect of the graceful Ionic order prevented somewhat, indeed, but not destroyed by the neighbouring presence of the towering columns of Du Cerceau's additions. Now there remain only such rare fragments of columns and pilasters as might be gathered by the hand.' The fragments now exist only in this Valescure garden.

I happened to be in Paris soon after Meissonier's death, and had the privilege of visiting his studio. His

method of working from small clay models rather than
from nature or memory explains much that is remarkable
in the style of his pictures. But what makes me recall
this visit now is that I saw there an exceedingly interesting sketch of the ruined Tuileries. It was an unusually
large water-colour representing the Salle des Maréchaux
after the fire as seen from the Tuileries gardens, with
Napoleon's battles still engraven on the cornice, bright
blue sky above, and the car and horses on the top of the
Arc du Carrousel appearing above a heap of ruins. Madame
Meissonier has since given this valuable historical record
to the Luxembourg.

At the end of about ten days I left my kind friends at
Valescure and moved on to Cannes, where I had not been
for many years. The whole place was changed beyond
recognition; the old olives are being cut down as not
profitable, and new buildings rise in all directions. The
yellow-flowering acacia (*Acacia dealbata*) has sown itself
everywhere. It began by being the pride of Cannes
gardens, and is now much encouraged by the peasants
for the exportation of its flowering branches for sale in
northern cities. To me it is a beautiful thing in the hand,
but singularly unbecoming to the colour of the landscape.
Of all the new plant introductions since my childhood
along the Mediterranean coast, the one that seems to me
to have taken good hold, and to be most suitable both in
growth and colouring, is the Australian eucalyptus. This
in a kind of way may take the place of the disappearing
olives. It is bluer in colour, and the growth is even more
weirdly picturesque. I often wondered how several trees
of it would look grouped together; for the present one
only sees single standards, some of them very tall. The
Eucalyptus Gunnii in my garden, which I mentioned in a
former book, has continued to stand our winter climate,
and flourishes only slightly protected by other shrubs.

I did not mention that I grew it originally from seed; many things adapt themselves better to a strange soil if grown from seed on the spot.

The orange gardens at Cannes have ceased to exist as they were in my day, when, being highly cultivated and manured, there grew beneath the trees carpets of Neapolitan violets, 'Violettes de Parme,' as they are called. Now everything is sacrificed to the imitation of English lawns, and in these are planted, very inappropriately, palms and bamboos, aloes and agaves, all dotted about, and producing a most unfortunate and ugly effect. From a gardener's point of view this is absolutely incorrect, for the grass requires much watering, whereas the poor aloes and agaves need a dry, stony soil in order to flourish and gain their beautiful blue bloom.

On the whole, the gardens that I saw at Cannes disappointed me dreadfully; but then I must confess it was a very backward year. Even the favourite of our English gardeners, the 'Old Glory,' or, as some put it, 'the Glory to die John,' was hardly in leaf.

As a happy instance of what the unlettered can make of a botanical name I have been told by a friend, who actually heard it, that as two old women were parting by a cottage-gate one said to the other, admiring her large laurustinus in full bloom, 'What a fine plant you have there!' 'Yes,' said the other, 'and such a beautiful name as it's got!' The first one, looking a little astonished and ashamed of her ignorance, said, 'And what is it?' 'Oh, don't you know? It's called "The Lord sustine (sustain) us"!'

In gardens where tiny streams ran through them, quantities of the lovely *Iris stylosa* were in flower. But beyond this there was little to attract attention. Not nearly enough care is given in England to this beautiful winter flower. Very large clumps are required, and with room to spread, all buds come out well in water. The only violet I saw in cultivation was the large single kind with long stalks and little smell.

On March 18 a German friend came all the way from Frankfort to go with me to see Mr. Hanbury's garden at La Mortola. The morning dawned with heavy, driving clouds, and by eight o'clock it was raining in torrents. She sent to ask what I intended to do. My answer was the irritatingly priggish one (which I should never have dared send to a man) that I never allowed weather to interfere with my plans. So we started for Mentone by rail, and then, after driving for over an hour in ceaseless rain, we arrived late for luncheon. Our longed-for treat of seeing this wonderful garden had to be enjoyed under umbrellas and holding up dripping petticoats as we climbed up and down the steps, the whole garden being on the slope of a steep hill. In spite of these unfavourable conditions, it far outstripped in interest any I had seen, and gave wonderful life to the charming and instructive book I mentioned before, 'Riviera Notes,' which is dedicated to 'Commendatore Thomas Hanbury, of La Mortola, Italy.' I do not know if this book has been reprinted, but I sincerely hope so for the benefit of all who can avail themselves of the privilege, so kindly granted, of seeing Mr. Hanbury's rare botanical collection. Everything in it is arranged in the manner best suited to the growth of the individual plants, coming as they do from all parts of the world and willing to flourish on his sunny rockeries. The garden is full of beautiful memories of the old, uncultivated Italian *podere*, and at the base, between the garden and the sea, runs the original, narrow, paved road made by the Romans, it is said, in the year 13 B.C., and called Via Augusta. The ordinary Cannes visitor is apt to be disappointed, I am told, not to find this garden the usual effect of turf and spring-bedding. The field of anemones that was in bloom the day I was there is far more beautiful than any bed of flowers that one can imagine. Of course, they had originally been planted, but their growth and general effect of colour in the field grass gave the appearance of wildness. There is still in the centre of the garden a most stately avenue of *Cupressus semper-virens*, the pyramidal cypress. It forms a beautiful feature in the landscape, and the people of the place say it recalls a long-forgotten cemetery, as the name 'La Mortola' is supposed to mean 'a place of burial.' The cypress-trees are sharing the fate of the olives and are disappearing from the fields and gardens along this coast. They are claimed by the church as being trees of mourning, and the peasants believe them to be unlucky round their dwellings and appropriate only to burial-grounds. I am very sorry for this, as they are a feature associated with all that seemed most paintable in one's youth, with their tall dark spikes against sunset skies. We are always trying to grow them in England, where they don't flourish, or to imitate them with shrubs of much less beautiful growth, such as *Cupressus macrocarpa* and Irish yews.

My friend and I moved on to join the rest of our party at Florence, where we arrived many hours after our time. The railway officials may have feared some danger from flooded torrents, as the train went very slowly. The weather was extremely bad and very cold, and the idea of going to live up on any of the hills outside Florence did not smile upon my party at all, but I confess that, to me, much of the charm of Florence disappears when I am boxed up in the centre of the town. It ended in our establishing ourselves in one of the hotels on the Lung Arno, and during the three weeks I remained here, we hardly had a fine day and only one that was really warm.

Among the first things I noticed were pots full of beautiful turban ranunculus. They never seem to be

grown in pots at home, which is a pity, as their colours
are so good. We had a happy time and did lots of sight-
seeing, but I said enough about Florence in my last book.
Towards the end of our stay, my son joined our party.
He had never been in Florence before, and one of his first
questions was : ' Where is the statue of David, begun by
Mino da Fiesole and finished by Michael Angelo ? ' I
was immensely surprised, as I had no idea that he knew
anything at all about Mino da Fiesole. He then quoted
from memory the following poem by Owen Meredith :

LOVE'S LABOUR LOST

In the old Piazza at Florence a statue of David stands,
'Tis the masterful work of Michael Angelo's marvellous art,
Yet a failure nevertheless; for it came to his master's hands,
Not a virgin block intact, but already rough-hewn in part.

And what Mino da Fiesole did to it, Angelo could not undo.
So the work was but half his own. It is finish'd, yet incomplete.
As that statue to Michael Angelo hundreds of years ago,
So are you at this moment to me : an achievement, and yet a defeat !

'Tis that others have been before us, of whose touch you retain the
 trace.
You are half my work, half theirs, thro' your spirit and flesh disperst
Is the mark of a love not mine, that my own love cannot efface.
 or you were not virgin marble when you came to my hands at first.

It is one of the many changes which annoy those
who knew Florence in its old days, that the David has been
removed from the place which Michael Angelo chose for
it. But, once it was found necessary to have a hideous
wooden shed over this statue to protect it from the weather,
it really was much better to put it indoors, and the David
gives immense dignity to the noble hall in the Belle Arti,
where it now stands in a splendid light.

About the modern alterations in Florence, a friend of
mine, who had not been there for years, writes : ' I go to

the galleries and churches and find changes everywhere. Like all old folk, I am disposed to say, " Changes for the worse " ; but this would be unreasonable—some are for the better. The great, and to me distressing, change is in the aspect of the city, the loss of its tall mediæval walls and the " improvement " (?) of its chief streets, which now have horrid shop-fronts. The place is no longer in sympathy with the great edifices which stand as if they had come to call ! As to the galleries, I suppose I must admit improvement, though the character of the Medici collections is gone, except in the Tribuna, which has been mercifully spared in spite of its very bad light. I had some difficulty in finding the Madonna del Calamajo, and that famous one of Lippi with the naughty boy angel. But when I did find them I confessed that I had for the first time seen them in a good light, and in a bad light it is not seeing at all. . . . I am angry with the Italian Government for transporting statues, &c., and thus taking their unique character from the streets of Florence. Everything is sacrificed to the idea of a " Museo " with catalogues and numbers and staring foreigners ! The object of these works is lost and ignored when thus removed, and modern horrors of poor Garibaldi who *was* picturesque, and V. Emmanuel, and above all that stout, podgy Manin, are everywhere. If artistic merit meant other merit, old Cosimo would beat all these moderns as he sits on his bronze horse, mercifully not yet inside a Museo. But two things are great gain. One is the superb portraits of the Portinari family in Hugo Van der Goës' great " Adoration." Sina Folco got it done ; it has been in the Spedale and hardly two years in the gallery. The other gain is the fine Botticelli of " Spring " which I never saw properly long ago. It was dirty, hung high and in an impossible light.'

APRIL

Arrival at Naples—Museum—English Society for the Prevention of
Cruelty to Animals at Naples—Slaughter-houses in England—
Art objects from Pompeii sometimes echoes of modern Japan
—Baiæ—Coleridge and Harrison on Gibbon—South of Italy
less affected by barbaric invasions than other parts—Aquarium
—Goats—Dr. Munthe on the housing of the poor—Mrs.
Jameson's picnic—Pompeii: smallness of the houses—Mr.
Rolfe on Pompeiian pins and matches—Cenotaphs and war
memorials—Pompeiian gardens as models for London—Sorrento
and Amalfi—Garibaldi on cremation—'Aurora Leigh.'

ABOUT the middle of April my son and I went on to
Naples, and the rest of the party went home. We did
shopping and sight-seeing all the morning of the day we
left Florence, saw our friends off in the afternoon, came
back and packed, left by rail at six o'clock in the evening,
changed into a sleeping-car at Rome at midnight, and
reached Naples at seven the next morning. We took
a carriage and drove three miles to an hotel high on the
hill to the north-west of the town, where we had great
difficulty in getting a front room. The back rooms were
close against the rocky sides of the hill, the consolation
being that the rock was magnificently clothed by a very
fine old wistaria in full flower. The view from the front
rooms in the early morning light was certainly most
glorious. Our first day was almost our only fine one.
Though well on in April, the weather was very cold, and
after the first day the view was so veiled that we were
only able to see Vesuvius twice. One of these times, as the

clouds rolled by, it was snow-capped and lighted from foot
to summit by the setting sun—bright crimson-red at the
top, graduating down to hazy purple, all the base bathed
in mist, the town standing out, yellow and red, in front
of it, and the smoke of the mountain rising up pale grey
against a coal-black storm-cloud. This was the one really
beautiful colour effect that I saw during my eight days at
Naples. The blue bay and purple Ischia never existed
for us. When we saw the island at all its volcanic peaks
were only silhouetted in a deeper tone against the grey
sky. On the whole I was disappointed with Naples; but
this is frequently the case on first visiting for a short
time a place that one has heard of all one's life and
thought one knew from pictures and descriptions. I was
disappointed in the one-sidedness of its bay, which curves
boldly to the east, but is cut off, to the exclusion of all
coast view, by the rocky feet of the high hills to the west.
These hide the sunsets, and prevent any distant effect on
that side of the town. In all other directions the suburbs
are huge, and getting out of modern Naples is almost like
getting out of London.

To return to our first day. In spite of our night
journey, we got a carriage and made at once for that
marvellous museum which I had longed all my life to
see. It is rich with every kind of beautiful object, in
which the traveller suddenly finds himself face to face
with the lost life of ancient Greece. The Greek invasion
of Southern Italy seems to have been some time later
than the building of Pæstum, of which the date has been
fixed with comparative certainty at about 650 B.C. I had
never been told of one small instance of conservatism
which I noticed in Naples. It is probably to be found all
over Southern Italy; anyhow, the same thing is to be
seen at Brindisi, so that it is known to many people who
travel for business and not for pleasure, and who may

never have visited any other Italian town but that one.
The horses are quite differently harnessed from anywhere else. They have no bit in their mouths, and are
guided by a tight bandage—with, I believe, nails in it—
over their noses. This and many other cruelties to
animals are very apparent in Southern Italy ; and I was
amused to see, by the prominent advertising of the
English Society for the Prevention of Cruelty to Animals,
that, with the usual characteristic of an English settle-
ment in a foreign town, they were keenly alive to the
faults of the people who surround them, while we at
home allow the continuance of the most frightful abuses,
content to be kind to our dogs and horses. I cannot
understand how this Society rests on its oars while the
laws at home respecting slaughter-houses remain inadequate, and even these are evaded, not only in our huge
London, but in every village and town in the land. The
animals in these slaughter-houses are allowed to get into
a state of frenzied terror before being killed, as they are
not protected against the appalling sights, smells, and
sounds which result from the slaughtering of their com-
panions. This state of fear produced in them before being
killed has a most injurious effect on the quality of the
meat. The curious thing is that the only agitation against
these abuses comes from the despised vegetarian, for
sentimental reasons, his own health being unaffected by
them, whereas to the meat-eater there are materially
harmful results in the fact that animals are slain while in
a state of absolute panic. There is, moreover, no proper
inspection—at any rate, in country districts—to ensure
against animals being sold for meat in spite of their being
actually diseased. Very little improvement can be expected while the butcher has to bear the whole loss of a
diseased carcase. I heard the other day, first-hand from
one of my friends, a tale which gives one cause to ponder.

N

In a consumptive sanatorium the patients who were forced by the doctor to eat kidney pudding were all taken more or less ill, as was also the doctor. The cook was questioned about the kidneys. She promptly answered, as if it was nothing very unusual, that several of them had stones in them, ' But I cut 'em out before cooking them ' ! My friend, having a natural dislike to ' innards,' had absolutely refused to eat the pudding, and her immunity from sickness led to the inquiry.

Much as I had looked forward to the Naples Museum, it far surpassed all my expectations. It was Garibaldi, when dictator in 1860, who proclaimed the museum and the territory devoted to the excavations to be the property of the nation. Our time in Naples was so short that, as we had just come from Florence, we resolved to avoid the picture gallery altogether, with its wealth of mediæval paintings, and give our whole time to the Greek and Roman art treasures to be found in the other parts of the museum. Good statues one has seen elsewhere, but the bronze objects used in daily life in Pompeii and Herculaneum, such as lamps, braziers, pots, vases, jugs, beds, &c., were an absolute revelation to me. One's astonishment is heightened when one realises that Pompeii was merely a seaside summer resort, about the size of, let us say, Worthing. The bronze braziers, beautiful as they are, must have afforded little heat to the inhabitants in the cold spring months, as most of them are so tall—some nearly six feet high. I suppose it is not known what was burnt in these braziers ; they may have had some artificial fuel, as is used in Japan to this day. The lamps are of endless variety of design, sometimes grotesque, sometimes quite plain, but all beautiful in proportion and balance. Placed on low tripod stands, they reminded me in form and general appearance of some of the best Japanese stands and vases for flowers.

Did the Greeks get it from the East, this wonderful feeling of balance, or did Japan gain from India a reflection of Greek civilisation ? Balance plays so great a part and gives so great a charm to many things. This is well illustrated by a tiny bronze statuette of Victory in one of the glass cases. The extended wings support her, and though one arm is gone, the sense of security is perfect, and there is no fear that she will slip from the globe beneath her foot. We spent two hours the first morning and two hours another morning in this paradise of treasures, but that was only sufficient to whet the appetite. All modern travelling is spoilt by hurry and want of time. One travels further and sees less, thanks to the rapid modes of locomotion. We returned to our little carriage with the bitless ponies, and, having lunched in the town, we resolved, with modern energy and want of faith that the weather would remain as fine as it was then, to drive at once to Baiæ. The delighted coachman never let out that the distance was fifteen miles. I think if we had known this we should have postponed the expedition to another day. But I was very glad we went then, for I believe we should never have done it otherwise, as the weather did get worse, and we should have had no time.

To go to Baiæ you drive through a tunnel which pierces the western range of hills, and emerge on to a beautiful open sunset space. The road is everywhere studded with Roman ruins. We visited, by order of the coachman, sulphur springs and eruptive holes along the road, and spent a delightful hour and a half at Baiæ itself trying to discriminate what were temples, villas, or amphitheatres. I found that my ignorance was complete with regard to all classic history or the development of its civilisation as shown in architecture. When in my youth I read Gibbon's ' History of the Decline and Fall of the Roman Empire ' it had dawned even upon my inexperienced mind that the instruction he conveyed was of an unsatisfactory kind, and in no way enlightened the reader as to the various influences of race, religion, or art, nor as to the evolution of one era out of another. It was, therefore, no small consolation to me to find in a little volume of Coleridge's ' Table Talk ' which I happened to have in my travelling-bag, his opinion as to the uninstructive superficiality of that immortal work. It may comfort others who have, and still more those who have not, read that history, so I copy it here :—

' The difference between the composition of a history in modern and ancient times is very great; still there are certain principles upon which a history of a modern period may be written, neither sacrificing all truth and reality, like Gibbon, nor descending into mere biography and anecdote.

' Gibbon's style is detestable, but his style is not the worst thing about him. His history has proved an effectual bar to all real familiarity with the temper and habits of Imperial Rome. Few persons read the original authorities, even those which are classical; and certainly no distinct knowledge of the actual state of the empire can be obtained from Gibbon's rhetorical sketches. He takes notice of nothing but what may produce an effect ; he skips on from eminence to eminence, without ever taking you through the valleys between ; in fact, his work is little else but a disguised collection of all the splendid anecdotes which he could find in any book concerning any persons or nations from the Antonines to the capture of Constantinople. When I read a chapter in Gibbon, I seem to be looking through a luminous haze or fog : figures come and go, I know not how or why, all larger than life, or distorted or discoloured ; nothing is real, vivid, true; all is scenical, and, as it were, exhibited by

candlelight. And then to call it a " History of the Decline
and Fall of the Roman Empire "! Was there ever a
greater misnomer? I protest I do not remember a single
philosophical attempt made throughout the work to
fathom the ultimate causes of the decline or fall of that
empire. How miserably deficient is the narrative of the
important reign of Justinian! And that poor scepticism,
which Gibbon mistook for Socratic philosophy, had led
him to misstate and mistake the character and influence
of Christianity in a way which even an avowed infidel or
atheist would not and could not have done. Gibbon was
a man of immense reading; but he had no philosophy;
and he never fully understood the principle upon which
the best of the old historians wrote. He attempted to
imitate their artificial construction of the whole work—their
dramatic *ordonnance* of the parts—without seeing that their
histories were intended more as documents illustrative of
the truths of political philosophy than as mere chronicles
of events. The true key to the declension of the Roman
Empire—which is not to be found in all Gibbon's immense
work—may be stated in two words—the imperial
character overlaying, and finally destroying, the national
character. Rome under Trajan was an empire without a
nation.'

This last sentence may be not without wise application
to the England of to-day. On my return home I referred
to that excellent book of lectures by Mr. Frederic
Harrison on 'The Meaning of History' (Macmillan,
1894), to read what he said about the 'Decline and Fall,'
and it is so supremely interesting and encouraging to see
what different views great men take that I must quote
what he says from the chapter called 'Some Great Books
of History':—

'It is no personal paradox, but the judgment of all
competent men, that the " Decline and Fall " of Gibbon is

the most perfect historical composition that exists in
any language; at once scrupulously faithful in its facts;
consummate in its literary art; and comprehensive in
analysis of the forces affecting society over a very long
and crowded epoch. In eight moderate volumes, of
which every sentence is compacted of learning and
brimful of thought, and yet every page is as fascinating
as romance, this great historian has condensed the history
of the civilised world over the vast period of fourteen
centuries—linking the ancient world to the modern, the
Eastern world to the Western, and marshalling in one
magnificent panorama the contrasts, the relations, and the
analogies of all. If Gibbon has not the monumental
simplicity of Thucydides, or the profound insight of
Tacitus, he has performed a feat which neither has
attempted. "Survey mankind," says our poet, "from
China to Peru!" And our historian surveys mankind
from Britain to Tartary, from the Sahara to Siberia, and
weaves for one-third of all recorded time the epic of the
human race.

'Half the hours we waste over desultory memoirs of
very minor personages and long-drawn biographies of
mere mutes on the mighty stage of our world, would
enable us all to know our "Decline and Fall," the most
masterly survey of an immense epoch ever elaborated by
the brain of man. There is an old saying that over the
portal of Plato's Academy it was written, "Let no one
enter here till he is master of geometry." So we might
imagine the ideal School of History to have graven on its
gates, "Let none enter here till he has mastered Gibbon."
Those who find his eight crowded volumes beyond their
compass might at least know his famous first three
chapters, the survey of the Roman Empire down to the
age of the Antonines; his seventeenth chapter on Constantine and the establishment of Christianity; the reign

of Theodosius (chaps. 32–34); the Conversion of the Barbarians (chap. 37); the Kingdom of Theodoric (chap. 39); the reign of Justinian (chaps. 40, 41, 42); with the two famous chapters on Roman Law (chaps. 43, 44). If we add others, we may take the career of Charlemagne (chap. 49); of Mahomet (chaps. 50, 51); the Crusades (chaps. 58, 59), which are not equal to the first-mentioned; the rise of the Turks (chaps. 64, 65); the last Siege of Constantinople (chap. 68); and the last chapters on the City of Rome (69, 70, 71).'

It seems ridiculous, but to me there was more life and realism in the pages of Sienkiewicz's novel, 'Quo Vadis' than in any chapter of Gibbon. Nowadays, it is the fashion to defend the character of the worst Roman emperors, making out, what is probably quite true, that they were not as black as they were painted. But Sienkiewicz's object is not to exonerate Nero; on the contrary, his line is a prejudiced one against the old world in favour of the new, and offends many people with its dogmatic tone.

As is usual with me when I get home, I find out all sorts of books that would have been a great help had I taken them with me. It is so different reading books in the atmosphere of the locality to which they refer, compared with merely seeking for general information. There are many books of this kind in Bohn's Classical Library, and the 'Lives of the Twelve Cæsars,' by C. Suetonius Tranquillus, translated by Alexander Thompson, M.D., is a classic which would vivify the old world to a great many. This account of these emperors by a contemporary—Suetonius was the son of a Roman knight who commanded a legion—seems to bring the old world as near to us as do Pliny's Letters, many of which were addressed to this same Suetonius, with whom he lived in the closest friendship. Lately, Macmillan has republished in the 'Golden Treasury Series' Cicero's two famous essays on Old Age and Friendship. It is a new English translation, by E. S. Shuckburgh. In the introduction he kindly tells us that these essays were written while Cicero in his old age (63) was travelling about with his secretaries. He says, ' With the palace of fame so laboriously raised tumbling about his ears, Cicero found consolation in two things—literature and philosophy. While moving from villa to villa on that enchanting coast' (which was the coast about Naples) ' he was incessantly reading and writing ; keeping his staff of literary slaves, or freedmen, so hard at work that they longed for the holiday to be over, and to return to the less fatiguing duties of city life.'

Our return journey from Baiæ was in the cool twilight, when our little ponies took us back to our hotel at Naples. I had had a pretty hard thirty-six hours, and reserved to myself the right of having milk and bread in my room instead of going to the table-d'hôte. I woke the next morning as fresh as usual.

Walking about Naples is full of interest ; for though the principal streets are like those of many other continental towns, this only shows up the remarkable contrast of the side streets and alleys, so narrow, so small, so picturesque, with their strip of blue sky above, the shops windowless—only large black holes in the wall. The goods displayed are half of them inside and half tumbling out on to the old cobbled roadway with the gutter running down the middle of it. So much in the Naples of to-day seems like Pompeii brought to life again. The very method of driving suggests a chariot behind the coachman instead of a fly. Everything seems to make one realise that the dark ages of barbaric destruction left Southern Italy comparatively untouched. Perhaps as a natural consequence of this, art, having lingered and clung in every direction to the traditions of the past, the Renaissance was here without life or originality, and the pictorial art which grew up in Naples in the middle ages permeated gradually from the North. This is interestingly shown by the couple of pages in Kugler's 'Handbook of Painting' which he finds sufficient to describe 'the school of Naples.' The chief interest of this 'school,' as he calls it, is that many of the pictures are original works by Flemish painters which served as models to the Neapolitan artists. Later they were affected by the Spanish school.

Oh! we had so little time for everything. What is a week in Naples for seeing all one wants to see, even in the most superficial way? The aquarium is, I believe, one of the best in Europe, not only for superiority of arrangement and lighting, but for the scientific interest of the exhibits. With a little friendliness to the custodian, he shows off for one's benefit some of the peculiarities of the inhabitants of the sea by feeding them. For me aquariums have a great fascination, though I have to harden myself against a feeling of extreme depression aroused by the cruelties of nature as there displayed.

Walking home in the evening, we met one of the flocks of goats that are driven about the better parts of the town to be milked at the house doors. Seeing goats in the streets of Naples first gave me the idea, which I work out earlier in the book, of what an immense benefit the keeping of goats might be in England.

I tried to find out where Lady Hamilton and Sir William lived in the Nelson time. Her letters in that most curious book by J. C. Jeaffreson, 'Lady Hamilton and Lord Nelson,' were generally written from what Mr. Jeaffreson calls 'her villa in Caserta.' The few people I was able to ask seemed never to have heard of the Hamiltons, or even of Nelson. But I saw no educated Italians. They would doubtless know all about it, for I fear hate lives longer than anything, and they will perhaps never forgive the terrible story which is recalled with great historical precision in the introduction to ' The Autobiography of Giuseppe Garibaldi,' finished by his friend Jesse White Mario. This book ranks to-day as almost an old one, having been published in 1899. It gives the Italian version of what I suppose is now universally recognised as Nelson's great crime, in helping the King and Queen of Naples to forsake their country, and despoiling it of its treasure and jewels, at the time of the French invasion in 1799. No anecdote in history sticks so much to its locality as that of the murder of the noble Admiral Caracciolo, who, when condemned on board the British flag-ship for not obeying the orders of his disloyal King, was hanged at his own yard-arm, having been refused the only favour he asked after his trial, which was to be shot. No one can look at the Bay of Naples without remembering how the Italian admiral's body, in spite of its heavily shotted shroud, rose to the surface and floated under the windows of the King's cabin on board the English admiral's flag-ship. Not content with this, we must never forget that the English banner waved over Naples in those terrible times when the early martyrs of Italian liberty and thirty thousand prisoners suffered inhuman cruelties, which Mario says ' defy description and surpass belief.'

The day we went to Pompeii was perversely unfavourable—grey, thick, and very warm. We thought we should be superior and do things in old-fashioned style by driving there instead of going by rail. This almost anyone could have told us was a great mistake. The rail is infinitely the better, though in fine weather I should think it would be possible to go by water—for those who like it. We drove through miles and miles of continuous slums, broken here and there by a handsome old villa. All picturesqueness of costume has disappeared—no lazzaronis, no red caps, no eating of macaroni in the streets. The street often runs close to the sea, but the double row of houses hides everything. Murray says the drive is interesting if you have time to visit the towns at the foot of Vesuvius ; but we had not, and we slightly resented his not having warned us against the badness of the road and the dreariness of the drive, though he does say that those who are pressed for time had better go by rail. In this I entirely agree. All along the drive I kept thinking of Dr. Axel Munthe's wonderful ' Letters from a Mourning City,' written at the time of the cholera epidemic in 1884. I have often heard Naples described as the hot-bed of every vice. But in making that no doubt true accusation, how few remember the sanitary condition of the place, which is worse and more crowded than in any other city of Europe ! Munthe gives the following description of the hygiene of Naples in 1885, and it is a great consolation to be told, as I was, that with the sacrifice of much of the picturesque, there is also a great improvement now in the sanitary housing of the poor. Munthe says : ' The laws of hygiene teach us how close a connection exists between the sanitary conditions of a locality and the density of its population. The history of the Neapolitan epidemic furnishes us with an example of this law concerning density of population. Upon a surface of eight square kilometres (amount of surface that has been built over), there dwell no less than 461,962 human beings. And according to the official statistics no less than 128,804 of these people inhabit underground dwellings and cellars. But there is something worse than these *bassi* and *sotto terrani* ; another step down the shelving ladder of society and we come to a still more wretched form of habitation—to the *fondaci*. You have often heard me speak of these places as the scenes of the most

appalling misery out here.   There are eighty-six *fondaci*
in Naples at the present moment; formerly they were
still more numerous, but more modern constructions have
done away with a good many. . . . A third sort of dwelling place consists of the so-called *locande*, where lodgers
are received for the night at two and three soldi a head.
I have even seen *locandi* where they are received for one
soldi a head, but there the people sit and lean their arms
and heads *against a rope* that is stretched across the
room from one wall to another, not a bad idea for accommodating a crowd.'   Dr. Munthe refers in such kindly
words to the superstition which means so much to
Italians and is often only an irritation to northern minds,
that I must make one other quotation :—

'. . . And those who sneered at their superstitions
and forbad their processions, what had they to offer them
in exchange for their obscure but rock-like faith?   Ah
yes, sanitary rules, veritable sarcasms on their poverty,
printed advertisements, which most of them were unable
to read and none of them were able to understand, recommending them to live in airy rooms, to avoid vegetables,
and take to meat, to disinfect constantly, either with
carbolic acid or with "corrosive sublimate," the most
effective microbe antidote according to Dr. Koch. . . .
And what has the obtuse brain of a lazzarone to do with
Koch and microbes?—he whose thoughts have never
crossed the bay beyond which the whole of the remaining world is "Barbaria" to him; he who knows a host
of Saints' days, a few prayers that he has been taught as
a child, the names of a dozen fish he has seen jumping
in the nets at Mergellina; he who can play at *morra* and
sing *Santa Lucia*, and that is about all!   How is he to
manage to "air the room"—he who lives with ten or
twelve others in one of those *fondaci*, into which the
light of day has never been known to penetrate, where

one of us is unable to remain for more than a moment
without going out to take a breath of fresh air? And he
is, forsooth, to choose his food; he whose expenses at the
best of times never exceed one or two soldi a day; he
who never in his life has had the chance of tasting meat,
and whom you may perhaps see standing in front of the
baker's shop, watching with the expression of a hungry
animal in his eyes the piece of bread which you have just
given your dog, and then fighting over the remaining
crumbs with a crowd of others such as himself.

'And as to disinfection! What does he know about
that, he who, alas! shows so little inclination to master
the first great rule of disinfection, the popular antiseptic
which consists in sometimes dipping one's hands and
face in water. . . . Just give a thought to all this, and
then it will no longer strike you as so wonderful that
these poor people should believe more in the censer's
clouds than in the sulphur's fumes, and more in holy
water than in a ten per cent. solution of carbolic acid.'

I felt thrilled on arriving at Pompeii, which all my
life I had so longed to see; but the dust, the dirt, the
crowd of tourists at the little hotel where we had to
breakfast were most destructive of all the higher emotions.
The old world faded away, and the degradedness of what
is called civilisation was terribly to the fore. But in this
motley crowd my son and I were really alone—a fact
which contrasted favourably with my recollection of Mrs.
Jameson's description in her ' Diary of an Ennuiée' of a
fashionable society picnic. How much happier and how
much more real is the enjoyment of anything we see with
one chosen companion than with a host of even intimate
acquaintances? Writing on March 30, 1824, Mrs. Jameson says: 'Yesterday we dined *al fresco* in the Boboli
Gardens; and though our party was rather too large,
it was well assorted, and the day went off admirably

The queen of our feast was in high good humour, and irresistible in charms. . . . Everybody played their part well, each by a tacit convention sacrificing to the *amour propre* of the rest. Every individual really occupied with his own particular *rôle*, but all apparently happy and mutually pleased. Vanity and selfishness, indifference and *ennui*, were veiled under a general mask of good humour and good breeding, and the flowery bonds of politeness and gallantry held together those who knew no common tie of thought or interest; and when parted (as they soon will be, north, south, east, and west) will probably never meet again in this world; and whether they do or not, who thinks or cares ? '

We finished our very indifferent breakfast as quickly as possible, engaged a guide—which is absolutely essential if one only has time for a single visit—and got a *chaise-à-porteur* and two strong porters to carry me about. The entrance to the town is now arranged with a turnstile, and everything you pass is known and explained to you by the guides, who seem to exercise their monotonous duties with great civility. Although it is absolutely right of the Italian Government to preserve their treasures against the modern inroad of tourists from all over the world, I think there must have been immensely greater charm in visiting Pompeii in the old free days when everybody could poke about and wonder and muse as he liked. Shelley certainly could not have said to-day as he did in 1820 in the ' Ode to Naples '—

> I stood within the city disinterr'd ;
> And heard the autumnal leaves like light footfalls
> Of spirits passing through the streets ; and heard
> The mountain's slumbrous voice at intervals
> Thrill through those roofless halls.

In those happy days before the Italian Government undertook the guardianship of the treasures at Pompeii, it almost seems that anybody could dig, buy, or rob at his own discretion. Even so late as Mrs. Jameson's day (1824), she tells how a little imp of a lazzarone clawed away the dust and dirt to show her a most beautiful aërial figure with floating drapery, representing either Fame or Victory. He quickly covered it up again that the other workmen might not see, and grinningly appealed for a few pence as his reward. Bulwer is, some say, little read now by the young; but I expect as long as the English language is spoken his 'Last Days of Pompeii' will remain to tourists at Naples what George Eliot's 'Romola' is to visitors at Florence. I wonder if it is more true to the spirit of those classic days than 'Romolo' is to the middle ages, though George Eliot is none the less interesting from being modern—not mediæval—in feeling.

The general characteristic of the Pompeiian buildings is their extreme smallness, and the houses must have been what we should think very cramped and uncomfortable, even in a seaside lodging, to-day. A wish for shade and dread of sunshine seem to have been the first considerations rather than warmth or ease. A cortile, or cloister, generally surrounded the little garden or yard in the middle of the houses. It was here that the Pompeiians evidently spent their days. In one of these I saw fixed tables placed in two or three different directions so as to be suitable for use at various hours of the day. One often wonders whether the climate of Italy is colder now, or if the people are much more chilly than they used to be from want of health. It is only in my lifetime that warming houses in Italy has been considered at all necessary, and certainly 150 years ago the clothes that women wore even in England were strangely insufficient compared with the furs and wraps which people hardly ever put away now.

There has been such immense improvement of late in the art of excavation, that now it can be done without injuring even the most delicate wall-decorations. So good have some of these proved that the houses are being roofed over in order to preserve them. This will make the future excavations much more interesting, as the treasures can now be left on the spot instead of carrying them off to the museum. The mural decorations were evidently not the work of great painters, like the frescoes of mediæval Italy. They are without much variety, they have no gradation, and are painted in a conventionally fantastic style rather than in the realistic or poetical. In fact, they rank as the upholsterer's purely decorative art. But, given this fact, the wonder is that so high an artistic level was attained, and they prove that the ordinary ' decorator ' of the day was essentially an artist in his own particular line. The patterns on the walls, even when only stencilled, are well drawn and in excellent taste. In spite, however, of their charm of a miniature kind in these little empty rooms, I do not think the decorations displayed on the black or red walls of Pompeii would bear imitation in our northern houses. The rooms evidently must have been comparatively devoid of furniture, which made the wall-painting their chief ornament ; as in Japan to this day, one hanging picture and one flower arrangement are thought sufficient decoration.

I only record one or two of my personal impressions, as Pompeii must appeal to everyone in his own especial way. I think people going to Pompeii would do well to get and read before they go there a book by our Consul at Naples, Mr. Eustace Neville Rolfe, called ' Pompeii, Popular and Practical : an easy book on a difficult subject.' This it essentially is. I did not see it till after I got back. As a specimen of the kind of interesting information he gives, I quote this one paragraph : ' It is unsafe to argue

that the Pompeiians did not have things, because we have
only recently invented them. The safety-pin, which is
quite a modern invention, was in common use in Pompeii.
Wire rope, which we look upon as a new discovery, was
known in those days, and a very fine specimen of it may
be seen in the Naples Museum. Martial speaks of sulphur
matches, which in our English kitchens replaced the old-
fashioned tinder-box scarcely fifty years ago, and the surgical instruments found in Pompeii were lost to science
for centuries and reinvented in our day almost in their
original form.' In various other directions besides these
Mr. Rolfe alludes to, it is clear that our civilisation has
not yet reached the level of the Greek long before the
existence of Pompeii. In a chapter on the ' Street of the
Tombs,' he has what was to me a very suggestive passage.
In commenting upon these tombs, he says : ' We must
not pass over the Cenotaph of Calventius Quietus, because
Cenotaphs represent an interesting phase of the post-
humous honours paid to the deceased by the Romans.
The word signifies "an empty tomb," and such buildings
were habitually erected to people who had passed away
without having burial rites. Thus if a man were drowned
at sea or killed in battle, a tomb was erected for him at
home in the belief that this would give his spirit rest, and
a safe transit across the Styx. The idea is pathetic, and
is frequently made use of by the poets, sometimes by
making the unburied corpse reappear, sometimes, as in
Horace's well-known ode, by making the spirit beg the
passing stranger to cast a handful of sand upon the unburied corpse, in token that rites of some kind had been
performed upon it, so that the wandering soul might have
rest.' This idea may seem heathenish to some, but surely
there is a dignity in the principle of honouring at home
those who die in the service of their country, no matter
where or from what cause. I think the soldier's grave

O

should be left to nature, as the sailor is buried at sea, and
it has seemed to me lately, with my mind constantly
dwelling on the sadnesses of the South African War, a great
mistake, though a very natural one, to try to make the
perishable imperishable by societies for the preservation of
soldiers' graves on the veldt. Even in ordinary church-
yards are we not too often reminded of a fact which Sir
Henry Taylor puts into his ' Philip Van Artevelde ' ?

> Pain and grief are transitory things no less than joy,
> And though they leave us not the man we were,
> Yet they *do* leave us.

The ideal way, I think, of honouring the dead and keeping their memory alive would be to have, let us say, in
each postal district of the place from which they came, a
stone raised on which would be written the names and
dates of those, from a general to the youngest private,
who have been killed or died in any war. The feeling is
almost a bitter one that those who die in what are more
or less ' fashionable ' wars, such as the Crimean, or the
recent great Boer war, should be more honoured and their
resting-places more cared for than those who have laid
down their life in the swamps of West Africa or the China
rice-fields.

At Pompeii there is an attempt, in one or two instances,
to restore the gardens. It seems to me the old gardens,
tiny as they were, are very good models of what town
gardens should be with the aid of pavements, brick paths,
tesselated pavements or mosaics, and constant renewal of
the actual plants, never trying to imitate the large spaces
of the country where things can be left to grow on from
year to year and where growth and size become a pride
and delight. I think in small gardens everything should
be kept very decorative, but very miniature. The difficulty
of flowering plants in London is the height of the houses,

which, in some circumstances, excludes nearly all sunshine. Nothing can stand this but ferns and ivy, but a
very charming, dainty little garden might be made with
various coloured ivies, euonymus, holly, privet, &c., provided
they were much pruned or even continually renewed. In
fact, like many other things, London gardens depend on
elbow grease—that is to say, a great deal of work, attention, washing, &c. There is much opening for individual
development in this direction. In a suggestive article
which has much interested me, by Mr. Whitmore, in the
'National Review' for July, 1902, he condemns me for
my objection to evergreens in London, and I quite see his
point of view as regards their use. But it must be in an
entirely different way from what has been generally done
in London suburbs, where the evergreens are left untended
from year to year till everything is black, except the new
shoot for a short time in the spring. I immensely like his
recommendation that no fair-sized London garden need be
without its blossoming fruit-trees. He praises mulberries
a little more than I think they deserve for towns, except in
cases of people having to live in London through the
autumn, as their leaf comes out so very late in the spring,
and they have no blossom to speak of. In these gardens
of miniature evergreens which I am picturing to myself,
colour might be introduced by buying the flowering plants
in pots as they come into season. In fact, the whole treatment of the garden should be as nearly as possible that
of an ornamented terrace, tubs, vases, pots, anything that
there is room for, and the instant removal of anything unhealthy or which reminds one in any way of the injury to
plant life of London smoke and climate. A clean well-
paved yard with one healthy plant gives me more pleasure
than a broad border with badly grown, unhealthy shrubs or
unsatisfactory herbaceous plants. People who are inclined
not to spend much money on their little yard or garden

should confine their expenditure to the original paving and a good hose. Then two or three plants, when they can afford to purchase them, will always look satisfactory. The visitor ought to be able to say, 'How pretty your little yard is!' instead of the rather bashful owner's apology, 'It's not bad for London.' We must never forget that in London all cleanliness is artificial and is only attained by giving the matter a great deal of attention.

On leaving Pompeii, we were tempted to send our carriage back empty and return to Naples by rail, but we decided to go home as we came, slums and all, and naturally found it less tedious, as we knew what to expect.

One more delightful day to record, which alas! did not include the much-wished-for visit to Pæstum. We went by train to Castelamare, then drove again along the eastern side of the Bay, to Sorrento, and part of this way was actually in the country. Had it not drizzled it would have been an enjoyable as well as a beautiful drive. The great hotel at Sorrento where we spent the night looked gloomy and damp, the rooms being well shaded and adapted in every way to keep off the brilliant afternoon sunshine. Capri was invisible, and I had to fall back again on my recollections of Robert Lytton's little poem to imagine how Sorrento could look on a fine spring afternoon. He sent me this poem in a letter from Sorrento ; it was afterwards published in the volume called 'After Paradise; or, Legends of Exile.'

SORRENTO REVISITED

(1885)

On the lizarded wall and the gold-orb'd tree
    Spring's splendour again is shining ;
But the glow of its gladness awakes in me
    Only a vast repining.

To Sorrento, asleep on the soft blue breast
   Of the sea that she loves, and dreaming,
Lone Capri uplifts an ethereal crest
   In the luminous azure gleaming.

And the Sirens are singing again from the shore.
   'Tis the song that they sang to Ulysses ;
But the sound of a song that is sung no more
   My soul in their music misses.

We went to bed early, with gloomy forebodings about
the weather for the following day. We woke the next
morning to find that, without being all we could desire, it
was very much better than the day before. Our little
carriage came early to the door, and the quiet solitude of
the long and beautiful drive lately made along the rock-
bound coast was as great an enjoyment as any day spent
this spring in Italy. Here I first noticed the extreme
beauty of flowering kales. They flower in England
when left to do so in somewhat untidy cottage gardens,
but this happens so much later than in Italy that they got
swamped amongst more showy spring flowers. I instantly
sent my gardener a postcard to preserve, so that they
might bloom, some of the winter kales in my garden.
Since then they have always been allowed to flower, and
make most beautiful pale yellow starry bunches to put
in water or send to London. They remain in perfection
in the garden for over a fortnight.

As we drove on, away from fields and gardens for
about two miles, the whole of the foot of a sloping hill
was covered with extra tall white asphodels (*Asphodelus
ramosus*). I never before saw them so tall or so fine. I
longed to get out and pick some, but my companion said :
' Oh, there'll be plenty more.' Once the two miles were
over, we never saw another during the whole day. The
slope of the hill, with its gentle moisture, protected from
east wind, probably exactly suited the plant, which, to my

mind, is not grown nearly enough in England in large open places, by sheltered streams, or by the sides of woods. I have had a plant or two in my garden for many years, but I have no space for the fine effects of massing plants of one kind together which are so striking in well-managed wild gardening, or what is *called* 'wild gardening.' To get a considerable quantity these asphodels must be grown from seed.

If anything could teach one the real value of protecting plants and vegetables, so little done in England, it would be travelling in Italy in the spring. The extraordinary industry displayed by these poor peasant owners over the lemon crops alone is a lesson never to be forgotten. Miles and miles of terraces, most difficult to reach, are covered over with rough protection made with poles, branches, and rough grass, woven together in the place of straw. Under this hangs the beautiful yellow fruit, only protected on the sides where cold winds from the sea might injure it.

Soon after twelve we arrived at Amalfi, where we spent two or three hours. We lunched at the old Capucine Convent, now an hotel; it is too well known to travellers in Italy to need praise from me. Amongst all its many beauties and interests nothing seems more worthy of notice to the gardener than the world-famed, vine-clad Pergola, which is illustrated in the second edition of Mr. Robinson's 'English Flower Garden.' It is an inadequate representation, and this no doubt is the reason that Mr. Robinson has omitted it from later editions. But how different is the sight of the real thing from any illustration or any description. The Pergola is in two parts, the old and the new. The old, as is so often the case, is infinitely the more beautiful, the more practical, and, needless to say, the more lasting. It is built *cornice*-like along the side of the hill. As you stand at the hotel

door and look down its long gallery of round columns,
snowy white, built of I don't know what, and plastered
over, you see a strong wall about four and a half feet
high on the left or hill side. This is wedged between the
columns, looking as if it were threaded through them,
to prevent the falling stones or slight landslips from
the hill above injuring the broad Pergola path, along
which the monks used to enjoy their daily winter walk in
the sun. On the top of this wall, where the gentle
summer moistures are caught as they drip from the hill,
is a bed of cultivated soil containing all kinds of flowers.
On the right or precipice side of the path, the artificial
construction is for the reverse object and supports the
long walk itself. The wall stretches for some way down
the surface of the rock, but rises only about three feet
above the level of the path, for the safety of those who
wander by moonlight, and to prevent any weakening or
slipping away of the level footway. On this side, a broad
flower bed, nearly three feet wide, is placed within the
wall and consists of a border of earth raised terrace-like
about a foot above the path. Here there is shelter from
the winds, and Madonna lilies and irises bask in the sun.
The roofing is of sapling trees—stems gnarled and rough-
hewn. These are placed, beam like, across the pillar-tops,
securely wedged by a deep rut in the plaster, and extending boldly for several feet beyond the column. The
length-way beams are tied haphazard, now under, now
over, these cross-bars, and overlap at the joins to give
strength. They are laid over and tied to the cross-bars
only on the inner side of the columns, leaving these un-
connected (except with the opposite side), like an avenue
of stone trees. The vines are planted in the beds within
the walls. Now that I am trying to describe it, I regret
very much that I took no measurements, as this was the
most beautifully proportioned Pergola I have ever seen.

Photographs of it, however, are easily obtainable, and it is a model which would help anybody to understand how beautiful Pergolas can be. In the newer part, though the walls are solidly built as the situation renders necessary, the supports are only made of rough saplings, round which the vines are twisted and tied. This part is more helpful than the older one for the arrangement of beds and the general planting of a Pergola on flat ground. The cliff is much less precipitous, so that the walls are lower and the flowers hang over the path from beds outside, where the vines are also planted.

All over the cliff was growing in great beauty the spiræa called by tourists ' Italian may ' (*Spiræa japonica*, known as *S. callosa*). Four or five hours' more driving brought us to La Cava station, and we returned to Naples by rail.

I had time to see only one garden at Naples. It lay under the shade of the western hills, and was a very beautifully wooded, tangled dell, more a shrubbery run wild than a garden, and it sadly wanted pruning. I was told that Garibaldi had lived and died at the villa belonging to this garden, down close by the sea. But this can only have been partially true. He may have lived there, but he died on June 2, 1882, in the Island of Caprera. It was his earnestly expressed wish to be cremated, but this neither his widow nor anyone about him had the courage to carry out. It is stated in his biography that, after being told by Captain Roberts of the burning of Shelley and Williams, Garibaldi said : ' It is the right thing, and it is beautiful and a healthy thing also ; you defy worms and corruption, you do not contaminate the air of the living. Only the priests oppose it ; it would hurt their trade.' I think this puts the matter well, though of course the sentimental reason against cremation is not alluded to, neither is there any acknowledg-

ment of the rather trivial objection that cremation renders
the exhuming of bodies impossible in cases of supposed
murder. This seems to me a very small evil compared with
the daily contamination of earth, air, and water for the living.
Moreover, from the sentimental point of view, there is an
almost unthinkable horribleness in the crowded ceme-
teries near towns, so different from the quiet resting-place
under an old yew-tree in the pure country air, where in
youth we most of us think we should like our bones to be
at peace when our end comes. Sir Henry Thompson, the
well-known surgeon, who has been the great promoter of
cremation in England, was telling me only the other day
how very, very slow is the progress it makes in our own
or any other country. To its advocates this is incredible—
it seems so obviously the most sensible form of burial.
Every year Sir Henry Thompson publishes the statistics
of cremation in the ' Lancet.' This year his article
appeared in the issue of July 5 (1902), and is very interesting reading to those who care for the subject.

April 23rd.—Our return journey from Naples was by
the German Lloyd steamer, and our finest two days were
the day we went on board and the day we got off at
Genoa. In old times, if one went from Leghorn to
Genoa by sea, it was by an Italian steamer. To-day the
fine ships along that coast are German. You get out of
the little boat that takes you from the Italian shore to the
steamer, and you feel almost as if you were in a great
hotel in Germany. I have travelled so little by sea that
I immensely enjoyed the sight of the Bay of Naples and
all the little scenes surrounding a big steamer, boys
diving for coppers, &c. Buildings, or the work of man
like Stonehenge, or the Mont St. Michel, are exceedingly
like their pictures, photographs, drawings, or paintings;
but Nature—in this instance Vesuvius—was far more
beautiful than any portrait of it ever conveyed to my

mind. Steaming along the coast was enchanting, and I enjoyed it to the full, with the sea so calm that the large vessel was absolutely steady. All the fine afternoon till it grew dark we sat on deck and enjoyed the beauties of the shores and islands. My mind flew back to Marion's description in ' Aurora Leigh ' :—

> . . . . I could hear my own soul speak,
> And had my friend,—for Nature comes sometimes
> And says, ' I am ambassador for God.'
> I felt the wind soft from the land of Souls ;
> The old miraculous mountains heaved in sight,
> One straining past another along the shore,
> The way of grand dull Odyssean ghosts,
> Athirst to drink the cool blue wine of seas
> And stare on voyagers.

Very few young people nowadays seem to read ' Aurora Leigh.' This is not surprising, as the cause of women as preached by Mrs. Browning has been more than accepted for some years. But I was interested to see, in a syllabus sent to me the other day by a suburban club for women's lectures to women, that one lecture was entitled : ' What do the women of to-day owe to " Aurora Leigh " ? '

Genoa was bathed in golden, bright, twinkling sunshine, and oh! the regret of toiling up to the hot station to take the train for the north.

MAY

Apology for more gardening notes—Journey to Ireland—New English Art Club—A modern landscape recalling Claude at his best—Spring in the West of Ireland—Glorification of flat garden by old yuccas—Persian ranunculus—Want of thinning out and pruning a universal fault—An East Coast garden—Cultivation of *Hydrangea paniculata*—' The Wild Geese '—Gardening letter from German friend—Two good spring plants—A sundial—Floating bouquets —The May horticultural show.

THE number of excellent gardening books that have been written of late almost makes one ashamed to write any further on the subject. But perhaps the personal experience of any one individual has always a certain interest for amateurs who are really fond of gardening. Not long ago, when listening to a discussion about these gardening books and their superabundant number, I heard one man remark, with rather a sad voice, for he had great possessions : ' No one ever seems to write with a view to helping people who have *large* gardens and woods.' So it came into my mind that I might write a few notes about large places I have visited during the last two years, and that, perhaps, any information I had gleaned as to any individual plants which I particularly noticed as doing well either at home or abroad, might be of some service to those who had gardens of no matter what size.

I have divided my notes into months, because gardening hints are so much simplified by taking the diary form , and in this way I can condense two years' experiences into

one. I have omitted the three winter months of November,
December and January ; for, though all months are equally
important at home, one gains but little knowledge in
winter from seeing the gardens of others. One's own
garden is never dull in the worst weather because one can
always picture what will be, and one can always think
over the errors of the past, as useful in gardening as in
other things, so long as it is not merely regretting the
past, but determining to do better. Maeterlinck says
somewhere that ' we are so constituted that nothing takes
us further or leads us higher than the leaps made by our
own errors.' If this is true of life, it is doubly true of
gardening, and winter is a time for planning and
reflection.

This late spring saw me journeying to Ireland instead
of Italy, as I did last year. The day I left home I break-
fasted, as usual, at eight ; I then gardened, wrote letters,
finished up the usual home-leaving business, and gave
orders. At twelve I dressed and caught the one o'clock
train to London, ate my bread and fruit travelling-luncheon
in the train, deposited my maid and luggage at my son's flat,
went to see a friend, and then visited two picture galleries.
On my return to the flat I wrote several notes, rested and
read for an hour, and dined at eight, on cheese, bread, salad,
and fruit. I travelled by night mail to Dublin with the
usual interrupted sleep of that journey. At nine the next
morning in the train I had some hot milk and bread and
butter, though sorely tempted to fall to excellent slices
of grilled fresh salmon, which everyone around me was
enjoying. I arrived at my destination at 12.50, and after
a cold bath and a change of clothes I felt as little tired as
if I had been at home and spent the night in bed. I
merely give these uninteresting details to prove that, after
nine or ten years of diet, my health and strength are better,
not worse, than those of most people of my age, or even

many much younger. This in no spirit of self-glorification, but for the sake of the cause which, I maintain, makes people stronger and more able to work, not less so. That is the whole point.

Before I go on with my Irish visits I must say a few words about one of the galleries I mentioned above, as it was the first visit I had ever paid to an exhibition of the New English Art Club. It interested me much, as it seems to represent to the young artist of to-day the very same kind of rebound that the early pre-Raphaelite brotherhood did to those interested in the new art in my youth, or, on broader lines, what the eighteenth century was to the seventeenth. So far as I could judge by this one exhibition, the school is a revolt against what is generally called colour, a determination to follow some special set of tones, and to arrive at a certain harmony and charm by pitching all bright pure colour, especially green, out of the paint-box. Is realism to be ignored in favour of a kind of *cinquecento* Titianesque harmony of tones ? Very decorative, very charming, as far as it goes, but only very exceptionally true to nature, because one so seldom sees those warm, blue-green and rich brown-yellow foregrounds in nature, and then only for a short time after a damp sunset. Does it not lead to mannerism to follow one special set of tones to express all moods of Nature ? The critics say the tendency of this club is towards a ' soberer tonality ' ; certainly this struck me most forcibly, for there seemed hardly a picture in the room with any of what I should call real colour in it. This school apparently seems to think that colour is tone, gradation, and values, and so it can be, but only every now and then. Perhaps the very reserve of these pictures would make them pleasanter to live with than what some of us would think better pictures in a gayer key. My chief interest in them is to try to discover the tendency of this school and its meaning. Titian's landscapes were almost always conventional backgrounds to figures. If we aim at a continuity of tone in painting, are we not limiting far too much the endless diversity of nature? Are we to go back to Claude and Poussin, or the Early English water-colourists? But the New English Art Club has ideas and objects that make one think, and I shall always now try to go and see their exhibitions. One picture, to my mind, illumined the whole collection; the subject exactly suited the school. It was by Mr. Roger Fry, and was called ' A Baroque Façade '—a quaint Italian palace, with figures on the top, all grey, standing up, mysterious and dark, against a most beautifully toned and graduated evening sky, streaked in all directions with delicate grey after-sunset clouds. The landscape stretched away to blue grey distance on the left, and there were indications of walled gardens, and orange-trees, perhaps, on the right. The picture caught just that moment before darkness comes, when one says, while gazing at nature, ' How beautiful ! ' knowing that in a moment it will be gone. I had just this feeling about the picture, yet knowing it would not fade. It is a long time since I have enjoyed a painted landscape so much, and it is certainly no reproach that, in a mysterious way, though in no sense an imitation, it recalled the most beautiful Claude I ever saw.

To return to my time in Ireland, which was very short, I paid a visit to a large, beautiful place not far from the West Coast. The weather was cold and wet and, even in the West, hardly a fortnight in advance of Surrey, but nothing could spoil the spring loveliness of the green grass, and the green woods, and the cowslips ! Such cowslips ! I had forgotten what cowslips could be till I saw them luxuriating in the moist Irish meadows with stalks a foot long. The garden was a large, flat, wind-swept, formal one, designed, I suppose, in the

middle of the last century, with the ordinary spring
bedding, wallflowers, anemones, tulips, and forget-me-
nots. But the formality was redeemed, almost glorified,
even so early in the year as May, by the unusualness of a
great number of old spreading yuccas. I never saw them
grown in this way before, with huge stems and young
ones increasing all around. They were not either *Y.
gloriosa* or *Y. filamentosa*, but *Y. pendula* or *recurva*. It
is a magnificent plant when old and established, and particularly adapted for the middle of beds, for the growth is
picturesque and yet symmetrical. The lower leaves sweep
the ground and the central ones point upwards as straight
as a needle, and they flower more frequently than the
other kinds. In this particular garden, when this host of
yuccas are in flower, it must be a rare and beautiful sight.
The whole thing can hardly have been planned, but was
evidently produced by one of those accidents of gardening
where Dame Nature takes the matter boldly in hand.
And how beautiful are these accidents in any old garden !
In this case it was leaving the yuccas alone to increase as
they liked, yet feeding them well, that had produced the
wonderful result.

In a sheltered corner, by a wall, was a very fine plant
of *Erica mediterranea*. This heath grows wild on the
West Coast of Ireland, and is supposed to have been
brought by the Spaniards, as it is a native of the South-
West of Europe. My friends gave me a good plant of it,
and I hope to induce it to grow in Surrey. Careful
pruning after flowering, for all kinds of hardy heaths,
seems the way to keep them healthy, especially in gardens
where dryness prevents their growing freely. I am full
of hope that there is going to be an increased cultivation
in good gardens of the hardy ericas. There was a fair
number of them exhibited at the Drill Hall. The 'Garden'
of June 28, 1902, had an excellent article on the hardy

heaths. In the later editions of Mr. Robinson's 'English Flower Garden,' he says, under ERICA (Heath) : 'Beautiful shrubs, of which the kinds that are wild in Europe are very precious for gardens. We should take more hints from our own wild plants, and bring the hardy heaths of Britain as an artistic element into the flower garden. Why we should have such things as the alternanthera grown with care and cost in hothouses, and then put out in summer to make our flower gardens ridiculous, while neglecting such lovely hardy things as our own heaths and their many pretty varieties, is a thing that would require some explaining. But very many people do not know how happy these heaths are as garden plants, and how delightfully they mark the seasons, and for the most part at a time when people leave town. A singularly pretty heath garden is that of Sir P. Currie at Hawley. In front of his house he has kept, instead of a lawn, a piece of the heath land of the district almost in a natural state, save for a little levelling of old pits. In such places the native heaths of Surrey and Hampshire sow themselves, and nothing can be more beautiful. Where, as in many country places, these heaths abound, there is no occasion to cultivate them, although we cultivate nothing prettier ; but certain varieties of these heaths are charming, and deserve a place in the garden or wild garden. In places large enough for bold heath gardens it would be charming to plant them, but a small place is often large enough for a few beds of hardy heaths. Once established they need very little attention. To some it may be necessary to state that most of our hardy heaths break into delightful forms, white and various coloured. The common heather has many charming varieties, also the Scotch heath. These forms are quite as free as the wild sorts, and give delightful variety in a heath garden, which need not by any means be a rocky or pretentious

affair, but quite simple; for heaths are best on the nearly
level ground. Though they grow best, perhaps, in peat
bogs and waste places, it would be a mistake to suppose
that only such soils can grow heaths well, because we see
them in Sussex in soils quite unlike those on which they
thrive in Hampshire, though certainly on heaths they
seem to form their own soil by decay of the stems and
leaves for many years. If rocky banks or large rock-
gardens already exist, choice heaths form often their very
best adornment, but such things are by no means necessary. Some of the best and most successful beds we
have seen were on the level ground, as in the late Sir
William Beaumont's garden in Surrey.' Then follows
in Mr. Robinson's book a long list of varieties of heaths.
At any rate in Surrey, why should we not have all the
hardy heaths in our gardens? They are lovely plants,
and, like many other things, not difficult to manage when
one knows about them, and very easy to propagate if the
little green tips are pinched off and put under a hand-
glass in sandy soil and in shade in July or August.
They are very much strengthened and improved by being
clipped back after flowering, some in autumn, some in
spring. I bought a good many a few years ago, but,
through my own fault and the dry seasons, they nearly
all died. What they really like is pure air and a moist
bottom, but with care one gets over many difficulties;
besides, these conditions are quite natural to many
gardens and woods. *E. australis* and *E. mediterranea*
and the Irish variety, I used to think, were none of them
quite hardy; but *E. mediterranea* grows at Kew, which
is damper and colder than many parts of Surrey. It
flowers from March to May; the typical plant is rosy-red,
and there is also a white-flowered variety. Then there
is *E. mediterranea hybrida*. It is the earliest of all the
heaths to flower. The hardiest of the taller ericas is

P

E. carnea. E. scoparia is not so pretty as some, but the species is quite hardy. At the Drill Hall I saw and took note of *E. cinerea*, which has many varieties. *E. vulgaris flore-pleno* is a double version of the heather of our moors, of which several varieties—*alba*—are very pretty. *E. vagans* grows wild in Europe. *E. Tetralix* is another wild variety.

All these ought to be much encouraged in some kinds of gardens, and especially so in heathy districts. I believe it is best to plant in spring. They are not easy to transplant, but they are worth any trouble. Mr. Barr told me he had lately made a good collection of the hardy heaths, and he sells them in pots.

To return to the garden in Ireland, in the kitchen garden was a fine and most successful bed of Persian ranunculus, planted in six inches of very light leafy soil. I have never succeeded well with ranunculus, though I have often tried to, so I said to myself, ' Here is a chance,' and I had a long talk with the gardener about them. He happened to be a Surrey man, and was most friendly. He told me he found it better not to plant the little tubers of ranunculus before the end of February or beginning of March, either in pots in a frame or out of doors, and that it was essential to very carefully open out the little roots and spread them in the soft soil, to keep them in the full sun, and well water them with care. With this kind of planting they grow straight away, and receive no chill or check, as they may do if planted earlier. I shall certainly try this method. He thought the Persian ranunculus did the best. For table decoration, this gardener had a most successful quantity of *Gladiolus elegantissima* (Veitch). Prettier things of the kind cannot be seen. Instead of flowers all up one side they flower both sides of the green stem, and are white with a red-purple stain. He said, with care, grow-

ing on their leaves after flowering and then drying the
bulb well in the sun, he kept them from year to year and
increased them. They should be potted up in October.
Diosma ericoides and *D. capitata* are lovely greenhouse
plants, but in my experience difficult to grow in crowded
greenhouses.

In every large place I go to, I find the same fault—
want of pruning. I do not mean cutting back, that is
done only too much, along walks especially ; but cutting
away, sacrificing great whole shrubs rather than let one
thing grow into the other, which makes hard green walls,
instead of showing off the growth of individual healthy
plants. The difficulty is that, in the case of evergreens,
severe thinning and cutting out must be done in spring
when owners are often away, and very few gardeners
dare take the responsibility of doing this by themselves.
All gardeners are also very busy in spring, and so the
beautiful shrubberies go on choking each other from one
year's end to another. A thin wood and a thin well
ordered shrubbery, turfed underneath, though only mown
twice a year, is an undeniably beautiful thing. Probably
it is so striking because it is so very rare. A straight
line cut through a wood is lovely at all times, but doubly
so if the edges are well cleared out and the strong pillars
(stems) spring straight from rough grass or bare brown
earth, as is the case with fir and beech. Few things will
grow under these, but the ground is warm and dry from
the fallen leaves or fir spines of scores of years.

From the wild West Coast I went back eastward to a
garden of another kind, beautiful to a degree from the
highest cultivation of years and much money admirably
spent—for money alone never makes a perfect garden.
Here, near the house, tall old yew hedges were glowing
almost a red brown from the splendid strength of their
spring shoot, and last year's clipping, but they would

never grow and colour like that in dry Surrey. These
yew-protected courts held all the usual glories of spring
bedding, well sheltered from spring winds, and at the
end of a long walk was an effect of blue as gorgeous as
a purple Apennine—a fine old plant of *Ceanothus azureus*,
unpruned, and a mass of bloom from top to bottom and
hanging forward on a tall garden wall—a very beautiful
effect, and one I have never seen here, as the early
flowering kinds of *Ceanothus* are apt to get killed by
spring drought if not pruned back hard early in April.
C. rigidus is another beautiful early flowerer. *C. Gloire
de Versailles* does here as a shrub, and should be treated
exactly like the most beautiful August flowering plant
there is when well cultivated, the *Hydrangea paniculata*.
Both these do admirably on the borders of damp woods
or shrubberies if not too much robbed or in too full sun.
All they require is not to be choked at their growing
time, and well pruned—indeed, cut down to within a foot
of the ground in April. The spring pruning and, in May,
mulching round the roots with leaf mould and well-
rotted manure are all they require. As time goes on, this
gives them almost a mound to stand on, which their
branches gloriously cover. The ceanothus will stand a
dryer place and more sun than the hydrangea, which
does not mind partial shade and a north aspect.

In this same garden a magnificent effect had been
gained by turning a large old walled kitchen garden into a
flower garden, preserving the old picturesque apple and
pear trees for the sake of their blossoms in spring and as
supports for various creepers in summer. It had all the
picturesqueness of a large, half wild Italian garden, and
all the beauty of each plant being healthy in itself, which
I have never seen anywhere so good as in the best of
our United Kingdom gardens. The retaining of picturesqueness can never be a certainty except where the

gardener and the master or mistress work satisfactorily together, as is eminently the case in this garden. At this particular time of year, this wall garden is reached by a walk cut through a field of young grass, full of beautiful long-stalked single tulips. There had been broad sweeps of fine daffodils, but they were nearly over. In the garden were some very fine autumn-sown calendulas —marigolds we used to call them—a Veitch improvement appropriately called ' Orange King' and ' Lemon Queen.' I am never tired of fine marigolds; they flower in spring, as well as autumn, only when the circumstances are favourable and the climate mild.

Ireland, with its beautiful ruins, its churches, its high civilisation in the Middle Ages, its later decay and misery, its connection with Spain, its mixed population, its poverty, its apparent content, its seething discontent, its hopeless emigration, its unsolved problems—all these things, when I am there, so rouse the keenest interest in me that I feel I shall never think of anything else again; but alas! new impressions stamp out the old. Oh, that statesmen and patriots would agree with President Roosevelt, who said the other day, ' Insistence on the impossible means delay in achieving the possible.'

Immediately on my return I came across Miss Emily Lawless's lately published volume of poems, ' The Wild Geese.' That which is perhaps the most serious of all the phases of Irish difficulties, enduring even till to-day in the form of emigration, is so pathetically shown in the following poem, that I rejoice in being allowed to copy it here. The book cannot be fully understood and appreciated by English people without the careful reading of Mr. Stopford Brooke's admirable preface, not only because of its historical teaching and explanation of the title of the book—and who, thinking of Ireland, can afford for one moment to forget its history ?—but because of his

most appreciative explanation of the various poems. Of
'Clare Coast' and the one I have selected, Mr. Stopford
Brooke writes: 'Then began, as I have said, the great
exodus of the Irish, and "Clare Coast" is the voice of a
handful of veterans who, sent over to collect money or
recruits, are leaving Clare with new exiles for the coast
of France. Those who speak are "war-battered dogs,"
and their cry reveals the temper and the soul of the Irish
Brigade. It is close to the reality. "The Choice" has
the same fruitful motive, but modified by the singer's
love for a woman, and by the offer made to him of high
place and prosperity by the Irish Government if he would
give up his religion and his patriotism—an offer made to
the leaders of the Wild Geese, and rejected by them all.
And the second part of the poem records the sick longing he has to see again his land, and his sweetheart,
whom he has left for ever. It, too, is close to reality.'
Mr. Stopford Brooke explains that the Wild Geese was the
name given to those self-exiled Irish soldiers who fought
as mercenaries for foreign nations. The Irish now fight
our battles and fight well. I wonder how many of them
still silently cry as they fall, 'Oh! that this was for
Ireland!' Apart from politics, I recommend this book
to all those with a love of poetry and feelings for
another's woe.

THE CHOICE

I

I who speak to you abide, with my choice on either side,
   With my fortune all to win and all to wear.
Shall I take this proffered gain? Shall I keep the loss and pain,
    With my own to live and bear?

For the choice is open now, I must either stand or bow,
   Secure this beckoning sunshine or else accept the rain.
Must be banished with my own, or my race and faith disown?
    Share the loss or snatch the gain?

Shall I pay the needed toll, just the purchase of a soul,
  Heart and lips, faith and promises to sever ?
Six centuries of strain, six centuries of pain,
    Six centuries cry ' never.'

Then let who will abide, for me the Fates decide,
  One road, and only one, for me they show.
There is room enough out there, room to pray and room to dare,
    Room out yonder—and I go !

II

    Heart of my heart; I sicken to be with you.
    Heart of my heart, my only love and care.
    Little I'd reck if ill or well you used me,
    Heart of my heart ; if I were only there.

    Heart of my heart ; I faint ; I pine to see you.
    Christ, how I hate this alien sea and shore !
    Gaily this night I'd sell my soul to see you,
    Heart of my heart—whom I shall see no more.

I cannot resist the pleasure of quoting one more poem which gives the picture from the point of view of those who are left behind :—

AN EXILE'S MOTHER

  There's famine in the land, its grip is tightening still ;
  There's trouble, black and bitter, on every side I glance,
  There are dead upon the roadside, and dead upon the hill,
  But my Jamie's safe and well away in France,
      Happy France,
  In the far-off, gay and gallant land of France.

  The sea sobs to the grey shore, the grey shore to the sea.
  Men meet and greet, and part again as in some evil trance,
  There's a bitter blight upon us, as plain as plain can be,
  But my Jamie's safe and well away in France,
      Happy France,
  In the far-off, gay and gallant land of France.

Oh not for all the coinèd gold that ever I could name
Would I bring you back, my Jamie, from your song, and feast,
  and dance,
Would I bring you to the hunger, the weariness and shame,
Would I bring you back to Clare out of France,
           Happy France,
From the far-off, gay and gallant land of France.

I'm no great sleeper now, for the nights are cruel cold,
And if there be a bit or sup 'tis by some friendly chance,
But I keep my old heart warm, and I keep my courage bold
By thinking of my Jamie safe in France,
           Happy France,
In the far-off, gay and gallant land of France.

On my return from Ireland, I found a somewhat dis-
consolate letter from my German friend who is a great
gardener. She seems to have forgotten while she was
writing that in many parts of England, too, the soil is heavy
and cold. As a rule, in the centre of Europe the spring
comes late, as she says, but when it does come it stays ;
while with us there are often glorious warm April days
which bring everything on, and May frosts which cut off in
a single night the azaleas in full bloom. All climates and
all soils have their trials, and it is in fighting these whims of
nature that the real interest of gardening consists How
dull it would be if all seasons were alike, and still more so
if all soils were the same and all climates temperate ! As
it is, everything makes a difference : a little clay, a mound of
earth, a piece of wall, a shelter, a little hole which collects
the water and so doubles the rainfall, and so on—endless
lessons, as my friend says, to teach patience and one sort
of philosophy—namely, not the resignation which says, ' a
visitation of Providence,' but the strength to begin again,
and the wits to find out what has been done wrong and
caused the failure. For those who dwell on soils in
some respects unkindly, I insert the letter, as I think

we gain strength from the difficulties and experiences of others, which is only perhaps a more amiable way of saying, ' There is something in the misfortunes of our best friends that is not altogether disagreeable to us.'

' Looking back upon, perhaps, the most barbarous and death-dealing spring-time in all my garden memories, along a very grey line of disappointments and failures, sowings of annuals repeatedly killed by cold and wet during this April and May, tender and half-hardy bedding things shrivelled and checked by frosts and torn by raging gales, there are very few bright lights to uplift the desponding amateur's heart and teach philosophy and patience. Years of apparent "love's labour lost" are suddenly crowned by a few of those haphazard successes by which nature seems to teach us the double lesson never to despair while we can strive, and never to forget that we are blind gropers in the dark. In my case, the object lesson is taught by a few humble little Alpines. You shall have the upshot of it at once, prefacing it by the fact, which must always be borne in mind, that, alas ! none of my experiences can be of use to you or any English friends, as everything that does well in your warm sand dies with me, and what will grow in my cold soil—limeless and slaty as it is—would probably suffer and die from dryness at the roots in yours.

' My chief joy has been the growth and bloom of that little gem *Campanula rupestris*. Tiny wee plants of this were sent me by Mr. Leichtlin of Baden-Baden. At first they filled me with apprehension, for with their thick woolly leaves and prostrate growth, and their fragile branches, emerging from a woody root-stock not half an inch from the ground, they seemed to me doomed to die in my heavy, wet, impermeable soil. But I knew how to manage the new arrival, thanks to my experience with another child of many anxieties, the *Campanula pulla* which I had tried in vain to grow for six years. This year it is full of growth, and has thrown up a dozen or more of its lovely, graceful, and delicate deep violet bells, as a result of planting it last year in a half shady nook of my little rock garden, in almost pure gravel mixed with a little leaf mould and sand. I planted it very firmly, and every bit of earth between its little crowns I entirely covered with small flat stones. The *Campanula rupestris* was given a little deeply-dug rock-pocket, in a rather more sunny and sloping place than the *C. pulla*. It had a little lime rubble mixed with the gravel, and I carefully avoided bringing the trailing woolly shoots into contact with the soil, laying a flat stone under each. The bloom was perfectly lovely, and I think by far the best and most interesting among the dwarf trailing species. The flowers are about an inch long, very constricted in the throat and deeply cut at the edges. The colour is a most delicate pale lilac with darker stripes going from the constricted part up the chalice. It seeds profusely, but the seed is not ripe yet, and I watch anxiously to see whether it will be fit to use, as it is the only way to propagate it. The little side-growths off the woody centre do strike in sandy soil in a cold frame, but they never make good strong plants. Another little success, after some trouble, has been the *Adenophora Potanini*, a crotchety thing at all times and horribly anxious not to be moved. It is perhaps not as good as the best platycodon, but desirable from the point of view of variety in the rockery's compound tribe. I have also had a fine batch of *Liatris spicata*, which I look upon with anxious eyes, as I lost all the *Liatris scariosa* last year, and fear my soil is too heavy, cold, and water-logged in winter for either of them. Rather a pretty group at the top of my rocks was formed by the *Sidalcea candida*, a charming little stiff white mallow not over a foot in height, several tufts of

Geranium argenteum, and a lovely, loose-hanging, white variety, of which, alas! I do not know the name. To complete the list of my little consolations in adversity, let me tell you that the *Ramondia pyrenaica* threw up such masses of their lovely flower stalks that the rosettes of leaves were well-nigh hidden. I have a new variety of this plant from Mr. Leichtlin with almost pure white flowers. I will go no further for fear of ending this with a wail, for the failures have far and away outnumbered the successes, and it has chiefly been to raise my flagging courage that I have apparently been praising what is my severest lesson in humility.'

May 26.—My home-coming was further welcomed by two new flowers in my garden which gave me great satisfaction, and which, if given the cultivation they require, do exceedingly well in our light sandy Surrey soil. One is a plant praised in gardening books and mentioned in catalogues, but which I have hardly ever seen growing in English gardens. I brought my first from Germany. There it grew in a rock garden in heavy, cold soil, facing north. This is conclusive evidence of its extreme hardiness. It is described in early and late editions of Mr. Robinson's ' English Flower Garden,' and figured in the second edition under the name of *Phlox divaricata.* The Germans add to this name *canadensis.* But in all the editions the colour is incorrectly described, as far as my flower is concerned, as ' lilac-purple,' which many amateur gardeners would decide with a shudder meant ' magenta.' In Nicholson's ' Dictionary of Gardening,' a most useful book, the colour is given as ' pale lilac or blueish.' This, though nearer the truth, gives no idea of its beauty. The corymb of flowers stands up from a bed of dark leaves on a stalk about a foot high. Its colour is a beautiful, real pale china-blue, more like the blue of the half-hardy Cape

Plumbago capensis than of any other flower I know. When picked it has the great merit of lasting well in water. I attribute its rare appearance in gardens to two causes, one that it wants dividing—gently pulling apart and planting out in half-shade every year after flowering —and replanting in October in a sunny spot where it is to flower, and the second is that, if left alone, the leaves are eaten during the summer by some unfindable insect, after which it dwindles and is worthless as a garden flower. So far as I can ascertain, slugs are its great enemy, and they seem very partial to its leaves.

The other flower is also a favourite with me and a great horticultural success. We all agree that, as a rule, single flowers are prettier than double, but the *Arabis alpina flore-pleno* is a very pretty flower. In appearance it resembles a miniature white ten-weeks stock. It is easy of increase from cuttings in June, which should be replanted in the autumn in full sun. It flowers more or less all through the summer, and though it grows quicker and stronger, perhaps, in good soil, it flowers more freely in poor dry situations.

After waiting for years to find a sundial to my taste and suitable for the centre of my green-paths (see 'Pot-Pourri from a Surrey Garden'), I happened to hear that a man in Kensington sold balusters from the old Kew Bridge, built in 1783 and demolished in 1889. I purchased one of these on the top of which was placed the face of a sundial, and I have felt it to be an immense improvement to the garden at all times of year. It is sunk into a small, square, stone base, and in spring a few yellow crocuses grow among the grass at its feet. It speaks for itself, and no motto surrounds it.

Mr. George Alison published in the 'Westminster Gazette' this charming, and to me original, version

of the least morbid feeling about a sundial and dark
days :—

> Serene he stands among the flowers,
> And only marks life's sunny hours.
> For him dark days do not exist—
> The brazen-faced old optimist !

I derive great pleasure at all times of year from a
way of arranging flowers of my own invention which I
call floating bouquets. I have the largest size of Green &
Nephews' 'Munstead Glass' bowl, the one without a
stand. This is filled with clear water to the very brim,
on which various flowers are floated throughout the year.
If in winter flowers are scarce, a bit of greenhouse fern
or autumn leaf may be added, but it is of great importance not to cover the water all over; a corner should
be left which reflects light as in a pond. The flower-
fancier may exercise endless inventiveness in this kind of
arrangement. Most people, I find, like it; others think
it absolutely ugly; for many plants it is practically
useful, as they live thus in water, whereas, if cut with
long stalks, they droop their heads immediately. This
is the case with Christmas roses, and still more so with
the later hellebores, which are such a joy about Easter
time; their soft greens and velvety plum colours look
lovely so arranged and picked short, the buds being left
on the plant to come out later. These flowers, if picked
in the ordinary way, even when the stalk-ends are split,
fade immediately in water. The forced Niphetos and
Maréchal Niel roses look most satisfactory so arranged ;
indeed, all the tea roses and China roses on this dry soil
are apt to hang their heads if the stalks only are in the
water ; floating, they rejoice in the cool water and show
all their full beauty looking straight up at you. Bunches
of blue *Plumbago capensis* do better so than in any other
way, also *P. rosea*, the pretty winter-flowering hot-

house species; it is the same with clematis, &c. Each flower-lover can find out endless new combinations, and which flowers flourish best on the bosom of the clear water.

May 28th.—I went as usual to see the Horticultural Show, which was more hopelessly crowded than ever. I suppose future generations will have large cool halls for these much-prized shows, instead of hot, crowded airless tents, which are as bad for flowers as for human beings. I did not notice much that was new. I think it such a misfortune that this great show is always at the same time of year. Some pure white plants of *Verbascum phœniceum* were very lovely, a *Rubus deliciosus* and a *Hydrangea stellata* (Veitch). Hydrangeas are useful plants for people with small gardens and greenhouses, as they are nearly hardy and so easy to propagate. *Tulipa persica* was new to me; it had several flowers borne on branching stems, inside brilliant yellow, outside golden bronze; also *T. pulchella*, with small glowing crimson flowers. How I do like going back to the type plants! they are so far more interesting, and lovelier too as a rule, than the others.

What I enjoyed beyond all in that day in London at the end of May was that peace—'coming peace'—was in the air. Blessed peace! Everyone seemed to believe in it. Will Mr. Laurence Housman, if he sees this book, forgive me for quoting his poem, which came out at about that time in some paper—I think the 'Spectator,' but am not sure?

THE WINNERS

We stand one with the men that died;
Whatever the goal, we have these beside.
Living or dead, we are comrades all,—
Our battles are won by the men that fall!

He who died quick with his face to the foe
In the heart of a friend must needs die slow:
Over his grave shall be heard the call,—
The battle is won by the men that fall!

For the dead man leaves you a work to do:
Your heart's so full you fight like two!
And the dead man's aim is the best of all,—
The battle is won by the men that fall!

Oh! lads, dear lads, you were loyal and true,
The worst of the fight was borne by you;
So the word shall go to cottage and hall,—
Our battles are won by the men that fall!

When peace dawns over the country side,
Our thanks shall be to the lads that died;
Oh! quiet hearts, can they hear us tell
How peace was won by the men that fell?

JUNE

Cuttings of double gorse and ericas—A gorse hedge—Gerarde on Solomon's seal—Preserving tulip bulbs after flowering—*Dictamnus fraxinella* in Wiltshire—The globe artichoke as food for man and goats—Peace—Glasnevin—Mr. Linden's garden at Brussels—Old wistaria and bignonias grown as shrubs—How to tell a good soil—Mr. G. F. Wilson's wild garden—How to grow Portugal laurels in boxes—Tamarisks and sea-buckthorns grown inland—The beauties of *Polygonum compactum*—London and the 24th of June—The Rose Show at Holland House.

June 1st.—I forgot to say last month that the best time to strike double and single gorse, and some of the ericas, is at the end of May. The little young shoots should be pinched out and planted in sand under a bell glass in the shade. In six months, or the following spring, they can be moved. It answers better to buy double gorse than single, which can be grown from seed, because the double is sent out in pots, and in this way a year or two is gained. A double gorse hedge requires nothing but a little cutting back, and an occasional mulching if the summer is very dry. When well grown a double gorse hedge is one of the loveliest and most effective I know, and quite impassable to man or beast. There is one in this neighbourhood that is such a glory in May that even cyclists stop to look at it. I am afraid I cannot say the same of motorists. It is best to plant the young gorse in a groove on a raised mound of earth, first of all because it protects the growth of its tender years from being trodden on, and secondly, the nature of

the plant with its weight of flower causes its branches
to fall downwards, thus increasing the beauty of its
wondrous glow. Even well-struck cuttings, if in an exposed position, where they do best ultimately, want a
little care—such as mulching, or covering with bracken
or straw—the first winter or two. Slightly nipping
back in spring helps the growth of all young shrubs. In
the case of laurustinus, large and small branches may
be cut out with immense advantage. Of course, this
does not mean cutting back in the gardener's hedge
sense, as that would remove all the bloom for the coming
year, while thinning out immensely improves the next
year's flowering. This applies to all spring shrubs,
lilacs, wistarias, spiræas, &c. In a corner of my
kitchen garden I keep a space for Solomon's seals, digging
them up for forcing as they are required. Gerarde in his
'Herbal' has a delightful allusion to this plant; he says
that Solomon's seal is good for 'bruises gotten by
women's wilfulness in falling against their husbands'
fists.'

June 2nd.—I find it very difficult to preserve tulip
bulbs when they are taken up out of beds in order to
admit of planting for the summer. The best way is to
take them up and put them, with their leaves still on them
into a pail of water for half an hour, and then lay them
in a shady open trench ; cover them up and water, mark
the place, and when ripe and dry, dig them up and put
them in boxes for autumn planting. To my mind the
most attractive are the type, or species, tulips. They
are beautiful both in form and colour, but they like warm
sunny situations, a rich loamy soil and to be left un-
disturbed for years. That is the great difficulty in small
gardens, as it means a great dry patch all the summer,
and no turning over of the earth for other plants   I saw
last year, at **Wisley**, *T. linifolia*, a bright scarlet, and, as

Q

the catalogues truly say, 'magnificent'; but at present
it seems to be scarce, for even in the cheaper Dutch
catalogues it is marked 3s. a bulb, which I consider very
expensive.

One of the June plants which gives me the greatest
pleasure, and which I strongly recommend for increase
in both small and large gardens, is *Dictamnus fraxinella*.
The white variety, *D. f. alba*, is much the most beautiful,
I think. It is a slow grower, and at one time I despaired
of making it do well. Now it is quite satisfactory.
Last June, in a schoolmaster's house in Wiltshire, the
whole of a large dinner-table was decorated with this
handsome sweet-scented flower, and the blooms were the
finest I had ever seen. As I sat down next my host I
thought to myself, 'Now I am going to find out the real
secret of cultivating the white dictamnus.' His answer
to my inquiry was, 'I really don't know; I think it was
here when I came twenty years ago. We will ask the
gardener to-morrow.' This I accordingly did, and he
said it had never been moved. He increased it by sowing
the seed directly it was ripe—the seed must be watched
or it quickly disappears, either by being eaten by birds,
or through falling out, bursting open in the hot July sun;
but he said the quicker way of increasing it is by taking off
pieces in the spring without disturbing the parent plant.
Both of these methods I tried on my return home, and
both answered. The moved pieces look healthy this
year, though they have not flowered, and I have a fine
crop of young seedlings. This is one of the many plants
worth growing, but with which quick effects are not
attained and patience is required.

For years I failed in what I call the satisfactory
cultivation of a favourite vegetable—the globe artichoke.
Of course, I grew them, but they were few and hard.
Many people said, 'Oh, they don't do in this light soil.'

I never like that answer, as the merit of a light soil is that it can always be made. I have now mastered the cultivation through the kind assistance of a French neighbour. The seeds came from Vilmorin, in Paris, and if sown in heat in January, and planted out in May and well watered, they will produce some artichokes the first autumn. The good plants are then selected and marked, or the bad ones pulled out, for the seedlings vary considerably even with the best seed. They require slightly protecting in winter with straw or bracken, and in March pieces are taken off the best plants, leaving only about three shoots on the parent plant, without disturbing it. The holes must be filled in, and the plants mulched round with good manure. The pieces taken off are planted in a row in good soil, and these produce a succession crop of artichokes in the autumn, the main plants bearing their crop in June and July. In dry soils and hot weather copious watering is essential. Another great charm of the globe artichoke to me is that its leaves are so beautiful and goats are very fond of them.

On Monday, the 4th of June, I came down and opened the papers in the ordinary casual way, and there in big letters was the long-expected and scarcely-to-be-believed-in news that peace was declared. It had been known in London on the Sunday, but no echo had reached us here. A large crowd went to cheer the King at Buckingham Palace; but, on the whole, the great news seems to have been taken soberly and quietly, very different from the reception given to good news during the war, when the London populace seemed to rejoice with savage rage and almost the lust of conquest. What the news must have meant in many homes one can well imagine. Here it could only accentuate for a time one huge personal sorrow and regret, so then I tried to think what Peace with a big P meant to the country at large—the coming back to happy homes, both rich and poor, of those that were left—and then to realise what it meant out there on the ruined and devastated veldt. However much one may blame the obstinacy and ignorance of the Boers in not understanding the temper of this country, and in not realising that the struggle for their freedom was useless and hopeless, one could not help pitying the contrast, and it was a relief at any rate to feel that now pity and sympathy were possible without the perpetual and silly reproach of being called a pro-Boer. It recalled to me the end of the Crimean war, and yet what a strange and different war this has been!—in a sense more like the conquest of Scotland and Ireland hundreds of years ago, and yet in another sense unique and modern. And, with all its failures, the effort certainly was towards making war more humane; for I suppose never before in the world's history have wives and children been sent into an enemy's lines for protection from famine and death as they were sent into ours. Living as I do so much alone, I have been struck by the amount of barbaric and sentimental superstition still afloat in the popular mind. The daily newspapers rang with the fact that a little pigeon had got into St. Paul's, and was noticed as an omen during the thanksgiving service for peace. A popular poet wrote of it :—

> When suddenly lifting my eyes
>   To the glooms half discovered above,
> I marked with a start of surprise
>   The white wings of a hovering dove.
>
> Blest messenger come to your home !
>   It is peace, blessed peace once again !
> And thou, spirit ineffable, come !
>   As at Pentecost, come and remain !

How little we know what the future of this peace will be, and what the gain will be to England or South Africa !

The best hope, and perhaps the most beautiful of possibilities, was expressed anonymously in the following lines in the 'Spectator' :—

THE SETTLERS

(A FORECAST ON THE DECLARATION OF PEACE)

How green the earth, how blue the sky,
　　How quiet now the days that pass,
Here, where the British settlers lie
　　Beneath their cloaks of grass !

Here ancient Peace resumes her ground,
　　And rich from toil stand hill and plain ;
Men reap and store : but they sleep sound,
　　The men who saved the grain.

Hard to the plough their hands they put,
　　And whereso'er the soil had need
The furrow drove : and underfoot
　　They sowed themselves for seed.

Ah ! not like him whose hand made yield
　　The brazen kine with fiery breath,
And over all the Colchian field
　　Strewed far the seeds of death.

Till, as day sank, awoke to war
　　The seedlings of the dragon's teeth ;
And Death ran multiplied once more
　　Across the hideous heath.

But rich in flocks be all these farms,
　　And fruitful all the fields that hide
Brave eyes that loved the light, and arms
　　That never clasped a bride.

Oh, willing hearts turned quick to clay,
　　Young lovers holding death in scorn,
Out of the lives you cast away
　　The coming race is born.

June 10th.—I think all garden lovers should visit, whenever they can, either at home or abroad, botanical gardens or the larger nurserymen's gardens. Glasnevin, near Dublin, is one of the most interesting botanical gardens I have ever seen, old and established, and planted in the truest botanical style—viz., picturesquely, and with the earth in parts thrown up into the mounds so essential for giving different aspects to plants. I was told by friends who visited Bruges this year— for, alas ! I did not go myself—that when tired of the wonderful art exhibition there, they found the nursery gardens round the town full of attractions and interest. At Brussels, Linden's garden has a very good show of orchids and Congo plants. The most remarkable feature in the methods of foreign nurserymen is the extraordinarily clean and clever way in which things are grown. At Mr. Linden's, orchids are only watered once a week owing to his system of saucers, which enables the plants to get their moisture in the nearest possible approach to their natural circumstances. Each pot rests on a little porous stand rising from the centre of a saucer, the space round remaining full of water. There is, of course, no hole in the saucers, which helps to keep the houses clean and free from the drip which makes the wood and ground of ordinary hothouses so damp, and helps to harbour all the pests. The floors of these Belgian greenhouses were covered with finely-chipped grey limestone. This description, which was given me by a friend, makes me long for some practical invention that would save the great trouble of the dripping of our own greenhouses in winter. It is most destructive and extravagant. If ordinary saucers are used for such plants, say, as carnations and pelargoniums, or indeed anything but water-loving plants, the result is damping off and general ill-health. I think zinc trays of various lengths and widths filled with shingle, or in some cases cocoanut fibre, as for bulbs which want more moisture, might solve the drip difficulty and enable shelves to be placed one above the other without injury, which is now impossible.

In my neighbourhood, Messrs. Waterer's garden near Woking has been established over 100 years, and, quite apart from their principal business, which is that of rhododendrons and azaleas, there are at these nurseries an immense number of interesting specimen plants well grown in a peculiar way. For those who are young enough, and who have large places, it would be of great instruction and interest to observe what planting for posterity means, though few people care to do that now. At Messrs. Waterer's they have wistarias and *Bignonia radicans* (*B. grandiflora* is the handsomer plant, and does best on a house) grown as shrubs, first supported on poles, but gradually grown, carefully pruned and twisted about, till they form self-supporting and most picturesque groups. In any private garden wistarias need never be cut down, as the old branches can be laid along the ground, and flourish equally well quite a long way off. I know an old plant in this neighbourhood that was so saved, and which now covers luxuriantly a new pergola. The suckers of wistaria are also easily layered in the summer, then cut off, and potted up and forced in a cool house in spring. The early horticultural shows this year had lilac and white wistaria so grown, and also the common laburnum as a standard, and very pretty they look in large conservatories.

It is impossible to give advice about a garden till one has seen it, and thoroughly taken in its aspects, its soil, its surroundings, and its natural advantages. I should love to begin garden-making all again, and to my mind perfection would be a woody slope, with a south and west aspect, protected from the north and east, and with a really good soil. 'What does a good soil mean?' I am asked. This can easily be judged by the wild flora of the neighbourhood. Woods where primroses and blue hyacinths grow well are not bad; but think of the woods in Somersetshire or near Petersfield where the wild orchid flourishes, the butterfly and bee orchid and the white helleborine, not to mention the more ordinary lilac kinds! The herb-paris tribe, with their quaint growth, Solomon's seal, wild garlic, forget-me-not, and violets, all of them grow in these districts, and later on they are choked and covered by stronger-growing things like campions, centaureas, and euphorbias, hiding the spring carpet of bloom. What a picture this is compared with the dry commons and miniature flora of Surrey. Unfortunately the soil where vegetation flourishes is one apparently less well adapted for the health of man, though if the air is bracing and people healthy, a rich soil will be considered not unwholesome in the future. The combination of good soil and fair aspect is perhaps rare, but I know two gardens in Surrey with these advantages. One is partly terraced and arranged in the most interesting manner down the side of the hill, and more or less divided into the four quarters of the world, Europe, Asia, Africa, and America, by planting the shrubs, trees, and flowers of each division as far as possible together. I had heard of a garden of this kind in Scotland, but had never before seen one. When it is a little older, it will be as beautiful as it is now interesting and instructive. The other garden was planted by Messrs. Backhouse, of York, and is down a sunny slope with large rocks and very informal terracing, and nooks and corners made by throwing up mounds and hillocks. Many plants will not grow at all if exposed to the summer sun all day long, and the breaking up of even a flat piece of ground in small gardens in this way is an immense advantage and improvement. All down one side of this garden was an artificial streamlet made with rocks and pools for various moisture-loving plants.

In making a new place, this very attractive feature would
not be difficult if the house were built on the top of the
hill, and the roof-water collected in a large tank. If the
water were allowed first to flow down the hill of itself to
show what its natural course would be, all stiffness would
be avoided and blocking would be made easy; and with
the natural rainfall encouraged into the channel a very
small dribble from the tank would maintain it.

The garden which of all others I most value and
admire, and from which I have learnt most of what I
know, is that of Mr. G. F. Wilson at Wisley. Alas! he
died this spring, and one wonders what will become of it
now that the directing spirit is gone. I believe it was
twenty-five years ago that Mr. Wilson first began with
about six acres to turn rough ground, part field, part wood,
into what is generally called 'a wild garden.' Here I first
learnt what could be done by thinning out a wood,
digging ponds, and throwing up the soil into mounds.
The drawback to this type of garden, which I watched
with interest, was that it seemed to be a necessity to take
in more and more fresh ground year by year, and now
there must be fourteen or sixteen acres under cultivation.
The plants and shrubs, as they spread and flourished,
were left alone to grow at their wild will, and new spaces
were planted with other specimens. It is partly wood,
partly slopes facing north, and partly flat ground, mere
ordinary fields, and it immediately strikes even a casual
observer how much better many things do in the partial
shade of the cleared-out oak-wood than in the open.
This is in a great measure owing to the protection afforded
in spring by the tall trees. Bamboos, for instance, planted
in this wood flourish amazingly, as do *auratum* lilies,
pernettias, all the tenderer heaths, andromedas, rhododendrons, azaleas, *Hydrangea paniculata*, &c., all of
which were finer specimens than I have seen anywhere

else. Needless to say, there are no laurels or any of the commoner coarse-growing shrubs. No one must think this kind of gardening easy, for it depends on endless and intelligent hand-weeding. The plants must be kept clear, or they never flourish. When Mr. Wilson made a bed in his wood and filled it with the soil best adapted to the plants he was going to put in, the place was always marked by rough stones from the field. That meant it would be kept perfectly clean till the plant grew strong enough to need no further care. At no time of year can anyone visit this garden without finding flourishing specimens of plants he has probably never seen before.

Experience teaches that a mound thrown up in windy exposed places, and planted with hedges or shrubs, is a far better breaker of the force of the wind than the highest and most expensive wall ; for in a gale the wind hits the top of the wall and dashes down on the other side, while the waving branches of shrubs break and soften its devastating progress—a wind-swept garden being, perhaps, the most hopeless and disappointing to the gardener. In Italy the garden walls are delightfully varied—high to the north, low to the south, broken in various places to suit different plants, tiled on the top here and there for protection, ramped on the top if the ground falls—in this way the picturesqueness of garden walls and their individuality as regards each place are immensely increased.

In all wild gardening there is much difficulty in keeping down not only weeds and strong-growing plants, but destructive animals like rabbits, hares, rats, and mice, which are the most injurious, while pheasants sometimes manage to do a good deal of harm. Very useful little shelters are made at Wisley by surrounding the small bed, till the plants get hold, with a band, about a foot high, of perforated zinc. Some people will think this sounds very hideous, but, to a real gardener, the sense of protection to the young plant does away with the very slight disfigurement such a shelter makes in the wild growth of a wood.

Wild gardening requires much more knowledge than any other kind, and is almost unknown professionally. I have never seen planting well done, from an artist's point of view, when left to even the best firms. At Wisley there are large quantities of the Japanese iris (*I. Kæmpferi*). grown in artificially made ditches in the open field, or on the dry slopes of the ponds, not quite close to the water, as *Kæmpferi* does not like damp, especially in winter, though it is glad of lots of water at the flowering time in summer. The ideal situation for it would be a dry ditch in full sun that could be partially flooded in summer.

The general impression seems to be that bamboos want a lot of sun; they are, on the contrary, perfect wood plants, if a space is cleared for them. Mr. Wilson was making a big collection of many kinds in his wood, and they are flourishing luxuriantly.

To come back to smaller plants, the Cape montbretias do excellently in woods, blooming, as they do, in the dull August time. There are eight or ten varieties of epimediums; planted together in a bed in a wood, they look charming with their early spring growth, but, like everything else, they must be kept weeded. In autumn their leaves turn a good colour.

Messrs. V. V. Gauntlet & Co. (Japanese Nurseries, Redruth) send out on application, and for inspection only, the most charming books, illustrated by Japanese flower-painters, of various Japanese plants—lilies, Japanese pæonies, irises, maples, magnolias, camellias, &c. The choice of Japanese *Iris Kæmpferi* is particularly ample. All the prices are marked in plain figures.

One of the gardening subjects I have oftenest thought out is what I should advise if anyone I cared very much about inherited a large house and a large, ready-made,

formal garden, with empty beds in winter, and filled in
summer with bright-coloured half-hardy plants.   The
formal garden, to my mind, is certainly adapted to the
formal house, though it often shares its ugliness.   In the
case of very beautiful old places I think what looks best
is either turf right up to the house, especially if the fine
timber has been spared, such as yews, cedars, oaks,
beeches, mulberries, &c., or if there is a garden round the
house, it should be in small proportions, paved, and planted
with roses, lavender, rosemary, carnations, &c., in large
masses.   The actual gardens for the picking of flowers
should be near the house, and yet apart from it.   Given
space enough, my idea of perfection for those who live in
their place all the year round would be to have various
gardens, of which some would be at rest and others in full
beauty ; say, a bulb garden, an iris garden, a garden for
early and late annuals, perennial borders connecting some
of these together, a double avenue of Michaelmas daisies,
another of lavender and China roses, and a straight walk
or terrace with a formal imitation of the way orange-trees
are grown in Italy.   The best I ever saw of this last kind
of thing was in a large Hertfordshire place last year : the
trees were Portugal laurels, apparently growing out of
square boxes such as the orange-trees are grown in
at Hampton Court.   My surprise was great at finding
the Portugal laurels had thick stems and were perfectly
healthy.   On examination, I found the boxes rested on
the ground and the trees had been allowed to root
through.   The effect was excellent, and equally decorative
in winter and summer.   Laurels pine in boxes from dryness, and bays die in boxes from frost in winter if not
housed.   I have never before seen this effort at Italian
gardening so successfully carried out in England.   To
put it in a few words, the only possible solution for decorative gardening near large houses is formal design with

informal planting and growth. Everything else can only be decided by the accidents of the situation.

As I said about Ireland, so I feel about almost every large place I have ever seen ; pruning, taking out of laurels, cutting out broad vistas in the woods, as they radiate from the house, wide spaces with shrubs and trees allowed to feather to the ground ; all this seems to me what the gardens of most big places really require. I have only known one place where the owner had the courage to root out every laurel, both common and Portugal, within half a mile of his house, replacing them with all sorts of unusual decorative shrubs. (See splendid list in ' Century Book of Gardening,' p. 423.) When actually growing in woods, the effect of laurels left entirely alone and un-pruned, except to take out dead wood, is, I think, exceed-ingly good, especially if they are kept in groups, and for the delectation of their owners in winter rather than summer. The slight protection of the trees saves them from the shabby appearance in spring so often caused by frost and wind in the open. Also I have seen one of the new kinds of large-leaved laurels, judiciously pruned, look almost as handsome as a magnolia nailed to the wall of a house or barn. In itself the laurel flower is very beautiful when seen close, but this is impossible when the shrubs are pruned, as they generally are, so as to flower only on the top. Bay-trees, which are hurt by frost even more than laurels, would do well in fairly moist woods.

It is curious that the tamarisk is always considered a seaside plant. Certainly it does very well there, but it does equally well inland if properly pruned. It is a picturesque grower and an amusing plant to play with, as it can be treated in all kinds of ways—as a hedge, in a group, cut down every year, or planted far apart from other things, and encouraged to grow old, when its effect is unique amongst shrubs. I only know one example of really old tamarisks, and these are supposed to have been brought from the East and planted by Lady Mary Wortley Montagu. The strong black stems and branches are very angular in growth, recalling Japanese plants, and their feathery tops wave delightfully against sky or distance. There is a kind not generally grown, which does well in my garden, and its cloud-like pink blossom appears in the autumn. Another kind which I first saw at Aix, *T. parvi-flora*, flowers in May, in tiny pink blossoms all along the branches before the leaves appear. This, too, I find well worth growing.

Sea-buckthorn thrives to perfection in inland places. At Kew they plant it near the pond, where it fruits well and gives a golden glow to the whole group of shrubs in early spring. Mine does not fruit, thus proving to me that the plant needs moisture.

The spiræas are innumerable in kind, and make beautiful, easily-cultivated ornaments for open shrub-beries and borders of woods. I think the reason one so seldom sees these shrubs is that they look shabby and untidy unless the old wood is cut out once a year. They seem to me to require nothing else, and yet it is never done, and gardeners seem to prefer doing without them altogether or to put them in a border where they are unsuitable. If a little ground is cleared round them, they send up endless suckers, which should be taken off in spring if a fine specimen is desired. In this respect the treatment is exactly the same as with all the polygonums, though these are herbaceous plants, not shrubs. *P. com-pactum* is one of the smaller varieties, but most desirable, for if planted on a bank in full sun and thinned out in spring, the wings of its seeds are more beautiful than its flowers—a lovely rosy pink at the point of every branch. *P. molle* should never be forgotten, as its large white flowers in October are very valuable and effective.

June 19th.—Not quite hardy bulbs are more mysterious in their wilfulness than almost any other plants I grow. You seem to treat them exactly the same, but the least little thing must affect them, for you get unexpectedly different results. This is perhaps the reason why so few gardeners grow them. The year before last I had a dozen or more of *Ornithogalum arabicum,* and they all flowered most beautifully. They were well worth growing, and have the usual merit of their tribe that every one flowers in water, and this kind has a distinct aromatic odour which is rare with ornithogalums. It is not quite hardy, and must be grown in a pot. It is a lovely plant with its pale cream petals and jet black eye. It is cheap to buy, and with a little care in the drying off ought to go on from year to year. This year I had a dozen new bulbs ; only two of them flowered, and those not very well. After a great deal of pumping, the gardener told me he thought the reason of this failure was that they came from Holland in a slightly growing state and were potted up later than usual. So it resolved itself into the fault being mine, as I had ordered the bulbs late. These plants are often seen now in London shops. They want a little protection, but no heat except the sun in spring. My *O. nutans* continues to flourish most satisfactorily in the grass, and it *is* a pretty thing !

I think those who have big places and pieces of water ought not to neglect planting the taxodium or deciduous cypress. It is such a beautiful tree, and in favourable circumstances must grow fairly quickly. Perhaps it does not live long in this country, for I think there used to be many more about in my youth than I see now. I suppose it was a fashion when it first came from America, where it grows best in the swamps of the Southern States. It must have been introduced before 1640, as Parkinson mentions it and says : ' Its seed was

brought by Master Tradescant from Virginia and sown
here, and doe spring very bravely.' It is not a true
cypress, of course, but a separate genus.

I have been singularly unsuccessful with the *Romneya
Coulteri*, having lost it two or three times over. Humi-
liating confession! I am now trying it again, and it
promises well. The finest I know in this light Surrey
soil is growing under a west wall and is slightly shaded
to the south by shrubs. It is a very fine specimen, throwing up suckers in all directions, as if it would say, ' See
how easy I am to grow! ' It requires a little protection
in winter with bracken, straw, or dry leaves. I killed
mine last year by stupidly ordering that ashes should
be put round it. Besides its beauty and uncommonness
—for one does not often see it—the buds come out in
water—a great merit. *Gaura Lindheimeri* is a perennial,
but I have never succeeded in keeping it here over more
than one winter. I think it is a plant well worth treating as a half-hardy annual sown early in January every
year, and planted out thickly in a sheltered place in full
sun. It is very attractive, and is much grown in the
gardens in Paris.

Tuesday, June 24th.—I was persuaded to go to London,
like most of the world, with the idea of seeing the Coronation. The day was foggy, hot, and heavy, and the tired-
looking crowds were already trapesing through the streets
as I drove from Waterloo about mid-day. The tawdry
decorations, the streets and thoroughfares blocked with
stands, the side streets shut off by dangerous barriers
such as I had never seen before, gave me an ominous
feeling of wonder as to what was going to happen.
I think London never looks so bad and so depressingly
like a mean but gigantic village as on these festive
occasions. Grosvenor Square was a mass of workmen
decorating handsomely and with feverish haste, and when

the news of the King's illness burst on the town like an unexpected blast from some evil genius, the almost magic arrest of all movement produced an effect which, in its own peculiar way, was one of the most remarkable experiences I have ever had. The contrast was so striking between cause and effect. The sickness of one man is, as a rule, so small an event, but in this case multitudes were instantly stupefied by the fact. By the evening the whole world was sympathetically affected by the sorrow and disappointment that had fallen upon London, and the proposed coronation was only celebrated on a few ships at sea, and in the islands of the Hebrides. It was, to many of us, rather a revelation of the methods of conducting weekly papers, that though the news of the King's illness was known on Tuesday, several of the weekly newspapers published loyal paragraphs about the Coronation, including detailed descriptions of facts which had never taken place.

I carried out my programme of the afternoon, which was to go with a friend to the great Rose Show at Holland House. As a show it was disappointing, for the season was a late one, and the roses were hardly out, and hung their heads in the intense heat. The crowd was gloomy and with bated breath spoke of but one topic. The beautiful building and gardens of Holland House remain a unique remnant of the many handsome houses and grounds that used to constitute the suburbs of London. Holland House, in spite of the black smoke-veil cast over it, is still one of the most splendid of the Jacobean dwellings. It was built by the well-known John Thorpe for Sir Walter Cope in 1607. Poor man! He had no son, and it went to his daughter and heiress. She married a smart Cavalier who lost his head in 1649, and Fairfax's soldiers were quartered in his hall. The after history of the house, with all its Whig associations,

R

in the time of the third Lord Holland, have been so much written about and are so well known that they need no repetition here. Even in my youth I can remember the huge ‘breakfasts,’ as they were called (really garden-parties), when many of the celebrities still haunted the place. In a delightful alcove behind the house there remains the inscription written by the late Lord Holland : ‘ Here Rogers sat, and here for ever dwell with me those pleasures that he sang so well.’ I fear poor Rogers would hardly be much known now but for the lovely Turner and Stothard illustrations which adorn his books.

The gardens have been much enlarged and improved by the present owners. I was anxious to see the several acres which used to be pasture and have been lately added to the grounds, and I must confess it was with a feeling of astonishment that I saw what can be done with care and cultivation in what is now virtually London itself. One had to look at the stems of the few remaining old trees to realise what was the amount of soot in the air, and what an atmosphere the gardener had to contend against. The roses looked as healthy as at Kew, and the Penzance briars and free-growing climbing roses ramped about as if they really believed they were in the country. The most interesting part of the new garden was an imitation of a Japanese garden. This is a very distinct feature which could easily be adopted where both money and water were plentiful and where the ground sloped gently. It was partly a small rock-garden, partly turf, and then a series of pools cleverly cemented and connected by an artificial rill which fed the various basins, some large and some small. In these were different varieties of aquatic plants, especially the new water-lilies (*Nymphæa*). It all looked very charming, and on the whole this was the best arrangement I have ever seen for growing aquatic plants. In a small way it could be imitated with tubs hidden by little rockeries, and placed one below the other with connecting tubes, also hidden. The whole of Holland House garden is a most interesting lesson as to what can be done in poor old London.

This week of the King's illness was the only really hot one we have had this summer, and during it the whole air was full of the most gloomy prognostications, the gloomiest emanating from the medical profession, and from certain headquarters of spiritualistic prophecy, all of which, as we know, happily came to nothing.

A volume came out this summer called ' A Little Book of Life and Death,' selected and arranged by Elizabeth Waterhouse, with a frontispiece from a painting by Mr. G. F. Watts. To those friends who liked so much the verses which were not mine in my other books I recommend this selection, for I think it so good as to be almost original. It is partly prose and partly verse, and from the former I take the following to give a taste of its general quality : ' We are evidently in the midst of a process, and the slowness of God's processes in the material world prepares us, or ought to prepare us, for something analogous in the moral world ; so that at least we may be allowed to trust that He who has taken untold ages for the formation of a bit of old red sandstone may not be limited to threescore years and ten for the perfecting of a human spirit.'—*Thomas Erskine of Linlathen.*

JULY

An account of lately bought gardening books—A lost poem by Milton
—Vegetable gardens and rotation of crops—How to easily cata-
logue a garden—More half-hardy plants suitable for large pots—
Carnations at Mr. Douglas'—Spanish rush-broom.

July 1st.—As at the end of the eighteenth century,
amidst wars and revolutions of all kinds, botanists and
gardeners and flower-painters went quietly on their way, so
it has been now, and since I published my ' Second Pot-
Pourri ' a great number of excellent gardening books have
come out, and, as time goes quickly and fashions alter, I
think it may not be uninteresting that I should give a slight
sketch of the books I have thought it worth my while to buy
in the last three years. In 1900 Miss Gertrude Jekyll
brought out her second book, ' Home and Garden.' Her
account of her home is most charming, and never did bird
build a more appropriate nest, and, as the Italians say, ' Ad
ogni uccello suo nido è caro.' Like her first volume this
one is excellently illustrated from photographs taken by
herself, which are very superior to the ordinary run of
photographic illustration. I feel sure that all Miss Jekyll's
books will be referred to again and again long after the
mass of present garden literature is of no more value
than autumn leaves. Next came out with indefatigable
energy from her pen ' Wall and Water Garden ' ; here the
illustrations from photographs begin to fall off and become
commonplace. They are evidently not taken by herself,
and have none of the individual charm so noticeable in
the earlier books. The letterpress, on the contrary, is, I

think, more useful, original, and instructive than perhaps
any of the others. Her advice about cutting up flat
ground by low walls, uncemented and with earth in between, is, I believe, entirely Miss Jekyll's own idea and
is most useful and beautifying. In wet heavy soils where
there are no stones, these walls could be made with clinkers,
or spoilt bricks, as they absorb the moisture, whereas stones
keep it in. I think anyone who casually looked at Miss
Jekyll's books when they first came out, will be surprised
on going back to them to find how much instruction there
is in them for all sorts of gardens—villa gardens, wood
gardens, field gardens, terrace gardens, and water gardens.
For those about to make a new garden, the 'Wall and
Water Garden' is most essential. I think tanks, both
big and little, especially oblong and square, are the most
beautiful additions to a garden. Unless carefully
watched the builders always place the waste pipe too
low down. What is really desirable is that the water
should be level with the edge, whether it be stone or
brick or merely grass, or, as in the case of dew-ponds on
the top of a chalk hill, the ground should slope gently
into the water. Last year came 'Lilies for English
Gardens, A Guide for Amateurs'; 'Country Life Library,
1901.' Most of these lily notes appeared in the 'Garden,'
Miss Jekyll being at that time co-editor of the paper.
I myself have made no progress in lily-growing in the last
three years. The lilies I buy flower and flourish for one
year or two, but the summers of late have been hot and
unfavourable. In spite of trying every kind of cultivation
and receipt the *Lilium candidum* is generally more or less
diseased with me, and the heads of bloom are not really
fine. The *Lilium croceum* lasted two or three years, but
now most of them have died off. I have one little white
Martagon lily which comes up faithfully year after year.
Lilium Hansoni was originally brought to me straight

from Japan. It flourishes very well, and stands re-planting and even dividing. That it has done well in this light soil through these last hot dry summers, flowering every year, should be noted as I have rarely seen it in other gardens. It blooms in June, and has bright green whorled leaves up its stalk and an orange flower slightly spotted with dark brown. Miss Jekyll says it is a lily that should be more known and grown. All the other lilies which were brought me from Japan at the same time some years ago have dwindled and finally disappeared. The late frosts this spring destroyed hundreds of auratums in Mr. Wilson's garden at Wisley. I think we may gather from Miss Jekyll's book that a lily garden means very frequent buying of bulbs. They are certainly more worth the money than many other plants, and in a woody dell near a house there is no more delicious effect than large clumps of well-grown lilies protected from cold winds and late frosts by refined and not too strong-growing shrubs and a certain number of overhanging trees. But half shade and moisture they must have, even if this is only procured by increasing the rainfall from planting them in a dell into which the sides drain. In this situation, and with the natural support of shrubs, I have seen *Tropæolum speciosum* flowering luxuriantly.

Last of all this year comes ' Roses for English Gardens,' by Miss Jekyll and Mr. Edward Morley. I was much disappointed with the illustrations of this book. I do not think roses growing or cut lend themselves to photography, but ' Roses coming over a Wall ' facing page 60 is a lovely exception. I had intended to make a fresh list of the roses which have done well here, but all Miss Jekyll's advice is so admirable that those who care for their rose gardens will be sure to have the book.

' Chemistry of the Garden : a Primer for Amateurs,' by Herbert Cousins.—This is a first-rate little book published by Macmillan in 1899, price 9d. Everyone should possess it. It is full of helpful, plain instruction as regards soils, manures, chemical requirements of plants and vegetables, and excellent directions for reducing blights and diseases that attack gardens.

'Small Gardens, and How to Make the Most of Them,' by Violet Biddel, another little 9d. book, familiar and chatty in style, and modern in feeling, names many plants which would otherwise be forgotten, and hints at many methods for gaining beauty of form and colour in small spaces. Best of all it preaches originality; for any garden, however small, which is the individual expression of its owner is of interest to every other gardener, who thereby sees variety.

'Flowers and Gardens,' by Mr. Forbes Watson.—This book, which I waited for for years, was republished in 1901, and prefaced very gracefully by Canon Ellacombe. It will always be loved and cherished by those who are fond of flowers. It curiously recalls some of Ruskin's writing. Forbes Watson approached his subject from a somewhat unusual standpoint; he was a great lover of flowers, a student of botany, and had the artistic temperament combined with marked religious convictions. We can hardly believe from the portrait in the beginning that Mr. Watson was only twenty-nine when he died. He practically wrote it on his death-bed. We now know who was the friend to whom the notes were entrusted when the pen fell from the author's hand, and who so well fulfilled his trust. The book has great charm, a charm which is no doubt due to the way in which it reflects the soul of the author, who was one of the old-school doctors with strongly-developed taste for natural history. Mr. J. B. Paton in his preface says : 'The papers published in this little volume were written to solace the languor of the last months of life, when a malady, which had crept by slow

approaches upon him, broke down his strength, and
arrested a professional career which had begun but recently. They betoken a mind gifted with quick, clear,
and delicate perception, independency of judgment, and
unsparing truthfulness. These were my friend's characteristic gifts. They are dimly mirrored in these pages,
but more clearly in the memory of those who knew him
well. To them this little volume will be welcome because
of him ; to others, perchance, it may be welcome for the
worth it has, because it tells of the beauty there is in
God's fairest, frailest handiwork in flowers, and bears
some trace of the rarer amaranthine beauty of a soul which
wore the white flower of a blameless life.'

'Favourite Haunts and Rural Studies in the Vicinity
of Windsor and Eton,' by Edward Jesse.—I so much like
Jesse's 'Gleanings from Natural History' that I bought
the above-named from a bookseller's catalogue, and did
not regret it. So much of the old world in the neighbourhood of London is disappearing that this book is of
especial interest to the antiquary and the garden fancier.
He mentions many old and exotic trees seldom now to be
met with, such as tulip-trees, catalpas, Turkey oaks,
deciduous cypresses, &c.—all these are remains of the
gardening craze at the end of the eighteenth century.
The book is illustrated with little old-fashioned prints of
cottages and churches, amongst them Upton Church, with
its 'ivy-mantled tower' of Gray. To those who like a
chat about things as they were in the middle of the last
century, I cordially recommend this book. Its out-of-date
language reads almost like a verse from Genesis which
gathers the whole firmament into those few words, 'He
made the stars also.' Jesse reminds us that Gray omitted
a beautiful stanza from his Elegy, thinking it made too
long a parenthesis at the end of the poem before the
epitaph ; but it is well worthy of insertion :—

There scattered oft, the earliest of the year,
By hands unseen are showers of violets found ;
The redbreast loves to build and warble there,
And little footsteps lightly print the ground.

'Life of the Fields,' by Richard Jefferies.—This is a darling little book by a kind of modern Jesse, full of the love of nature and distinguished by that rare spirit of unworldliness which, clothed as it is in very different form, has made a popular favourite of another book, read and deeply cared for by a very different class of reader from the one that appreciates Richard Jefferies. It is the peasant and artisan who make a sort of text-book of Robert Blatchford's 'Merrie England,' of which the motto is, 'Words ought not to be accepted because uttered by the lofty, nor rejected because uttered by the lowly.' Its teaching may be summed up in the following extract : 'Love of truth, love of knowledge, love of art, love of fame are all stronger motives than love of gain, which is the only human motive recognized by a system of political economy supposed to be founded on human nature.' Let all young Englishmen who dread failure read this book.

Mr. Peter Barr, who has lived for some time in South Africa, was kind enough to send me two lectures he delivered at Sea Point Horticultural Society, Capetown : one was on 'The Daffodil,' and the other on 'The Lilies of the World.' As very few people here can have seen these lectures, I give one interesting extract from the lily pamphlet, and, in passing, would also recommend the Lily Conference, reported in the 'Journal of Horticulture,' December 1901, which gives a most comprehensive account of all the varieties of lilies now grown. The more I read about them the more I feel sure that, like hyacinths, for ordinary people they represent a nurseryman's bill, as one has to buy afresh every year, But the subject is interesting enough for a real garden-lover to devote his whole attention to lilies and lilies alone. No doubt the climate of the Cape—and Mr. Barr gives a long list with historical notes of those most likely to succeed there— would prove much more favourable than England, especially for the Japanese varieties.

Under 'American Lilies,' Mr. Barr writes as follows :— ' *Sulphureum* (syn. *Wallichianum superbum*). When in Auckland, New Zealand, 1900, I saw in Mr. Ball's garden this really handsome lily, and a noble sight it was. Height, 5 to 6 feet, with a coronet of seven long white flowers surmounting a sturdy stem. I had a letter recently from Mr. Ball, informing me that the lily this year had fifteen flowers. At another garden in Auckland I saw the same lily behaving in a similar manner, and with a considerable progeny of all sizes. The owner expressed a desire to sell some of them, and I undertook to sound the lily's praises as I travelled this progressive and aggressive wonderful country, " the home of the brave and the free of the Southern Hemisphere." I had a letter of thanks recently from the owner referred to, informing me he had sold well on to 20*l*. worth of the bulbs of *Lilium sulphureum* to the persons to whom I had re- commended it. Now, this man a few years previously had bought a bulb for 5*s*. 6*d*. So say no more about the gigantic profits of the Capetown Cold Storage Company. Here is a record-breaker for you in lilies, and he expects to reap an annual revenue from this original investment of 5*s*. 6*d*. He let me into the secret of increasing the bulb, so pardon my not taking you into my confidence, as I promised not to spoil his market by letting others know how he worked his diamond mine.'

The other lecture is even more interesting to the general public, as it is on the cultivation of the favourite daffodil. It is, of course, addressed to those who live in South Africa, but it is full of information of the most fascinating kind for everybody. He says : ' Double daffodils, as far as I can make out, originated in a wild state, but when, and where, and how, are unsolved questions. No one in modern times has added a double daffodil to the existing ancient forms ; indeed, I am not quite sure that we know as many double daffodils as Parkinson cultivated in his garden in Holborn, London. Some names of double daffodils have been changed, it is true, but a change of name does not make a new flower.' Then he goes on to describe the wild double daffodils of the world. He gives an interesting account of the various people who worked on daffodils, and how he himself took up the study of the daffodil early in the sixties, working on the older forms known to Parkinson. Describing how they came into fashion, he says, ' I do not suppose that Oscar Wilde knew anything about daffodils, but there is no doubt in my mind that the great public are much indebted to him for the revolution in taste caused by his lectures on æsthetic colours. He broke down the prejudice to yellow. The artists followed him, and the public followed the artists.' Both lectures teem with interest, and I have no doubt they could be procured at Barr & Sons', King Street, Covent Garden ; for if the demand was sufficient, Mr. Barr would probably reprint them.

I always think books devoted to one kind of plant very interesting. I bought last year ' The Clematis as a Garden Flower,' by Thomas Moore and George Jackson, of the Woking Nursery, 1872. At that time the clematis was very fashionable, now it is not grown enough or given the attention it deserves. In Messrs. Jackman's catalogue for 1902, they give an excellent description of how clematis should be cultivated. They also mention that the book I have just named has been republished, no doubt with newer information, for the price of 2s. 6d.

The suggestions as to the various uses to which the types of clematis may be applied would be very useful to the amateur. I mention, elsewhere, a notable example I saw in Suffolk. I must say the illustrations in my old edition are in the worst form of mid-Victorian colour-printing and drawing.

Whenever I see in any old book catalogue a work by one of those indefatigable gardeners, Mr. and Mrs. Loudon, I always buy it. The last one I got is by Mrs. Loudon, and is called the 'Lady's Country Companion, or how to enjoy a country life rationally.' It is choke-full of information, also full of her bad taste in garden planting, has excellent teaching about bread and biscuit making, cream cheeses, &c. Her experience of goats seems to have been narrow, as was natural in the days before railways had drained the milk-supply of the country for the big towns, and also before the breeds of goats had been improved by foreign blood. We find them gentle, affectionate, and highly intelligent animals. Her information is so often picturesque as well as enlightening. For instance, she says of cabbages : ' The cabbage tribe is very much improved by cultivation, but the plants contained in it require a great deal of manure and frequent watering to make them succulent and good. It seems strange that such different plants as broccoli, cauliflowers, cabbages, Scotch greens, and savoys should all spring not only from one genus, but from one species, *Brassica oleracea*, which is, in fact, a British plant, and which, I have no doubt, you have seen growing on the cliffs of Dover, though you have no idea that a tall straggling plant with alternate leaves and very pretty flowers could be the wild cabbage.'

' The Century Book of Gardening : a comprehensive work for every Lover of the Garden ' (George Newnes, Limited).—This is a true title, and I think the book will

prove most useful, helpful, and suggestive to anyone
wishing to learn quickly and easily about gardening.
I bound my copy in two volumes, as its bulk and weight
are an objection. The photographic pictures, as usual,
exaggerate size and distance, but they give a good impression of many gardens and help people to know what
they want. The two long lists (one of shrubs and trees
at p. 387, the other of herbaceous plants at p. 69) alone
make the book worth possessing, and for those too idle
to make their own lists as I recommend, marking these
lists would be a way of cataloguing their garden.

After resisting the temptation for some years, I purchased the 'Illustrated Dictionary of Gardening : a
Practical and Scientific Encyclopædia of Horticulture for
Gardeners and Botanists,' edited by Geo. Nicholson, A.L.S.
—My copy is in eight volumes, which makes it easier to
handle, but it is extraordinary how puzzling the alphabet
can be when one wants to refer to a book of this kind. It
is an endless joy to me, and no evening can be dull or
lonely with one of these volumes to look at under one's
lamp.

'The Romance of Wild Flowers,' by Edward Step,
1899, was strongly recommended to me, and I got it.
The appropriateness of the title strikes me every time I
look into its fascinating pages, which have indeed much of
the romantic lure which makes a reader forget everything
else—even hunger—while the mind is held by a favourite
book.

I have been reading lately, with profound interest, the
little 'Mémoire' of Grant Allen, which came out in 1900,
when I had not the courage to send for it, because I
wished so terribly that I could have sent it to my son, who
had always been fond of Grant Allen's writings. It is a
very genuine, well-told account of a most interesting
character. To all who know his work, even partially like

myself, Grant Allen is a personality, and his biographer quotes, with regard to him, the story of Henry III. and Montaigne. 'I like your essays,' said the King. 'Then, Sire, you'll like me. I am my essays.' So might Grant Allen have replied. My son and I knew nothing whatever of the man, and of his fiction I still know nothing, but in his other writings he brought home to the ignorant in a wonderful and charming manner his own long and deep study of the natural sciences. A letter of his from Jamaica ought to be deeply interesting to all Englishmen who have to deal with the great difficulty of how natives can be justly ruled. I am always being told this is done better by Englishmen than men of any other nation. But have we not also made huge mistakes, and have we not much to learn in the matter of just application of laws which cannot be equal between white men and black? He is interesting, too, about the much-vaunted flora of the tropics. 'The fruits of which one hears so much are stringy and insipid; the flowers don't grow; and the "tropical vegetation" is a pure myth.' This may be slightly exaggerated, but I have always suspected something of the kind. He says, further, 'The tropics are the norma of nature—the way things mostly are and always have been. They represent to us the common condition of the whole world during by far the greater part of its entire existence; not only are they still, in the strictest sense, the biological headquarters, they are also the standard or central type by which we must explain all the rest of nature both in man and beast, in plant and animal.' All Allen's life seems to have been hampered by bad health and poverty, but the latter probably made him all he was. And when he was dying his last words to his son were: 'I want no memorial over my remains; tell those who care for anything that I may have done to buy a copy of "Force and Energy."'

The whole of this 'Mémoire' to me is extraordinarily touching, interesting, and inspiring. To have so little time left in which to learn is the true sadness of old age. On the first page, in Grant Allen's lovely little miniature writing, is the following : ' I don't know that any phrase or quotation has ever been of much use to me in life, but the two passages most frequently on my lips are probably these : What shall it profit a man if he gain the whole world and lose his own soul ? and

> To live by law,
> Acting the law we live by without fear ;
> And, because right is right, to follow right,
> Were wisdom in the scorn of consequence.
> *Tennyson.*'

I never saw Grant Allen, but Mr. Clodd's book makes me intensely wish I had known him, and I think the book is what such a book should be, ' a faithful portrait of a most fascinating personality.'

July 10th.—In that very charming work, by Mr. Rider Haggard, the ' Farmer's Year Book for 1898, with illustrations by G. Leon Little,' he gives an interesting anecdote which touches indirectly on my favourite subject of diet. Under date September 8th, when speaking of the obscure causes of sunstroke, he says, ' Here is a curious instance of the power of an English sun. In a Norfolk village with which I am acquainted, lived a man, a retired soldier, who, when serving in India, had married a native woman and brought her home to England. This woman, while working in the fields at harvest-time, was struck by the sun and died. Certainly it seems strange that she, who had passed her youth beneath its terrific rays, should have fallen a victim to them here in foggy Britain.' It is strange, but may not the cause have been the great change in the woman's food ? To pass suddenly in mid-life to an English labourer's diet from the Indian peasant's food of rice and vegetables and fruit, would be likely to set up an unwholesome condition of the blood which would result in mischief. And are not our own people who take much alcohol much more affected by sun than other people? I use the word *charming*, to describe Mr. Haggard's book, for two reasons; first, because it is rare to have a book on such a subject from a cultured point of view, and secondly, because it is equally rare in these days to have a cheap book illustrated from an artist's sketches. In this instance the illustrations are often very good indeed—truthful, well-drawn, with a human touch, especially in 'The Dead Foal,' but, personally, I wish they had been reproduced in black and white instead of brown. They are all, to my mind, very superior to the usual machine-made reproductions which make the book illustrations of to-day so tedious, though I quite own that now and then, at long intervals, the photographer, by the happy selection of his subject, raises himself almost to the level of an artist. A most conspicuous example of this is to be seen in Sir Harry Johnstone's important new book about the Uganda Protectorate in East Africa. Early in the first volume there is a picture called 'An Andorobo drinking as primitive man drank,' which in its grace of limb, freedom of action, and extraordinarily lovely balance is, to my mind, fit to shake hands with some of the best of the world's statues! I wonder, by the by, if in nature the bow was held in the left hand, or whether it appears in reproduction reversed as on the lens? Anyhow, the picture is a marvellous representation of its subject.

July 12th.—In an old 'Pall Mall Gazette' I have found an account of a lost poem, by John Milton, which an Irish correspondent sent to the paper as a literary curiosity. As it is not included in any of the later editions of the poet's work, and as I am always impressed by the excellence of much in the daily papers which, from their

ephemeral nature, must needs be soon forgotten, I include
it here on the chance of introducing it to a few more
people:—

I am old and blind,
    Men point at me as smitten by God's frown,
Afflicted and deserted of my mind,
    Yet am I not cast down.

I am weak—yet strong!
    I murmur not that I no longer see.
Poor, old, and helpless, I the more belong,
    Father supreme, to Thee!

O Merciful One!
    When men are furthest, then Thou art most near;
When friends pass by, my weakness shun,
    Thy chariot I hear.

Thy glorious face
    Is leaning towards me, and its holy light
Shines in upon my lonely dwelling-place;
    And there is no more night.

On my bended knee
    I recognise Thy purpose clearly shown.
My vision Thou hast dimmed that I may see
    Thyself—Thyself alone!

I have naught to fear;
    This darkness is the shadow of Thy wing.
Beneath it I am almost sacred—here
    Can come no evil thing.

Oh! I seem to stand
    Trembling, where foot of mortal ne'er hath been;
Wrapped in the radiance of Thy wondrous hand,
    Which eye hath never seen.

Visions come and go!
    Shapes of resplendent beauty round me throng:
From angel lips I seem to hear the flow
    Of soft and holy song.

It is nothing now,
    When Heaven is opening on my sightless eyes,
When airs from Paradise refresh my brow,
    That earth in darkness lies.

S

In a purer clime,
  My being fills with rapture ; waves of thought
Roll in upon my spirit : strains sublime
  Break over me, unsought.

Give me now my lyre !
  I feel the stirrings of a gift divine ;
Within my bosom glows unearthly fire,
  Lit by no skill of mine.   *John Milton.*

The tone of sublime joyfulness in this poem reminds
me of the well-known lines on ' The Celestial Surgeon,' by
R. L. Stevenson :—

> If I have faltered more or less
> In my great task of happiness ;
> If I have moved among my race
> And shown no glorious morning face ;
> If beams from happy human eyes
> Have moved me not ; if morning skies,
> Books, and my food, and summer rain
> Knocked at my sullen heart in vain :—
> Lord, Thy most pointed pleasure take,
> And stab my spirit broad awake ;
> Or, Lord, if too obdurate I,
> Choose Thou, before that spirit die,
> A piercing pain, a killing sin,
> And to my dead heart run them in !

It is now nearly twenty years since Mr. Robinson
wrote his preface to the English edition of Mons. Vilmorin's 'Vegetable Garden,' an invaluable book in every
way, now out of print. I mentioned it before, but return
to it here because, although I have been about a good
deal, I have never seen a kitchen garden, old or new,
planted and managed as Mr. Robinson recommends, and
which seems to me admirable for anyone making a new
vegetable garden. For the sake of those who have not
the book, I quote the following : ' One point deserves
the serious consideration of every owner of a garden, and
that is the " muddle " method of planting the kitchen

garden with fruit trees and bushes, and so cutting up the surface with walks, edgings, &c., that the very aim of the garden is missed. It is quite a mistake to grow fruit-trees over the kitchen-garden surface. We cannot grow vegetables well under them, and in attempting to do so we destroy the roots of the trees. This induces canker and other troubles, and is the main cause of our poor garden-fruit culture. One-fourth of the space entirely given to vegetables, divested of walks, large hedges, old frame grounds, old walls, rubbish, and other impediments, would give a far better supply. Such a spot well culti-vated would be a pleasure to see. It is not merely the ugliness and the loss of the mixed garden which we have to deplore, but the troubles of the unfortunate gardener who has to look after such a garden in addition to other work. How is he to succeed with many things so hope-lessly mixed up? Put the fruit-trees in one part—the higher ground, if any—and the remaining part devoted to vegetables, cultivating the ground in the best way, and having it always a fertile green vegetable garden. The vegetables, too, would be more wholesome from continual good light and air, for shade from ragged and profitless trees and bushes and hedges is one of the evils of this hopeless kind of garden. The broken crops, too (for the most part sickly patches) are not such as one can be proud of. Separation of the two things, complete and final, is the true remedy. There should not be the root of a fruit-tree in the way of a vegetable-grower.'

I agree with every word of this, but many of us have bought or inherited old gardens planted on the wrong system, and then it must be a compromise. The greatest difficulty is that under-gardeners will dig with a spade, and so cut the roots of the fruit-trees, which greatly injures them when they are young, and perhaps already making too little growth. I think fruit-trees, when once established, for this reason do best with the ground under them turfed, so long as there is a bare circle round the tree so that it can be manured, and mulched, in dry hot weather. Years ago Mr. Robinson advised me to sow pips of apples and pears, and one of these seedling pear-trees has fruited well this year and looks stronger and healthier than any other tree in the garden. In size and flavour the fruit also surpasses all the rest.

I have always found great difficulty in growing rasp-berries here. Every now and then, in a wet year, I get a certain number, but then they are more or less spoilt by the excess of rain. I am more and more convinced that, as with roses, they want better treatment than I have ever given them. In heavy soils, mulching with lawn-mowings is sufficient. Here, that is no use. I am sure they want heavily mulching with strong manure three times a year—in the early winter, at flowering time, and at fruiting time; though, perhaps, at the fruiting time, liquid manure would be best. In spite of what Professor Owen used to say when asked why he did not protect his fruit, ' They are the salaries of my orchestra, the wages of my choir,' I have wire-netted in a portion of my kitchen garden for the protection of the small fruit.

A kind neighbour and very clever gardener has written out for me a rotation of crops in the kitchen garden. Farmers pay great attention to this, but in smallish gardens it is often too much neglected.

' Divide garden into five equal parts or sections. In No. 5 perennials should be grown, such as asparagus, globe artichokes, herbs, strawberries, &c. . . . and no rotation applied.

' In the four remaining parts or sections the rotation of crops should be as follows the first year :—

' First Part, or Section A.—Generally such plants whose produce is *stalks*, *leaves*, or *flowers*. But cabbages, endive, lettuce, cauliflower, spinach, and (exceptionally) early potatoes, celery, and celeriac.

' This section requires very heavy manuring, applied as follows before or during winter :—

'The soil to be dug straight down, and the soil scattered on the manure, which must be laid on the inclined surface of what has been dug. (See diagram.)

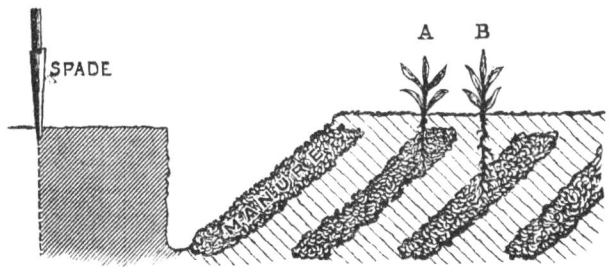

' The spade must be inserted as upright as possible, and be driven in entirely till the foot is level with the soil to be dug. The soil thus dug will show only soil on the surface, but below the surface each layer of manure will occupy an inclined surface, and these inclined layers will overlap so that any plant put in will surely reach the manure at a depth varying from one (A) to six (B) inches.

' Second Part, or Section B.—Such plants as are cultivated for their *roots* : Carrots, onions, turnips, salsify, scorzonera, chicory. The soil, having been prepared as above the previous year, will be found full of rich decomposed mould. It should then be trenched, taking care that the upper spit is not mixed with the lower spit. It will then require no additional manure, root crops being best without it.

' Third Part, or Section C.—Such plants as are cultivated for their *seeds* : Peas and beans. For these any addition of artificial manure or ashes may be applied (mulching after the bloom has commenced is recommended, not before).

'Fourth Part, or Section D.—To receive odds and ends: Seed beds or hotbeds; gourds or marrows grown on specially prepared places for their culture; also the reserve of flowers for autumn bedding. As the fourth section will become first section and grow A in the following year, the heavy manuring and the planting of biennials (broccoli) should here be anticipated.

'Each of the four parts of the garden will receive the proper culture for A, B, C, and D in regular rotation. Thus the first section, which grows A the first year, will grow B the second year, C the third year, D the fourth year, and A again the fifth year, and so on.'

July 15th.—An excellent way, which I regret not having followed, for keeping the lists of all you have in your garden—a plan as useful as it is instructive—is to get what are called visiting-books, with the alphabet marked on the pages down the side. It is better to have two of these books—one for shrubs, which would include fruit-trees, the other for perennials and annuals of all kinds and greenhouse plants. The very earnest student might add a vegetable book, with remarks as regards success or failure of experiments.

July 18th.—I have continued to experiment with the cultivation of plants in large pots. The Pelargonium 'Pretty Polly' is shown to full advantage grown in this way, and its name is appropriate, for it is *very* pretty. Any of the hardier zonal pelargoniums are satisfactory. One of my favourites is 'William Gladstone,' a beautiful rose-pink. A great improvement in the last few years has been made in these zonal pelargoniums. Mr. Cannell has a very fine collection, but in writing for them it is desirable to note that none should be sent with that

peculiar blue shade, verging on magenta, which kills all
the other colours. None of the tenderer pelargoniums,
though they live well out of doors in the summer, show
to perfection, except under glass. Of these 'Enid' is one
of the best for flowering all the winter. Even out of doors,
I think, the ivy-leaved ones do best against a wall if left in
ordinary six-inch pots with only moderate watering and
feeding. Inside, in this way, if the greenhouse is sunny
they are smothered with blossoms and not too much
leaved. They simply spend their life in flowering. 'Lord
Derby,' which one can only describe as a salmon-scarlet
with a large silver-splashed eye, is a very good one. This
reminds me of the old, well-known catalogue description
of a verbena—' " Lady B."—a good bedder with straggling
habits.' ' Lady Mary Fox ' is another of the old-fashioned
pelargoniums that grows prettily in a large pot, and
' Lucrece ' is one of the best pink zonals. Of my recent
discoveries in large plants for these pots I think *Cassia
corymbosa* is one of the most effective. It is of shrubby
growth, with handsome yellow flowers, which continue
till the frost comes. The plants can be housed in any
rough place where actual frost is excluded. Its fault is
that it has that rather tiresome tendency to go to bed
very early, and when your friends come to tea its leaves
droop and the petals turn back, and it looks shabby;
but the sun of next morning makes it triumphant and
bold. I can quite recommend it.

July 28*th.*—I went to-day to pay a visit to Mr.
Douglas, the well-known carnation-grower at Great
Bookham. It is always so interesting to see a large
number of plants of one kind grown to the greatest perfection. His carnations were lovely, and many were his
own seedlings. Once potted up, he never waters them
with anything but plain water. I am sure success or the
contrary must much depend on careful watering. He

layers all his carnations in their own pots in the glass-houses, gives them quantities of air and sun, and never puts any out of doors at all. They root quickly, and by keeping his plants in his sunny, low greenhouses, their seeds ripen.

This wet July has made *Ligustrum sinense floribundum* flower profusely, and when it does so I think it is the best of all the privets. It should face north, or be in partial shade.

In one of my many happy visits to the East Coast in July, I came across, in quite a small garden on a high bank facing south-west, against a bright blue sky, one of the most beautiful results of cultivation I ever saw in any garden. It was the Spanish sweet-smelling rush-broom (*Spartium junceum*), one sheet of golden bloom. It is one of the clearest and brightest of all the yellow flowers, and keeps for days in water, retaining all its colour at night. I instantly asked how it had been treated, and found the bank had been well planted with a few young plants far apart eight or nine years ago, and that the only cultivation it had had was cutting back hard with shears every year late in the summer, after it had done flowering, not allowing the seeds to ripen. Certainly the result was most satisfactory and well worth any one trying, whether near the sea or away from it. Even individual plants do far better and are hardier if well cut back. I used to lose them in winter, now I do not. This rush-broom planted on the southern side of the double gorse hedge before mentioned would carry on the band of golden blossom later into the year. *Genista tinctoria* planted in a *very* dry place on a wall and cut back hard becomes almost a dwarf plant, but flowers in great profusion. I think nothing so amusing as success or failure in varying the cultivation of certain plants, particularly in small gardens.

I think the cimicifugas much better worth growing

than seems to be the general opinion, for I seldom see them. I have both *C. racemosa* and *C. japonica* in the middle of a large herbaceous bed. They flower at quite different times, require no staking, and, being well fed, attract considerable attention in my garden. The tall thalictrums, *T. flavum* and *T. aquilegifolium*, I have placed near them, and the two together make a very pretty quiet background for brighter flowers. I have a specimen of sanguisorba which grows like a cimicifuga. Most people think them both spiræas. Sanguisorbas do excellently by rivers or ponds, and have a charming growth.

Bartonia asteroides is an autumn flowering plant that appears to ordinary people to be a Michaelmas daisy, but its leaf and its habit are quite different. It does best not divided, likes a dry sunny place, is more refined and graceful, and lasts longer in water than any of the asters I know.

One of the handsomest of the shrubs as a single specimen, or in front of other shrubs, is the Chinese Guelder rose (*Viburnum placatum*). Pruning back after flowering, and an occasional mulching, is all it requires. Its flowers are most striking in spring, and its foliage tints beautifully in autumn.

AUGUST

Cultivation of various plants — Outdoor fig culture—Rhubarb in France—Effects of *Nicotiana sylvestris alba* —Potatoes in succession—Colonial branch of Swanley Horticultural College for Women—'Animal life'—Letter about monkey's food— Hampton Court garden and the old railing—The motor and Bramshill--Building a house—Rose planting—Cooking receipts —Autumn work in a German country-house kitchen—Household receipts.

August 1st.—I have this year thoroughly conquered the difficulty of growing and flowering the single *Datura cornigera (Brugmansia Knighti)*. It is beautifully figured in the sixth edition of the 'English Flower Garden.' This picture would convince anyone how well worth growing it is. In a windy and exposed situation it is no use trying to grow it. What I do with it is this: It is kept all the winter with other half-hardy plants in a pot; in the spring all the lower suckers are taken off and the stem is bared. If you want to increase your plant, you leave two or three shoots at the bottom of the plant earthing them up, and in the autumn you take them off. When you have potted up in the spring it is well, if you have room, to leave the plant growing on in a cool greenhouse. It is planted out at the end of May. If the head is too thick, a little judicious pruning improves its shape. The hole must be well dug and well manured, the earth left as a cup round the plant, liquid manured later on as the flower-buds form, and copiously watered

in dry weather. By this method I believe anyone can
have this plant growing and flowering during the whole
summer in a way that any gardener would feel proud of.
The sweet Italian double datura I use as a wall plant in
a large pot, and have a plant so old that a large knob has
formed just above the ground, which, I suppose, is
Nature's way of retaining moisture in its unnatural
condition of being left in a pot. I have another of these
double daturas planted in a bed in my greenhouse, and
cutting it hard back once or twice a year makes it also
flower twice. In the cultivation of cannas, too, I have
made great progress. The common kind have luxuriant
foliage, but the flower is poor. Two years ago I bought
twelve good ones from Mr. Cannell. Before the frost comes
they are taken up and put under the shelter of some
shrub with dry earth thrown over their roots. This
makes them die down naturally. They are then put into
a box with dry earth and wintered in the cellar. In
January they begin to show signs of growth ; they are
best then taken up and very much divided and put into
separate small pots. Some should be kept and potted on
for greenhouse use, the others planted out in rich soil at
the beginning of June. The great object is that they
should be good-sized plants ready to flower at the end of
June. Some heat may be used short of making the
plants weak. The better sort of cannas never show to
advantage if they flower too late. A canna called
' Alphonse Bouvier ' makes a most splendid pot plant on
my wall. It stands plenty of liquid manure when in good
growing condition. Most plants stand high feeding when
they are at their strongest and making their flowers.
Over-manuring constantly kills young plants.

We keep our dahlias also in the cellar. They, too,
form better plants for a good deal of dividing when
planted out. The shoots should be thinned out as they

grow, and all the lower leaves removed in June or July. This immensely increases their flowering capabilities.

Œnothera taraxacifolia has been a most useful plant here this wet summer. We have to treat it as an annual, as it does not stand our winters. The vagaries of seasons have to be much taken into account. All my dry-soil, sun-loving plants, which generally do so well here, have utterly failed this peculiar season. Another Cape perennial which has done splendidly this year is *Venidium calendulaceum*. It is of dwarf spreading growth and the most showy, orange, marigold-like blooms. Cuttings can be taken in August and kept in a greenhouse through the winter, but I think growing from seed in a hotbed in spring is the better way.

Francoa ramosa, when grown for the greenhouse, should be potted on and *not* divided. Its wealth of bloom, if this point is observed, makes it a handsome plant for a greenhouse.

Many people say to me, 'Do you get figs off your trees?' I answer, 'Yes, since I have known how to treat them.' There is no difficulty in growing good figs out of doors, provided—first, you have a wall with south aspect. Second, select the right sort of fig, such as (in mild situations) 'Bourjassotte Grise,' 'St. John,' 'Osborn's Prolific,' and 'Brown Turkey'. It is of no use to try 'Castle Kennedy,' 'Negro Largo,' or many others. Third, if the shoots grow more than 18 inches long, the ground is too rich; in this case, cut the roots freely and pinch the points of the roots. Fourth, no shoots must be shaded; if they are, you may get figs, but you do not deserve them. The first frost will destroy the unripe figs. I always pick mine off. This information was given me by a man who is a most successful fig-grower. It is better to prune figs in the spring because some of the shoots may be damaged by frost if done in autumn.

Two or three years ago, visiting in a French country house in August, to my surprise, the most excellent rhubarb tart was served hot for the English guests! I immediately thought, ' Oh! a new and refined kind which tastes just like plums; how wonderful these French are!' On inquiry, I found the plants had come from England, and that, the family being away, the French gardener had been bored by their enormous growth and had cut them down in May. Their new green growth made this delicious autumn variety. I recommend it to all, as it comes in most usefully when the small fruit is over. The plants must not be forced or cut for eating till May.

August 3rd.—In a beautiful old garden close to the Thames, I saw a simple but really lovely effect of garden planting. In a small, inner, walled garden were some old yews, sombre and dark, and in front, instead of the ordinary mixed border or the eternal autumn yellows, was an enormous bed entirely planted—and very well done, not too close together—with *Nicotiana sylvestris alba*. Its beautiful long white flowers stood out against the dark background, and in the pale evening light the effect was magical. One great advantage of this plant over the old *Nicotiana affinis*, is that it does not close by day, but, in spite of this, *affinis* is the best for small gardens. The other must be massed and have such a lot of room to look really well. Gardeners are rather afraid of it, as some years it fails and only goes to leaf. I think this is greatly owing to treating it too well in the way of rich soil.

Against autumn and deciduous trees and shrubs in the front of park-like scenery, nothing looks so handsome as very large clumps of yellow flowers shading up to orange, letting montbretias be the highest touch towards the red. The endless varieties of perennial sunflowers—

helianthus and harpaliums, rudbeckias, &c.—make these easy and successful, and they can be varied and improved year by year, as they are all the better for replanting late in each autumn. Mr. Robinson gives such a splendid list of the varieties of these sunflowers that I need not repeat. The great thing to note is the variety in their heights. Lately, a double rudbeckia, ' Golden Glow,' has been added to these yellow autumn flowers and is most effective and useful. In the latest of autumn sunshine these yellow flowers look beautiful against autumn leaves.

For mixed borders, a low-growing white phlox—' La Reine '—is a very good one, I find. If cut down, it flowers again in autumn.

All the hardy statices (sea lavender) do well in this light soil and, had they more space than I can give them, would form a beautiful patch of blue-grey colour in this bad month for bloom. They are easily increased, as they are what is called root-plants. You dig them up in-tending to move them ; probably, the bit you left behind produces a finer spray of flower than the large root you take away. That means that every piece of root will grow if in a sunny, dry situation.

For those who like unshowy plants with various charms of their own, the hardy euphorbias (spurges) are a fascinating family to cultivate. The well-known Cape spurge (*E. Lathyris*) is a plant with a distinct habit and considerable beauty of foliage and demeanour when well grown. It sows itself in this garden, and I merely pull up those that are not wanted. *E. pilosa* and *E. amyg-daloides* are very attractive in spring from their yellow flowers, which look like leaves when little else is in bloom.

I have grown for years *Veratrum nigrum* merely for the pleasure of the growth of its handsome, green, crinkled leaves in spring, and this wet summer brought me the surprise of seeing it throw up its unusual tall flower-spike with numberless blackish purple blossoms. There is a white variety which I shall now feel encouraged to get. They are handsome alpines, rarely seen, for they do not thrive in a dry rockery. *Cineraria maritima*, that distinctive perennial with grey leaves, I used to lose year after year, when I first lived here. Then it struck me to plant it on the southern side of a moisture-absorbing shrub, and ever since it has flourished year by year, and is one of the joys of late autumn.

I think most gardeners plant their spring potatoes all at once ; I find it a great advantage to plant a few rows every month up to the beginning of August, as this continues much longer the delicacy so prized by many people of little waxy new potatoes. Seakale beet is a vegetable which does not seem to be generally known. I have grown it for some years, and, though it is not extraordinary in any way, it makes a pretty dish as described in receipts.

One of the things that my readers never seem to have forgiven me was my condemnation of the Virginia creeper and the *Ampelopsis Veitchii* in my first books. They never seem to understand that it had nothing to do with dislike of the lovely plant, but with the way it was misused, and I am as convinced as ever that, like ivy, it does spoil beautiful brickwork. This year I have been haunted by someone quoting me as an authority as to when the Virginia creeper was first grown on London houses. When I put it as late as 1841, it was a great misstatement. J. C. Loudon, in his 'Arboretum et Fruticetum Britannicum,' published in 1838, says, 'The Virginia creepers grow freely in the smoke of cities and in London, and it was introduced into England in 1629.'

I have lately received papers with regard to establishing a Colonial branch of the Horticultural College at Swanley, Kent. This with a special view to the immediate demand for competent women in South Africa, in which country there is a present deficiency of 70,000 Englishwomen. The proposal seems to me well worthy the consideration of those interested in emigration questions. The Horticultural College, in its women's branch, seems to have been suffering considerably from want of funds.

Last month came out a new periodical called 'Animal Life.' I spent a delightful five minutes in front of it at a railway bookstall. It begins with a picture of Sir Harry Johnstone's new animal, the Okapi. I am sure many—especially young people—are pining to see this new creature : the stuffed specimen is at the Natural History Museum at South Kensington. The first article is by Professor Garner and called ' Monkey Land,' with such pictures of monkeys ! I said to myself, ' I must buy this new magazine. First numbers are always the best.' It was only 7*d*. The next number had an article on the great apes, by Sir Harry Johnstone. That finished me, so now I take it in. I wrote to Sir Harry to ask for a few more details about the food of the apes in Africa. He very kindly answered as follows : 'About apes. They are to me a profoundly interesting subject. I look upon them as quite half human. Their food consists of the fruit of several species of *Amomum* (a plant allied to ginger and to the banana) ; of the core or heart of some palms or of other trees ; of mushrooms (most abundant and succulent in the great forest) ; of the plums of the parinarium-tree, and the fruits of many other forest trees ; of roots (in the case of the chimpanzees, of young birds, rats, beetle grubs, and other animal food) ; and, finally, of plunder from the forest negroes' plantations. The great apes are *mainly* vegetable feeders, but certain chimpanzees readily devour flesh.'

The amomum (from *a*, not, and *momos*, impurity; in reference to the quality of counteracting poison) referred to here is described in the ' Dictionary of Gardening ' as a stove herbaceous perennial, chiefly aromatic, and formerly used in embalming. I wonder whether the fact of the chimpanzees' taste for flesh points to a possibility of their evolving through a meat-eating phase into humanity? Sir Harry Johnstone calls them the most human of all the apes.

August 9th.—I find in my note-book of last year that I went to Hampton Court and never saw its gardens in such great perfection. It was as beautiful as could be, bathed in soft golden sunlight, not foggy or misty, as, alas! it so often is from smoke, but with clear pearly distances. The flowers were really gorgeous, and one saw in perfection the modern kind of bedding. The last of the carpet beds have disappeared. In not one bed could the earth be seen, no pains and no expense having been spared. A lovely arrangement of red and white bouvardias scented the cool, moist air, and everything had as much water as it wanted. I have never seen a more successful public garden. I heard afterwards that all this meant a new gardener, and I note it now as an encouragement to those who are depressed by failure in their own gardens, for this year, the weather being unfavourable, the whole thing was utterly different. The herbaceous borders had gone to leaf and many of the beds had failed. The restoration of the old iron railing, near the river, at the end of William and Mary's garden, is a great public benefit. It appears that this railing had been carted about to various places, some reaching as far as Edinburgh, to save the expense of buying handsome gates for public or Crown property. Its restoration to its old

T

foundation does much credit to Lord Esher's management during his term of office as First Commissioner of Works.

This year, for a short time, we had a motor-car, and I could not help thinking that my agonising fears on first driving in it must have been exactly what our grandparents felt when they first used railways. When I meet these things along the road—which, alas! I do very often— what with their dangerous pace, their horrible dust, not to mention the smell that poisons the country air, the only expression that comes to my lips is, 'Beastly things, how I hate them!' But when you are in one all is changed. There is a sense of power and independence, and almost an exultant feeling of rushing through the air and covering long distances without effort. This is, I confess, an enjoyable experience. Besides, there is the great pleasure of enlarging one's area, and of seeing in comfort places and towns one could only reach before by two drives and a railway journey. The distant visit I enjoyed most was to Brams Hill, the famous Jacobean house in Hampshire. Such a lovely place and house, quite unspoilt by restoring and changing! It is close to Kingsley's old home at Eversley, and is so admirably described by Lucas Malet in her novel, 'Sir Richard Calmady,' that I felt the thrill of a *human* history when shown the hall where Sir Richard's father died, and the panelled bedroom where the agonised mother pressed the naked baby to her bosom after making her terrible discovery of its deformity. Rocking herself to and fro in a paroxysm of rebellious grief, she cried, 'God is unjust! He takes pleasure in fooling us. God is unjust.' From the white-panelled old bedroom on the first floor one walked straight into a large sitting-room, from the windows of which one could see the broad lawn, with the summer woods sloping away behind it, where Lady

Calmady's brother shot the horse. No one could have chosen a more perfect background for such a story. The house was, I believe, built by the same architect who built Holland House, and was intended for the residence of Henry, Prince of Wales, son of James I., who died young. Its position is my ideal one, and Kingsley says of it, ' It stands high, looking out far and wide over the rich lowland country from its eyrie of dark pines.' The guide-book says, ' The Scotch firs in this park are among the oldest and finest in England.' I suppose Scotch firs were the fashion when Scotland's king came to rule over England. We were pressed to return by the kind owners of Brams Hill, and I should much have liked to do so, for one never could tire of its beauties, of the old treasures it contains, and of the unprofaned air of antiquity which surrounds it. But, alas ! the motor departed with its owner, and I have had to go back to my ten-mile radius.

On the homeward journey we passed a large open space planted with rows and rows of oak-trees, which was done by the Government at the time of the Napoleonic wars, in a panic that there would be no oak left in England wherewith to build ships. Though a hundred years old they are still quite small, and as the exact date of their planting is known, these straight alleys will be of interest to future generations. After driving through the glorious old nature-planted woods of Brams Hill, it was curious to note the contrast in this dry carrying-out of an officially ordered planting. Even to the future generations who see these oaks in their old age, they will never recall a wood such as, in its wild beauty, suggested the French proverb which Miss Una Van Artevelt Taylor adopts in the following poem—one of a charming series written by her for the ' Westminster Gazette ' :—

NOUS N'IRONS PLUS AU BOIS, LES LAURIERS SONT COUPÉS

The path is there—the hedge of high hornbeams,
 Grey-stemmed, brown-leaved, and then the gate, and then
The little wood we called the Wood of Dreams :
 That wood our hearts will never find again.
  Nous n'irons plus au bois, for well we know
  Les lauriers sont coupés, long, long, long ago.

In that green place the woodland sunlight streams
 On windflowers white, and every April day
Paints bluer violets. But our fair-faced Dreams
 As shadows came, as ghosts have passed away.
  Nous n'irons plus au bois ; nor here nor there
  Our dreams abide, and all the wood is bare.

The path is there—the sudden silver gleams
 Of birch-boughs through the dusk. New eyes will see
June foxgloves blossom ; to the Wood of Dreams
 New dreamers and new dreams will come—but we—
  Nous n'irons plus au bois, ah no, ah no ;
  Les lauriers sont coupés—even, even so.

I heard the other day of an American couple visiting
one of the most beautiful and typical of our south-country
places, with broad, soft, velvety green lawns. They
were wild with astonishment and admiration. At last
the wife exclaimed, 'Oh ! Jack, however do they get
these lawns here ? I can't understand it. *We* can't do
it.' Her husband replied, ' Well, my dear, I guess we
can't have the two centuries of mowing which these
places have had ; that may have something to do with it.'
He was, in a sense, right : some of our old places have
had over three centuries of mowing. But the climate,
too, is answerable for a great deal. The cold winters,
even of the centre of Europe, entirely prevent grass from
growing as it does with us. We could get a better lawn
in three years than they could ever have. There is no
doubt that the British Isles have a quite wonderful

climate for gardening, though it is little suited for sitting
out, which many people seem to think is the chief use of
a garden.

As I am always planning and planting imaginary
gardens, or thinking how I should alter the gardens of
my friends, so I am always building imaginary houses.
Nothing I see satisfies my ideas of what a modern
' health house ' should be—that is to say, a house which
is not only a home, but a model-dwelling for acquiring
the highest degree of physical health compatible with our
modern life. In small houses and great houses, in
villadom and in hall, the first thought seems to be that
the outside of the house should be pretty, or ornamental,
or picturesque. Now, no one admires beauty more than
I do, but the first law I should lay down to myself would
be that the *inside* of the house, both as regards comfort
and health, should come before any structural external
beauty. I do not deny that every house must be adapted
to its owner, and, of new houses, the one that fulfils this
best of any that I know is the one which Miss Jekyll lives
in, which she describes in her first two books. My
house would of course have to be quite different. The
outside appearances would have to be very simple and
very plain. White walls with a red-tiled roof for modest
houses such as I am speaking of seem to me most suited
to the English climate. My house should be low, only
two storeys, but covering a good deal of ground. Beautiful brick chimneys, and many of them for the sake of
having my fireplaces in the part of the room best suited
to its requirements, would be the principal additional outside expense. By beautiful chimneys, I mean such as
stand out good in form against a clear sky, like those
which were built to farmhouses three hundred years ago.
The roof over the windows might have eaves, but there
must be no eaves where creepers are to be grown. Eaves

were used before gutters were invented to carry the rain-water from the roof and away from the walls. In those days, the drip fell on the roots of the plants; now there is no drip, so the plants must receive the rain as it falls if they are to be healthy. The absence of eaves being ugly, the difficulty might perhaps be solved by having a slightly perforated gutter which would allow enough water through for the needs of the plants below, but would carry off the bulk of the rain-water, but I fear this would make the walls damp. But could my little white house ever be pretty, whether it cost much or little, seeing that it would have to be nearly all windows—enormous windows, tall and broad? I hear most people saying, ' How frightful—and how uncomfortable!' It is not only a matter of air, as that can be got more or less by small windows if intelligently placed and made to open easily, but to catch all the sun possible in our sunless climate. In the sitting-rooms the windows must be in part glass doors reaching to the ground; for to live in the country and not be able to watch nature, the birds pecking, the flowers growing, &c., when sitting in a room, seems to me a great drawback. Not only must it be possible for me—

> While safe beneath the roof,
> To hear with drowsy ear the plash of rain,

but I must *see* it : I must also, without getting up from my employment, be able to see the great storm-clouds roll up from the horizon, and the sun rise and set—all this with firm big windows that do not rattle in the wind and that a child or an invalid could open or shut easily. What have I ever seen that comes nearest to these ideas? All old English cottages have small windows, and the tall windows of the beautiful old Jacobean houses are generally very narrow, and the leaded panes do not open enough. Perhaps some of the Georgian houses built quite at the end of the eighteenth century, full of the recollection of their simple Dutch origin, are most like what I want, but even these have not enough window space for me. I think I should like my drawing-room something like the old Queen Anne orangeries, only not of course nearly so high or large, and the morning-room might have circular windows like some of the old Brighton houses. The woodwork would have to be strong, and there comes in either initial expense or endless painting. I am even Philistine enough now and then to like a large pane of plate glass over a fireplace let into the wall, like a window without a frame, provided the view is good. The flue is led up on each side.

For ordinary windows, twelve by nine inch panes do very well for not too heavy a framework. For exposed situations, where windows opening to the ground are liable to catch the full force of the west wind, a plan by which the heavy plate-glass door slides back in a strong iron groove, top and bottom, with a handle inside and outside of the framework to move it easily, is far the most serviceable and convenient that I have ever seen. In bedrooms where the whole of one side of a wall is window, there may be a window-seat across, but then the glass should come down to within a foot of the seat; but I think shelves under the seat with sliding doors are better and more useful than the customary lockers. Doors that are cut in half like old cottage and stable doors have been too little used of late.

My dining-room should always face west, and I think it a great objection to have the fireplace in the middle or at the window-end of a dining-room. It should be at the far end, on one side, so that no one has to sit with back to the fire; the table being as near as possible to the window. I should never allow hot-air pipes in any house of mine, though a stove or large fireplace in the

middle of the house is almost essential for keeping it dry.
I do not like the usual central hall as a sitting-room,
because I do not come into the country to sit within
prison walls, and these halls are generally lighted from
the top, or from some dark courtyard shut out by opaque
glass, or the windows are so high up there is no seeing
out of them. To come in on a fine afternoon, with the
views perhaps at their very best, to have tea in a darkened
room, seems to be a non-appreciation of the best the
country has to give us. Here, again, I know most people
will disagree with me. I have heard these halls more
praised than anything else in modern architecture.

I suppose it would be difficult to manage, but I think
the best bedrooms ought to have small balconies to
facilitate air-baths night and morning, or even for sleeping outside in fine weather. The modern fashion of
roofed alcoves with open sides and glass to the north
makes very delightful open-air rooms. Paved yards, too,
are full of capabilities for sitting out, or for miniature
pot-gardening. What I like best is sitting in my own
room with all the windows and doors wide open. This is
what most people dread as a full draught, but I have
never found any harm from it. Perhaps a safer plan in
bedrooms is to have two windows high up towards the
ceiling, *both* of which are kept open, as a single window
in a room, however wide open, does not properly venti-
late it, because it creates no draught.

I think the rich man has yet to arise who will build a
house and furnish it well and in good taste according to
the ideas, inventions, and manufactures of his own day,
or perhaps twenty years before. A glimmer of the
beautiful uses that might be made, for instance, of glass
and iron was to be seen in the Petit Palais of the French
Exhibition of 1900. Imitation of old things is in no sense
true evolution.

In the building of houses I think the greatest care and
thought are necessary in order to prevent the builders
from throwing out the earth of the foundations in a way
that produces a flat space in front of the house. Given
that the house is on the slope of a hill, or even on the
flat, I would rather throw up the earth into a large mound
on the north or north-east and plant this with any kind
of shrub liked by the owner, than in any way make a
terrace in front of the house. Even in the case of making
terraces, they should be laid out *after* the house is built,
and are far prettier if they slope away from the house
than if they give the appearance of the house being in a
hole. Having decided where the foundation earth is to
go, and how it is to be planted in a manner that will
cause the greatest protection with the least exclusion of
view, one must try to make up one's mind what one
really wants most. I came here to a ready-made
villa garden which I knew could never be really beautiful or picturesque; therefore I decided that what I
wanted was to grow the greatest number of plants and
flowers for picking and giving away which could be grown
healthily in a small space. This can only be done by
growing an immense variety of plants, as constant succession is the only method by which I could gain my
object. One must always reckon that in certain years
whole families of flowers fail altogether, so that, besides
the seasons, one has to provide for wet and dry years.

The beautiful combinations of flower effects which
come in all gardens as if by chance, come to me also,
though sometimes, I confess, by design; but if I were to
build my imaginary house it would be on an uncultivated
piece of ground—say, wood, common, or field—and then
I think my object would be to keep the character of the
ground as it was. In some ways the flat field would
be the most difficult to manage, and then an effect of

wildness would have to be given by breaking up the
ground and making miniature mountains and valleys.
In a small space this would be impossible, so that a
small cottage garden must be straight in design, a
mixture of flowering shrubs and herbaceous plants.
The kitchen garden and the orchard would be kept
apart, as I said before. Endless small hedges would
have a charming effect on flat ground—spring beds
divided by well-clipped hedges of *Ribes sanguinea* ; the
Japanese loniceras, which flower three times as well
if grown on short railings of bamboo-sticks instead of
against a house and pruned twice a year ; the Penzance
sweet-briars—in fact, the combinations for cutting up
flat ground are innumerable and would occur to any
good gardener. Long lavender-bordered walks would
also look very well. I was told this year that the spikenard of the Bible was made from oil of lavender. This
may have been an adulteration of the ancients ! for Anne
Pratt says it was made from the spikenard-grass, nardus.
She adds : ' When an army rides over the plains of
Persia, covered with this tall grass, an almost overpower-
ing sweetness, arising from its stems and roots, fills the
surrounding air. The ointment which takes its name
from this grass was used among the rich Jews at their
baths and public feasts. . . . Its value among the
ancients may be inferred from the mention made by
Horace, that the quantity contained in a small box of
precious stone was regarded as equal in worth to a large
vessel of wine.'

I mentioned before the difficulties of wood-gardening.
The heather and gorse ground which I mean when I say
' common,' is so beautiful in itself and so easily enriched
by plants that require little but what they find, that it
would be undesirable in any way to spoil its nature, and
all plants that would not grow without moisture would

have to be relegated to the kitchen garden. Even there it would be better to keep the varieties for each season apart, as no real gardener minds large bare places where plants are resting, any more than a mother minds her children being asleep. The toil and trouble of a garden close to the house is that its beauty is expected to be perennial and successive through every month in the year. This is of course impossible, but even striving for it entails immense care and labour if no month is to be without outdoor flowers.

Everybody tries to grow roses, and in many soils they succeed without trouble ; but for the best I ever saw in the first year of planting, the bed was made in a way worth recording. The earth was dug out two feet deep and replaced by carefully prepared soils, but instead of mixing them all together, or laying the clay at the bottom as is usually done, the three soils were laid diagonally one on top of the other, clay, richly-manured loam, and the top spit. In this way the roots of the freshly planted roses found that which they required, at different levels.

August 26th.—My experience in cooking has been very different from what it used to be, for I seldom go out, and my own food at home is of the very simplest. Cookery books of all kinds have become so inexpensive that I shall content myself with naming just a few books which have come to my knowledge of late, and only give in addition such simple cooking hints as experience has recently taught me, and which concern health rather than gastronomy ; and some still cheaper receipts will be found in the chapter called ' Wholesome Food at 3*s.* a Week.'

The opposition to simpler foods and the dislike to the simple receipts I now give are so widespread that my old sympathy with the other side of the question makes me give the following little newspaper story of an

American Judge's opinion of vegetarianism : ' The Courts have granted a divorce to a wife in Cleveland, Ohio, whose husband was not only a vegetarian himself, but insisted on her following the same *régime*. She was not aware of his dietetic eccentricity until after marriage, but she soon found she would have to eat vegetarian dishes or go without eating altogether. At last she packed her trunks and went off. The Judge ruled that in denying his wife the food which to her was a necessity of life he was guilty of neglect.'

' Leaves from our Tuscan Kitchen, or How to Cook Vegetables,' by Janet Ross. (Dent & Co., 1899.) There are now innumerable vegetarian cookery books, but I find this one most useful, perhaps because it is rather a nation's cooking of vegetables than a vegetarian cookery book. It is by the same Mrs. Ross of Poggio Gherardo, Florence, who sells such excellent olive oil made on her own estate. I used to think much change was very essential in housekeeping. I now think if people are well they like things best dressed most days alike, or nearly so, with only the change the seasons bring.

' Hilda's Diary of a Cape Housekeeper ' is a book everyone going to the Cape should get, as it would be very useful. It gives many hints besides cooking receipts, and an account of the Cape climate every month in the year.

Her former book, ' Hilda's Where Is It of Receipts,' was much appreciated. (Chapman & Hall.)

A good little book for humble households is to be got for one penny, from the School of Cookery, Colquitt Street, Liverpool. It is cheap cooking, but not vegetarian.

An excellent Vegetable Soup.—Take two carrots, two onions, two turnips, a little spinach, lettuce, endive, and sorrel. Tie up together a sprig of every sort of herb you have in the garden. Boil all in water. When the hard vegetables are cooked, take out the sweet herbs, rub the whole through a fine sieve, make it not too thick by adding the water the vegetables were boiled in, add a little butter, pepper, and salt, and serve hot.

Cheese Salad.—For a small luncheon dish, cheese salad can be quickly made by putting thinly-sliced or grated cheese in the middle of the dish, surrounding it with lettuce, endive, or cress, and covering all with a good salad dressing.

Globe Artichokes may be economically treated by boiling in the usual way, taking out the choke without disturbing the leaves, making a *purée* of the bottom with butter, pepper, and salt, and serving it hot on fried toast cut into rounds. The leaves are kept together and served cold the next day standing upright in the dish in a little mixed oil, pepper, salt, and lemon-juice. Watercress cooked exactly like spinach makes a useful early spring green vegetable. Landcress and chervil mixed, cooked in the same way, make an equally good autumn *purée*.

To Keep a Cake Fresh put an apple with it in the storing tin. This keeps it soft for several days on the same principle as the well-known receipt for keeping tobacco moist by enclosing a few potato peelings in the jar with it.

Chocolate Toast.—Take some pieces of the very best French chocolate, melt it in a little hot water to the thickness of gooseberry-fool. Pour this hot on rounds of fried toast.

A New Jam, and the very best I ever tasted, is made of morella cherries mixed with raspberries.

Onions and Sauce.—Peel little onions, and boil them tender. Make a sauce apart of butter and flour cooked a light brown coffee colour, and mixed with water. Stir the onions into this, and serve. Another way is to fry the small onions in butter till a nice brown, then braise till tender, and serve hot.

Chou Braisé.—Take a nice spring cabbage, split it, and wash in salt and water; put it in a saucepan of boiling water for ten minutes. Take it up, drain well on a sieve, put it in a casserole pot for one hour to braise with a little butter, and pepper and salt to taste, but no stock or water.

Chou à la Crème is made when cabbages have come to their full maturity, and are hard of heart and large of circumference. Boil till quite tender, chop roughly, and then pass through a hair sieve; add thick cream fairly liberally, and serve 'piping hot' and well seasoned, with golden-brown strips of fried bread leaning up against the cabbage all round.

Pommes à la Caramel.—Four large apples, $\frac{1}{4}$ lb. lump sugar, 2 oz. butter, half a cup cold water. Make a caramel of the sugar, water, and butter by boiling carefully till it is like a thick brown cream; flavour it by putting in the peelings and pips of the apples. Pass the syrup through a sieve, and put it into a saucepan with the apples whole or quartered. Stand on the stove to simmer for one hour, when the apples should be nice and soft, and look a golden brown floating in the syrup, which is nearly reduced to a jelly.

Gnocchi.—Bring $\frac{1}{2}$ pint milk to boiling point, and stir in two tablespoonfuls of semolina, 2 oz. grated cheese, 1 oz. butter, a little onion. Boil for fifteen minutes in double pan, or constantly stirring, turn out, and when cold make up into cakes, which flour, and fry a golden brown.

Purée of Potatoes.—Boil in salted water, drain well, put back on the fire in stewpan with butter and cream, and stir till there are no lumps, and it is about as thick as bread sauce. It must not boil.

Green Sauce.—Taking a handful of parsley and one

of chervil, and some leaves of tarragon, scald with hot
water, put into a mortar; do the same with a few leaves
of spinach, add two gherkins and two spoonfuls of capers;
pound all together with a piece of fresh butter, then
pass the whole through a fine sieve; mix this into white
béchamel (see ' Dainty Dishes ') sauce at the last moment.
Stir and warm up—must not boil. Pour this over the
vegetable the moment before serving. Boiled fish,
poached eggs, turnips, carrots, branches of seakale, beet,
oxalis, Jerusalem artichokes, celeriac, waxy potatoes, are
all good with this sauce.

Macaroni and Portugal Onion.—Soak some
macaroni in a little milk: cut some very thin slices off a
Spanish onion. Cook the onion a little first, then add
the macaroni, and boil till tender. Drain off any milk
left, and serve dry.

Apple Meringue.—Stew some apples as for apple
sauce; when very hot stir in the yolks of one or two
eggs. Whip the whites and throw over the apple; brown
in the oven or in front of the fire.

French Way of Cooking Endive.—Boil the leaves
in lots of salt and water, just as you would spinach or
cabbage. When tender, pour the whole thing into a
large sieve, and when the hot water has run off put
under a tap, and let the cold water run over it in quantities. This applies to all green vegetables boiled in
salt water—cabbage, sorrel, cos lettuce, cabbage lettuce,
&c. Then put the endive on a chopping-board, or, if
required very smooth, pass through a fine hair sieve;
in both these cases return it to the fire after having
first put a pat of butter to dissolve with one spoonful
of Vienna flour. Do not put the vegetables in before
the butter and flour are well amalgamated. When this
is achieved, stir the vegetable well up with the butter
and flour; let it simmer another fifteen minutes, add a

little cream or even milk quite at the last moment to make it soft and pretty. It must not be thicker than a thin *purée*.

Waxy potatoes, if wanted plain, are best first boiled, and then put in the oven to roast on a tin without fat or butter.

Cut raw well-bleached winter chicory leaves as thin as possible, and mix with hot beetroot for a winter salad.

Seakale Beet (Côtes de blettes), large ribbed leaves.—Cut the white ribs in lengths, like a finger and as thick as a little finger, boil in a little salt and water till tender, having left them in floured water while peeling, which keeps them white. Dress with a white béchamel. The green parts of the leaves are passed through a fine sieve and dressed exactly like spinach. For serving you put the white ribs with their sauce in the centre and the green all round.

Potato Salad.—Boil the potatoes in their skins at least two hours before dinner. Cut them in slices a quarter of an inch thick, put them in a china bowl with one or two soup ladles of hot broth or vegetable stock, and let them soak and remain tepid till dinner-time. They should so absorb the broth that they look dry. When served mix with a thin mayonnaise sauce, and surround with young lettuces previously dressed with oil and vinegar. They can be dressed with beetroot, cucumber, tomatoes, or herrings.

Milk Soup.—Cut one onion into small pieces, put them in a saucepan with some butter, let it fry until it is chestnut colour, then add the milk. Let it boil for some minutes, then pass it through a strainer and serve. Fry well in butter small pieces of bread and serve separately.

To Bottle Fruit.—Put clean cold water into the bottles, and sink them in hot water till the water boils in the bottles ; cover up top and let it get cold. Put the fruit fresh picked into a hot oven for a few minutes, take out and cover up till cold. Then drop it into the water in bottles; no sugar, nothing; cork and seal down at once.

Barley Water.—4 oz. pearl barley, 2 quarts water. Thoroughly wash the barley, add the water and boil till reduced to 1 quart. Strain through hair sieve (previously scalded) and press through some of the barley to thicken. Time to reduce three or four hours.

Semolina Pudding.—Boil some milk and sugar to taste (best rather sweet), sprinkle in some semolina to thicken it, flavour with vanilla if liked. Leave it to cook slowly by the fire till as dry as bread-sauce. Then butter a pie-dish *well* with fresh butter and pour in the mixture. Bake in an oven till thoroughly brown.

Sauce Sevillane (enough for two or three people).— The grated rinds of one orange and one lemon, the juice of both, two dessert-spoonfuls of red currant jelly, a little cayenne pepper or not. Mix together, and pass through a hair sieve. Serve cold in a sauceboat. Good with cold meat, or tomatoes cut in slices. Heated, with the addition of a tablespoonful of port wine, it is good with roast duck.

Boiled Cheese.—Take a piece of the best English cheddar, not Canadian, and melt it. Add a tablespoonful of white sauce made with 1 oz. butter, 1 oz. flour and ½ gill milk, all well cooked. Serve in dish with lamp under it. Eat with dry toast.

Chestnuts as a Sweet.—Peel and boil some large sweet chestnuts in water. Prepare a syrup apart, and drop in the chestnuts as it boils, and let them get cold together. Serve with or without whipped cream.

Stewed Figs.—Many children are fond of stewed figs, but I think they will be found better for everybody if, instead of being stewed, they are steeped overnight in

U

enough cold soft water to cover them. A few drops of fresh lemon-juice is a grateful addition to grown-up palates

How to Make a Dry Curry.—These directions are for 1 lb. meat, raw preferred, cold will do. Cut two onions and one apple in thin slices from top to bottom, fry them in a good lot of butter in a china-lined frypan till they are golden brown, drain them quite dry from the butter, and put them into a Nottingham jar (the round, flat shape with lid is the best). Put the butter back into the frypan, cut your meat into large dice (no fat or skin), fry the meat till it is brown, drain it from the butter, and put it into the jar with the apple and onions.

Put the butter back into the frypan, mix one dessert-spoonful of Mrs. Atkinson's curry powder smoothly with it, and add two tablespoonfuls of stock ; boil it for half a minute, and then empty into the jar with the other things, and stir well with a wooden spoon kept on purpose. Stand it on the hob and let it simmer slowly for six hours, stirring occasionally. The great secret is the slow cooking. Should it get too dry, add a little more stock or milk. Taste the curry, as some meat absorbs more powder than others ; if not flavoured strongly enough, add a little more powder just before serving. A little fine-grated cocoa-nut added is a great improvement. Serve with rice and chutney.

The above curry powder and chutney are to be got wholesale at 13 Church Street, Windsor, or retail at any of the London stores. I never eat these things myself now, but I am told this kind is exceedingly good, and I give the receipt for the many who like curried vegetables.

Dàl Bhàt (Indian breakfast dish).—Get from the chemist ½ lb. ground turmeric, ½ lb. roasted and ground coriander seed. Mix and keep in a bottle.

Soak a pint of lentils in cold water overnight. In the morning shred two or three onions, brown them in clarified butter, and add a tablespoonful of the mixed powder. Strain the water off the lentils, and put them in a saucepan with some stock or milk, and the browned onions, &c. Put on the lid and simmer gently till the lentils are quite soft. If they get at all dry, add a little more stock or milk. Serve with a separate dish of boiled rice as for curry. The cooked Dàl can be kept in a basin, and warmed up when required. Pepper and salt may be added if liked. The Dàl should be about the consistency of thinnish porridge.

Partridge, Grouse, or Young Chickens, à la Crème.—Half roast a grouse or partridge in butter in a sauté pan, with an onion cut in quarters, then take the bird and put it in a little casserole. In the sauté pan, where the bird has been cooking, put a liqueur glass of vinegar and reduce to half, add ½ pint of cream, and boil with salt and pepper and a little cayenne. Strain it over the bird in the casserole, cook slowly for ten minutes in the cream, and serve.

Fig Pudding.—6 oz. of chopped suet, 6 oz. of figs cut in small squares, 6 oz. of breadcrumbs, 3 oz. of brown sugar, two eggs, little milk, pinch of salt; mix all together, put into a buttered mould, and boil 2½ hours. Serve with any sweet sauce hot.

Sabajone (a Venetian sweet dish).—Make some sponge cake and toast it dry and crisp in a slow oven; cut it into convenient pieces, allowing two for each person, and spread thickly with good quince marmalade or jelly. Take ten lumps of white sugar, and about two tablespoonfuls of water to dissolve them. Whisk eight yolks of eggs in a bain marie, and then add to them the boiling dissolved sugar syrup. Whisk all hard for ten minutes or longer in a bain marie in a pan of hot water. Have the hot cake pieces nicely arranged in a *silver* dish for choice, or a white china one, and pour over the frothy sauce just before serving. Sponge rusks will do instead of sponge cake, and a little liqueur or wine can be sprinkled over them before adding the sauce, and this improves the sweet ; but do not let them be too damp and soft—they should be crisp under the froth.

The best **Gingerbread** I know.—1 lb. flour, 1 lb. black treacle, one dessert-spoonful of ground ginger, one dessert-spoonful mixed spice, ½ lb. brown sugar, ½ lb. fresh butter, ½ pint milk, ½ teaspoonful carbonate soda, four eggs, a little finely-chopped citron peel. Mix all the dry ingredients together, warm the milk, and dissolve the butter in it ; beat up the eggs, then add the milk and treacle. Bake three-quarters of an hour in a square cake tin, not very deep, or flat sauté pan. The mixture should be quite a running consistency before baking ; add a little milk if necessary. This is a moist, black gingerbread.

Cold Lemon Soufflé.—An economical and excellent sweet. Take the yolks of three eggs, the juice of three lemons, and grated rind of two. Add ½ lb. loaf sugar, about a tablespoon of gelatine dissolved in milk. Stir all together over the fire till it thickens. Whip ½ pint of cream and beat the whites of the three eggs up stiff, and stir in gently to the custard mixture. Pour into the soufflé dish or deep china dish, and sprinkle with grated crumbs and sugar, or with chocolate powder, or, for choice, with some chopped pistachio nut, and serve. If in summer stand in ice.

To Cure two Tongues at a Time.—These are much better than bought ones. 2 lbs. common salt, 1½ lb. black treacle, 1 lb. bay salt, ¼ lb. saltpetre. The salts finely pounded and mixed with the treacle. Rub two large fresh ox-tongues with the mixture, and turn them every day for four weeks or three and a half weeks, letting them lie in the pickle. Send them to the nearest curer to be smoked for two days. This last is not indispensable, but

improves them. When cured boil with plenty of vegetables, herbs, a few cloves and peppercorns ; garnish with
home-made glaze and chopped aspic.

Cats' Tongues.—A good French ice or dessert
biscuit. 2 oz. sugar, 2 oz. pastry flour, mixed with
cream to a soft paste, adding, perhaps, a little milk and
flavouring with vanilla or lemon. Put in a forcing-bag
and force out on to greased baking-sheets in rounds or
oblongs. Bake in a quick oven, serve fresh.

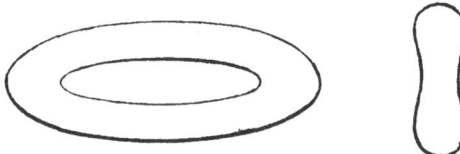

Orangeade as made in Paris.—Cut the rind of
two oranges and two lemons very thin, place in a jug,
pour in 1 pint of boiling water, add very little ginger (to
taste). Squeeze six lemons and six oranges, add the
juice of these and a syphon of soda-water. Strain, ice,
and serve. Blood oranges, when procurable, are best,
and two or three drops of cochineal improves the colour.

The best **Orange Jelly** I know.—½ lb. loaf sugar,
sixteen or eighteen oranges, two lemons, 1 oz. gelatine.
Boil the sugar to a syrup, pour it, boiling hot, on the
thinly-pared rind of two oranges, squeeze the juice of all
the oranges and lemons, pass through a fine sieve. Add
the dissolved gelatine and syrup and a few drops of
cochineal ; ice, and serve. This may not be stiff enough
to serve in a shape, but it is much nicer than when
stiffened with more gelatine.

Sand-Torte (a very good light foreign cake).—Clarify
1 lb. fresh butter ; when cold beat it to a cream, add
12 oz. white sugar, 1 lb. potato flour passed through a
sieve, four whole eggs and the yolks of two more, the

best of one lemon and a few drops of lemon or vanilla essence. Beat the whole for at least one hour, and until it makes bubbles. Bake in flat round or square, shallow mould, buttered and finely breadcrumbed, in a moderate oven. Dust over with fine white sugar—icing sugar, for choice—before serving.

Cold Chocolate in Glasses, as in Paris.—Quantity for about six custard glasses. Take 6 oz. of *good* chocolate, not less than 2s. per lb. Dissolve it in hot water in a clean copper over the fire—just enough water to melt it to a paste. Add a teacupful of milk, and let it boil up. Pour the mixture on to five yolks of eggs beaten up, and put it all on the fire again, stirring it till it just boils. Take off quickly and leave it to cool, or in summer stand it on ice.

Bread.—People are constantly saying to me, ' One of the chief difficulties of your diet is the good bread you recommend. The cooks say they have no time to make it, many members of the family won't eat it, the bakers won't change their methods, and the mill is either far off or the flour is all ground by the new method of china rollers, which is especially designed to make the flour very white.' My answer is that all these drawbacks are easily overcome by a little trouble and conciliation. When the cook finds that the bread-making is less trouble than cake-making (see Receipts), and that her mistress cares more about it than about made dishes, she soon falls in with her ideas. Millers I have found most obliging, and even interested in the matter. Our own old water-mills here at Cobham, by the Mole side, have been lately bought by Henry Moore & Son, of Leatherhead, and on being written to they would send to any part of the kingdom the various flours I use, and also the fresh bran (see receipt for Bran Tea), which is not easy to get in London. I asked them for a specimen of English wheat ground through stones, and received a most civil answer, in which they say they also send a gallon of whole meal with the bran taken out, adding : ' We hardly agree with this ourselves, as we think it much better with the bran left in.' This is not the first instance I have had that millers themselves appreciate the value of bran in wheat meal. Unfortunately, many modern digestions cannot stand whole meal with the bran left in, and for these it is better that they should have the bran sifted out either by the miller or at home, in which case they can get the phosphates of the bran by using it for bran tea.

I have heard of another mill at Dorking, kept by Mr. Atlee, where the wheat is ground in the same way with the old-fashioned stones.

My letter from Mr. Moore concludes : ' If we can do anything in the way of experimenting for you, we shall be very pleased to do so.' Thus proving that millers who have to grind to please the bakers, whose flour cannot be too white, are equally ready to please private customers when they have the chance.

For those living in London it may be a convenience to know that at the Haymarket Stores they can get the finest wheat meal, ground by the well-known Great Barton Mills of Mr. Scott, of Ipswich ; also at Bax & Sons, 35 Bishopsgate Street Without, E.C. When the bread-making at home is an insuperable difficulty, a good deal can be done by a little love-making to the baker. He is, naturally, anxious to please his customers, and will soon make wholesomer bread if he finds the public really demands it. People who have once had pure wheaten bread from a baker will often get him to send it to them wherever they go. The poor in their own way are very particular about bread, often getting it from a more distant village, in the belief that the whiter and puffier it is, the better it is.

Directions for Preparing Good Unleavened Graham Bread (according to Louis Kuhne).—Take 5½ lbs. of unbolted wheat meal (Graham flour), or the unbolted flour of any other grain (in tropical regions maize meal with wheat meal or rice grits, &c.), in a pan, pour over it about 1½ quart (2 wine quarts) of cold water, and mix it thoroughly. I consider cold water preferable to warm, as experience shows that warm water sets the bread in fermentation more readily than cold, and this may render the bread somewhat lighter, but less nutritious and pleasant to the taste. Now divide the dough into three or four equal parts, mould a loaf of each, lay each upon a dry tile (not brick) sprinkled with Graham flour, wet the loaves well on the top with water, and place each with its tile upon an empty flower-pot in the quick stove oven or brick oven. No other articles or pots should stand in the oven at the same time. The heat in the oven must be kept up by a steady fire, and should not be immediately intense. In half an hour, during which the oven must not be opened, turn the front side of the loaves to the back. A quarter of an hour later see whether the upper crust is well and firmly baked, and then turn the loaves over, as they are usually still soft on the lower side. The loaves must now bake until they sound quite hollow when tapped in the middle with the finger; this usually takes half an hour longer. One may then feel sure that the bread is well baked and the crust not too hard.

Directions for Preparing Graham Gruel.—To obtain a plate of gruel stir up a heaping tablespoonful of Graham flour with a little cold water till a liquid pap is formed. Pour this into boiling water, and let it boil some minutes, stirring continually. Salt very sparingly, or not at all. This gruel tastes very good when sprinkled over with raisins. (From Louis Kuhne's ' New Science of Healing.')

A Perfect Baby Food.—Cut thick slices of Graham, or whole meal, or white home-made bread, and bake them to hard rusks in a slow oven. Break up the rusks and put them into a jar, a Gourmet boiler, or an earthenware-lined milk-saucepan with cold distilled water enough to make into a thick jelly when cooked. Stew slowly for three to four hours, and strain through scalded muslin or very fine hair sieve. Mix with warm milk as required.

The rusks may be made twice a week and stored in air-tight tins, but the jelly must be freshly made each morning.

This food, by the long and thorough cooking of the cereal, contains all the elements necessary for the formation of strong and healthy flesh and bone tissue, good blood, hair, nails, teeth, skin, &c.—in short, everything required to keep an infant in the best possible condition. The Graham or unleavened bread is by far the best, as yeast or powder raised breads are likely to upset a child's digestion. Neither constipation nor diarrhœa need be feared on this food. It is valuable also for invalids and the old.

On the very important subject of bread Mr. Albert Broadbent, F.R.H.S., read an admirable paper last autumn at the annual meeting of the Vegetarian Society, in which he referred to certain remarkable statements made by Mr. T. G. Reed at the meetings of the British Dental Association to the effect that modern bread causes the destruction of tooth tissue, and that wherever roller mills are in operation carious teeth are found instead of sound ones. Mr. Broadbent did not think it was necessary to return to the old stone mills, if we take care not to disturb the natural balance of the wheat berry constituents by separating out the starch and making our bread of that to the exclusion of other valuable parts. He went on to say, 'Wherever we find a race of men retaining primitive milling customs, or living on uncorrupted grain food, we find their teeth strong, and there is an absence of decay. Mr. Albert Carter, surgeon dentist, related, in "Natural Food" some few years ago, some investigations he had made in this connection. He found the natives of Punjab and North-West Provinces— whether Hindoos, Sikhs, Punjabees, Afghans, or Goorkhas—had splendid teeth. He went to the banks of the Ganges and examined the Brahmin skulls. He failed to find one showing general dental decay such as he was acquainted with at home. In Ceylon he found that, while the native Cingalese had good teeth, the children of European parents had dreadfully bad teeth. In Australia it was the same; the aborigines had splendid teeth, while the Melbourne dentists were hard at work repairing disease in the Colonial-born portion of the population. Edwin Cox, Licentiate in Dental Surgery, R.C.S., in "Degeneracy and Preservation of the Teeth," ascribes teeth degeneracy to the use of white bread. This is undoubtedly a serious and important subject, and must continue so while we consume more bread than any other one food. Bread should be of the very best kinds available, and when it is made from the best kinds of wheat it forms an almost perfect food. For many years I have observed the beneficial results that have followed the use of good whole meal bread; with it there is scarcely any toothache or neuralgia, and constipation is scarcely possible. The teeth are made stronger and less tender, owing to the proper supply of lime and phosphates contained in whole meal.'

The following poem is by Mr. A. J. Munby, a contemporary and well-known Surrey poet, and belonging to the days that are gone. I read with great pleasure one day this spring this spirited description of a class that has almost disappeared; brass cans and lively ponies rush now along our lanes :—

THE MILKWOMAN

She was tall and strong, and she walk'd along
　　With a firm substantial tread,
Like one who knows that wherever she goes
　　She is earning her daily bread.

Her frock was print, and there was not a hint
　　In the whole of her simple dress
Of that milliner's touch which adds so much
　　To a lady's comeliness.

Yet she is aware that her face is fair ;
　　But she also understands
That the best of her charms are her stout red arms
　　And her strong hard-working hands.

' It's them,' says she, ' as has work'd for me,
　　Wherever my work has been ;
And as for my face, why it's no disgrace,
　　For I reckon it's always clean.

' Well, there's Jack, I know, he bothers me so—
　　But what do I care for him ?
I'll ha' nothing to say to a lad that's gay,
　　So long as I've life and limb !

' Such chaps may do for a wench like you,
　　As is fond of an easy life ;
But if I get a man, I shall do what I can
　　For to make him a working wife.'

She smiled as she spoke, and she settled her yoke
　　On the back of her shoulders broad,
And she stoop'd to her pails by the area rails,
　　And harness'd herself to her load.

Then she went on her beat through bustling street
　　With a step like a martial man's ;
A step that suits her iron-shod boots,
　　And the weight of her clanking cans.

For her cans and she had the bulk of three,
　　And deftly as she might steer,
'Twas the silent might of her strength and her height
　　That kept the pathway clear.

There were many who eyed her stately stride,
  As she moved through the yielding crowd,
With her hands on her hips, and a smile on her lips,
  And a look both calm and proud.

But none, or few, of the gazers knew
  The worth of her humble trade;
And beauty alone may never atone
  For the lot of a milkman's maid.

They could not see what was clear to me—
  That the loftiest lady there
Might envy the part in Dame Nature's heart
  Which is owned by Kitty Clare.

When first people begin strict diet they are very apt to think they will die without change. It has been a great humiliation to me to find how much more easily the young take to living on nothing but bread, fruit, potatoes, vegetables, and cheese than their elders do. A contemporary said to me in such a sad voice the other day, 'Is it not distressing to find out how one pines sometimes for some of the old foods, which at the time we ate them we hardly knew how much we enjoyed?' I earnestly caution people against not only bought breads, but also bought biscuits. I am sure that they are not the same help to health as things made at home, and biscuit-making, which by universal consent is now left entirely to manufacturers and bakers, takes even less time than boiling potatoes; for a skilful cook can make and bake a piling plate of biscuits or wafers in fifteen minutes. The exception to my mind is Lemanns, 28 St. Swithin's Lane; especially his thick captains.

One's great wish when practising diet is not to make the cook feel she is no longer wanted, but that her skill can be adapted to all sorts of new requirements. In staying away at a friend's house lately the cook was quite miserable because she felt she had nothing to do for me.

In another house I was unfortunately half-poisoned by the home-made bread being ruined by much salt, and the vegetables tasted as if steeped in brine.

From a German friend I received in late autumn last year a list of what the work had lately been in her kitchen. The winter climate is so severe in mid-Germany that nothing can be left in the ground and storing away from frost is very difficult. Everything is so easily brought to our shores all the year round now that we have almost forgotten the old days when preserving went on in our own country houses. Now most people say it is cheaper to buy; so it may be, but it is not so good. My friend complains that she cannot get twice-ground whole-meal flour in Germany, and she adds that she cannot get bread-making done at home, as her cook has such heaps and heaps of preserving to do. My answer would be that her family and her household would be in far better health with good sound bread to eat and much less preserved fruit and vegetables. She asks : 'Would you like to hear all we do?' and then gives the following list : Thirty or thirty-five tins of strawberry compote, two dozen marmalade and jams in pots (this does not mean orange marmalade, which would be out of season); ditto clear fruit juice to drink with water, also thick juice to use for puddings, sauces ; cherry compotes in tins, ditto stoned in glass pots ; marmalades and jams, half-fermented juice to drink with water, fifty or more pounds of apricots dried in oven for winter, servants' compotes, mirabelles, the same as above. Cucumbers, two barrels, filled and preserved in salt; cucumbers, small green, pickled in vinegar; mushrooms dried in oven and pressed in tins, ten quarts of tarragon vinegar for winter use, three dozen tins of flageolet beans, three dozen tins of shelled and preserved green beans, two barrels cut beans preserved in salt; choucroute, one barrelful preserved in salt ; blue plums peeled and preserved in bottles, ditto not peeled, preserved in vinegar and sugar ; tomatoes preserved whole in tins to use for ' garniture,' artichoke bottoms the same way, and many bottles of tomato *purée*. A good deal of this kind of luxury it seems to me desirable to reduce as much as possible, but of course everything depends on the size of the household.

I must not fail to mention one of the most useful articles for domestic purposes that has come out to my knowledge in the last few years—a powdered soap in tins, not expensive, to be bought at Harrod's Stores, or others, and called Sapon. Its *especial* use is for the washing of all nursery woollen things, Shetland shawls, under-clothing, blankets, &c. Everyone I have recommended it to has been delighted with it.

Another very charming but expensive luxury has been imported from France by Messrs. Goode & Co., of the china shop in Audley Street, and if I were going to begin life again I should certainly have it—copper stewpans of all sizes lined with silver beaten into the copper, in the old Sheffield-plate manner, before it is made up, and so thick that nothing will wear it off. This saves all the bother and expense of re-tinning, and the risk to health of copper pans being neglected, and not tinned soon enough. The price is not at all prohibitive, considering the goodness of the pans.

For Drying Flowers.—Take cotton wool and tissue paper. Lay the wool on the flower and the fine paper over it, and put this between the pages of any old folios or books. Tie a string round the books to keep all in place, and put them under an impromptu press of a board with a stone or weight on it, in the sun if possible, or even near the fire. The cotton wool absorbs the moisture and the heat dries the flower quickly. The wool prevents the delicate petals from being crushed, and so the character of the flower is preserved.

Corns.—So many people who refuse to believe that one of the minor merits of the simpler foods as ordered by Dr. Haig is the cure of corns, still suffer from these inconveniences, that I think the following letter too good to lose :—

Sir,—I trust it may not be beneath the dignity of the 'Westminster Gazette' to permit a humble country doctor to add his little seasonable mite to the gaiety of nations. The one fatal word which is the title of this letter represents the most common, perhaps, of the small ills of life. Once upon a time 'an old wife' in the village of Carisbrooke said this to me (and her endearment of address would have been above suspicion had my readers seen her), 'My dear, you may know a lot about Anna Tummy' (I thought this, at first, some hideous gastric surname, but I realised almost at once that it was an orthoepic liberty only), 'but you dunno nothin' about " karns "—I do.' Well, I happened to know that this good lady had a reputation in the country-side for the relief of the *hard* variety of these elementary drawbacks to easy labour. On her death-bed she told me her secret. Here are her *ipsissima verba*, and Heaven forbid that I should lessen the force of her teaching by any grammatical paraphrase. She said, 'You takes beeswax and you deeps it hot right on the karn and covers 'un ; then you puts on a bit o' swealed rag and lets 'un set for four days. Then you pulls 'un out after you've a-soaked your foot in water hot enough for you to bear. And when 'e's out you'll see a big hole where he was.' Now, I don't the least care if a 'thousand and one gems' of correspondence reach you saying the remedy is old. So is the Bible old, but very few people know much about it !

Your obedient Servant,
GEORGE W. R. DABBS, M.D.

Shanklin, I.W.: December 18.

Chilblains.—Far the best cure for chilblains is a very simple one. Rub the feet twice a day, when well warmed by a fire, with some soft soap. Chilblains are half a gouty symptom, and no doubt the magic is the strong alkali in the soap.

Personal.—The following suggestion is worth knowing for anybody who has a bilious headache and is obliged to make a speech, or any great effort. Put a whole tin of Colman's mustard into a large hot bath, stay in it ten or fifteen minutes, lie down after it for half an hour. The brain will then be far clearer and better than in ordinary health. This is a remedy only to be used for an emergency. It is, of course, rather a severe trial for the circulation, but less likely to be injurious than the drugs which are so often taken for headaches.

SEPTEMBER

Visit to Northamptonshire—Peterboro', Fotheringhay, and Kirby Hall
—Iris in pans for spring flowering—A last year's autumn letter
from Germany—Kew and the smoke curse—Pruning back of
shrubby plants to imitate sub-tropical gardening—Japanese
anemones in shade—Sunflower seeds as a possible farming
industry.

September 1st.—Last year I went to stay in Northamptonshire. On my way down, having an hour at Peterboro' station, I visited its magnificent old cathedral—the largest and most important Norman building that I have ever seen. I know nothing really about architecture, but I suppose it must have been built just at the Transition period, for all the arches are round, except just across the nave, where they are slightly pointed, probably the better to carry the weight of the tower. The whole effect of the church is most remarkably simple and dignified. When I got home, I looked it up, and found that Peterboro' Cathedral took more than half a century to build—from 1118 to 1193. The eastern aisle, which is Perpendicular, was begun in 1438 and not finished till 1528. Katharine of Arragon was buried there, and her remains are there to this day. Peterboro' stands on the old North road, but though I remember that my mother, in speaking to us as children of her many journeys from London to Newcastle, often talked of York Cathedral, I do not remember her ever having mentioned Peterboro'. This perhaps means that, although an important place, they only changed

x

horses there in the old posting days; just as we now, in
trains, pass the most interesting places, only staying three
minutes at the railway station. During my visit to Northamptonshire, I stayed in a most beautiful old Jacobean
house, not much spoilt and well maintained. My kind
hostess drove me about to see the sights of her neighbourhood, and we went first to the village of Fotheringhay.
The name remains, though the castle has ceased to exist
where Mary Queen of Scots ended her days. The village
has evidently shrunk in importance; the church is very
large, and alone retains a certain old-world magnificence.
A farmhouse is pointed out as being one in which the
soldiers were lodged during the time of the trial. We
walked a short distance down narrow lanes, between
farm-buildings, to see the site of the old castle. Even
when we have read of a certain place in a novel, and
afterwards visit it, the scene becomes alive with the
imaginary characters of the book and the whole story is at
once a reality, as I mentioned before when I visited Brams
Hill. How much more is this the case when we stand
on a spot where the prominent figures in history lived,
and looked with their eyes on what we now see to-day!
The general effect of nature never changes; a stream
winds as it has always wound, and the long shadows of
evening are cast by the sun in the same manner at the same
time of year as they were cast three hundred years ago.
I had this feeling at its highest pitch when visiting
Fotheringhay last autumn. All trace of the castle is
absolutely obliterated, except for a huge grass-grown
mound, verily a grave. Standing on the mound, it is
possible to trace the double moats which surrounded
Mary's last prison, and one huge mass of conglomerate
stone lies on the surface of the field below. The landscape is very fair—a lovely, winding river crossed by a
stone bridge of a more recent date than the one erected

by order of Elizabeth after her stay at Fotheringhay and which fell into ruin. The broad green meadows, the cattle grazing, everything is just as it might have been when Mary arrived, full of forebodings, in September 1586. As the meadows flooded, the fogs of autumn arose and surrounded her, her case became more hopeless, and, crippled with rheumatism, she began to lose heart, poor thing! In February came the cruel end, just after her forty-fifth birthday. The story told in so many history books that her son James razed the castle to the ground seems to be a fiction ; for, although his filial affection caused him to move her body to Westminster Abbey from Peterboro' Cathedral, where she had first been buried, it did not extend to the destruction of the castle, as there are documents to show it was standing and in good repair after his death. Its owner, Sir Robert Cotton, removed the woodwork of the hall, where Mary was executed, to Connington Hall, in Huntingdonshire, where it still exists. Froude gives many interesting details of Mary Queen of Scots' last days, but I think it is clear he never visited the place himself, as his description is not quite correct. He says that the village was nearer the river than the castle was ; this is not the case. He leaves out the second 'h' in Fotheringhay, and calls the river 'Nen' instead of 'Nene.' Froude, though so notoriously Pro-testant in his views, gives Mary a splendid testimony when he says : 'In point of form and grace, Mary Stuart had the advantage of her rival everywhere. Elizabeth, with a general desire to do right, could con-descend to poor and mean manœuvres. Mary Stuart carried herself, in the midst of her crimes, with a majesty that would have become the noblest of sove-reigns.' Froude, in the twelfth volume of his history, gives a most interesting description of Mary's dress on the scaffold. It must have been prepared beforehand

and required much thought. True woman to the very end, she made the most of her last opportunity of creating an effect. Froude says : 'The lawn veil was lifted carefully off not to disturb the hair, and was hung upon the rail. The black robe was next removed. Below it was a petticoat of crimson velvet. The black jacket followed, and under the jacket was a body of crimson satin. One of her ladies handed her a pair of crimson sleeves, with which she hastily covered her arms ; and thus she stood on the black scaffold with the black figures all around her, blood-red from head to foot.' Froude describes how, when she placed her crucifix on a chair, it was seized by one of the executioners. He was made at once to replace it, and everything she had worn was burnt at the huge hall fire before the spectators left the room, every precaution being taken to prevent the keeping of relics. This adds immense interest to the fact that, in the year (about) 1830, an old man digging in the castle grounds found Mary's ring bearing her initials and Darnley's tied with a true-lover's knot. This ring is now to be seen in the Waterton collection at the South Kensington Museum. The supposed explanation of the finding of this ring is that it dropped unperceived from Mary's finger at the time of her execution and was swept up and thrown into the moat with the bloody sawdust. The moats are dry now, but in winter the Nene still floods the lowlands. It is said that many medicinal plants are still found about the castle grounds, and a peculiar thistle is known locally as ' Queen Mary's Tears.'

Another of the places in the neighbourhood which I went to see was the well-known Kirby Hall, the fairest and most Italian of Elizabethan houses. It has shared, in these latter days, the fate of Fotheringhay in the time of Charles I. The owners, I suppose, finding it expensive to keep in repair, removed all the valuables, including the

panelling and chimneypieces, and allowed this beautiful
Elizabethan gem to fall into decay. Oh, the pity of it!
In Jones's 'Views of the Seats, Mansions and Castles of
England,' published in 1830, he gives two views of Kirby
Hall, taken, in my opinion, from the least beautiful sides,
and in no way doing the house justice. But these prints
represent the house as well-roofed and showing no signs
of ruin. No doubt fifty years ago a comparatively small
sum would have kept it weather-tight and preserved
it. Now thousands of pounds would hardly restore it.
This magnificent pile of buildings was founded by one of
Elizabeth's favourites, Sir Christopher Hatton. The
guide-books attribute the early part of the building to that
somewhat mythical genius, John Thorne, who gets the
credit of having designed all the best houses in England
for about a hundred years. However much he may have
been consulted about the planning of this house, it is
very different and far more remote and Italian in style
than either Holland House or Bram's Hill. A later
owner called in Inigo Jones to 'modernise and improve'
it. Sir Christopher Hatton died a bachelor before Elizabeth, in 1591, and was buried in St. Paul's Cathedral.

I wonder in what the wealth of England consisted in
Tudor times that enabled people to build such splendid
residences. In Suffolk, I am told, there were rich
clothiers and woollen merchants. Perhaps it was the
same in Northamptonshire, where the pastures are very
rich. Or shall we dare to think that the money was
mostly ground out of the wretched peasantry of the
country? I noticed, driving about in Northamptonshire,
several large old pigeon-houses like those that existed in
France, and which were part cause of the Great Revolution. The pigeons fed on the grain sown by the peasants,
and the peasants were paid in kind by gifts of these same
pigeons.

September 10th.—I have in flower a Cape bulb called eucomis, which is seldom seen, but which I think very handsome and well worth growing. There are three or four kinds ; the one I have is, I believe, called *E. undulata.* I have had it unmoved for several years, so it is quite hardy. The flower spikes are about two feet long, and the upper half densely arranged in a cylindrical manner. The flowers are not showy, but they look exceedingly well picked and arranged in a vase alone or with some spikes of *Lobelia cardinalis.*

I have flowered this year for the first time a most attractive little rockery plant, with quite a weak-looking body and a fine big showy campanula-like flower. It is called *Cyananthus lobatus.* It should be grown on a damp rockwork ; this I have not got, but as we have only had one fine dry week this year it did not matter, and the plant evidently thought it was in a damp spot.

August, or early in September, is the time to plant in pans, just like hyacinths, the small bulbous irises. They thoroughly repay growing, and should be planted and put in a frame under cocoanut fibre all the winter, and only be brought into the greenhouse just before flowering. *Iris reticulata* is the best, as it is very pretty and very sweet. *I. Bakeriana* is another excellent one.

September 17th.—I copy here a letter lately received from my German friend. Our gardens are a constant interest to each of us, as what does easily and well with her does badly with me, and she is then quite jealous of the sun-dried little successes of my light soil.

' We are nearly drowned, and to hear you talk of a dry garden sounds like distant myths of El Dorado. For a solid fortnight now we have had sheets of water, day and night, and everything is decaying and reduced to pulp ; one's very bones creep with the damp and feel mildewed. It is impossible to work the soil or to go

on the grass without wearing indiarubber boots, like
waders. You ought to see it once like this to convince
you of the numerous difficulties of my gardening—the
parching, cutting east winds and droughts all this spring
and summer, and then those indescribable masses of water
that rot and spoil everything, so that when the frost *does*
come everything is so saturated and water-logged it starts
the roots and bulbs from the soil, the water turning to
solid ice. What makes it a worse grievance than other
years is the fact of its beginning a month or six weeks in
advance of the usual time, and thus ruining the whole of
my painfully-built-up scheme of a pretty September
garden. The terrible havoc of last year's winter, and the
heavy snows and frosts of the latter half of March, had
so effectually ruined all prospects of a good spring garden
that I set to in April and May, so as at least to ensure a
good display for the summer and autumn. And, with all
due humility, I believe I had succeeded; but now it is
all battered and beaten into jam, and one could almost
long for snow to cover it up mercifully. These are the
woes of strong heavy soils, where things grow luxuriantly
and drought does but little harm, but where rains do
untold damage, and a wet autumn means destruction to
all tender things and disaster among the bulbs. Yet even
so calamitous a month of September bears some lessons—
one of them not to delay the gathering of seeds of the
best annuals, the other to push forward planting and
transplanting of perennials, so as to get them well established in the first days of the month. This exceptional
year proves that the usual rule of counting upon the
whole of September for all the transplanting and re-
planting is not a safe one. For the first time this year,
driven by necessity and by the rows and rows of corpses
which I discovered on coming out to Cronberg in April, I
have really had a good show of annuals and half-hardies.

The single branching larkspurs have been quite lovely, both in pots on the terrace and in big masses by themselves in the little paved flower-garden. There were lavender-coloured ones, and a batch of quite lovely bright pink coral ones, besides the usual tall white kind and the different shades of blue. I had never yet got them to thrive in my stiff soil, and I was determined to have them, so I deliberately took out eighteen inches of my soil in more than half the flower-garden, only leaving those beds that were to grow dahlias, asters, pyrethrums, and Michaelmas daisies, and replaced the soil by fine compost made of sand, road-scrapings, turfy loam, and a *very* little decayed manure. All the annuals sown in February and March in pans, or in the soil of a cold frame, were pricked out into this good light soil at the end of April, and some were sown in place as late as the first week in May. The latter were portulacas, a few patches of *Nemesia strumosa*, *Convolvulus minor*, *Viscaria oculata*, and the lovely sky-blue viscaria. A large square bed of *Nigella damascena*, sown in a mixture with *Omphalodes linifolia* and well thinned (it makes a perfect mixture and is most satisfactory), two big beds of salpiglossis of the largest strain (Haage & Schmidt at Erfurt), and smaller patches of the dwarf orange-coloured eschscholtzia, as well as the pink and white, both very pretty. Large patches of Shirley poppies and tall French double ones severely thinned out, the giant Machet mignonette, and the pretty bright red Stuttgart variety, then the lovely but very ephemeral *Phacelia campanularia* and the beautiful *Bartonia aurea*, the charming *Linaria bipartita* and *L. Cymbalaria*, the delightful tribe of Chinese fringed pinks, the purple cornflowers; all these did beautifully sown in place, and would have lasted until frost if they had not been battered and smashed by these terrible weeks of deluge. The frame-raised annuals were the single delphiniums, the rhodanthes and acrocliniums, the tall self-coloured antirrhinums; these last I can never grow except as annuals, nor can I the white, the bright yellow and bright red, and *very* dark red, nearly black—the *Chrysanthemum carinatum*, the tall, beautiful *Cosmos bipinnatus*, *Commelina cœlestis*—which began to flower six weeks after sowing, when it was only two inches high, and is now still in full lovely bloom and two feet high, and a gorgeous gentian blue—the annual *Lupinus Hartwegi* and *polyphyllus*, the whole tribe of tall and dwarf orange and pale yellow autumn marigolds (tagetes), and the best strains of scabious, a large patch of very big-flowered pale lilac, exactly like *S. caucasica*, being excessively useful, besides the bright coral pink, the self-white, and the ruby-coloured ones. My paved flower-garden was all arranged on the principle of a market garden, each kind in a bed, or half a bed, by itself, with the exception of two mixed ones, one of which I kept nearly entirely blue with lupins, *Phacelia viscida*, *Viscaria oculata cœrulea*, the Swan-River daisy, single-branching larkspur, *Commelina cœlestis*, cornflowers, nigellas, dark-blue salpiglossis, dark heliotrope, and on the edges tufts of dark violas, broken by patches of mignonette and the above-mentioned precious *Omphalodes linifolia* as the only exceptions to the general blueness. That bed was lovely; it was about fifteen yards long by two and a half yards wide. I forgot to mention tufts of the dark-blue *Salvia Horminum*, and a tuft or two of lovely self-sown *Veronica Hendersoni*, with long blue spikes. The Swan-River daisy above named is properly called *Brachycome iberidifolia*, and is lovely. The other bed, where I had several annuals mixed and not one kind planted alone, was a long narrow bed—only five feet wide and about twenty long—and here I planted all the shades of orange and terra-cotta zinnias, the white single larkspur of the tallest sort, the yellow antirrhinums, and the white feathery " Comet " aster, and then a large batch of brown, yellow, and orange *Helichrysum macranthum* (both these and the single zinnias have been very large and fine and satisfactory, and last so wonderfully in autumn and defy rain and drought and every fatality of weather) ; also a very pretty pink rudbeckia, about two feet high, with a lovely mahogany-coloured inside and broad zone all round. I found it in flower in a friend's garden in Hesse, and gathered some seed.' (It is not an annual, I think. Same as *Echinacea purpurea.*—M. T. E.) ' I forgot to mention among the annuals I have used largely for fringing and filling up the edges of my beds of dark-red hybrid perpetual roses is a good strain of tiny pink petunias, and *Phlox Drummondii* in all the shades of red, from salmon-pink to nearly black, barring all magenta, yellowish, or white. They did beautifully and are so pretty and luxuriant now ; also the fringed *Dianthus Heddewigi* and the *Viscaria oculata cardinalis.* By keeping to tones of red or pink-red, I really got rather a nice harmony of colour low-growing in the bed among the hybrid perpetuals ; but, alas ! in spite of the book, " Chemistry in the Garden," I have again had mildew and rust, though the latter, perhaps, to a lesser degree than other years, owing to painting the plants with brimstone early in March and frequent syringings in the growing season with the mixture that book recommends. But the cutting, horrible east winds we have here, and the changes from very hot to almost freezing which are so common in spring and after August, will, I am afraid, always be a terrible evil to fight against amongst my roses. To fill the gaps created by the winter frost among my rose-beds (especially tea roses), I had quantities of self-raised little plants on their own roots (cuttings), but they are small and make no show yet, so I introduced patches of an early-flowering and quite lovely outdoor chrysanthemum, apricot coloured with a touch of pink; it is called "Gustave Grunewald," and flowers from August profusely. It has all the characteristics of a Japanese chrysanthemum, and not the ordinary quite hardy out-of-door flowering kinds. When the weather becomes too rough, and the lovely pink chrysanthemum flowers become spoilt, I take up all the plants and pot them, and they continue to flower well into October when the others come on.' (Any rose-grower would say, 'No wonder her roses don't do, if, for the sake of effect, she plants so many things among them.'—M. T. E.)

'Now to tell you about my half-hardies; the number of them is of course very great, and the labour of potting and housing increased tenfold to what it is in England, from the extreme cold we have, which makes it impossible to leave out of doors any sort or kind of veronica, or *Verbena citriodora*, rosemary, cistus, hydrangea, *Lobelia cardinalis*, tritomas, *Campanula pyramidalis*, or the better sort of carnations. The few just named mean hundreds of little pots to be got through the winter, if I am to have anything like an adequate supply for next year. The pentstemons and best snapdragons, too, I keep in pots, then a few large plants of the glorious red salvia and the other salvia with large pink blooms, the white Paris daisy large and small leaved, the yellow marguerite, the calceolarias brown and yellow, the pretty blue *Agathea cœrulea*, the best kinds of bright pink verbenas to be kept to make cuttings from in the spring, and the best double fringed petunias, then all the Cape pelargoniums, the tuberous begonias, of which we raised a big batch from seed, and had some really splendid shades and sizes. I used these largely to produce patches of bright colour in the autumn garden and to fill up places from which the early flowering phloxes, tulips, English irises, hesperis, *Aster alpinus*, and *Campanula persicæfolia* had been removed to the reserve garden after flowering. These gaps were filled by red salvias broken by scarlet tuberous begonias, by Paris marguerites with yellow and white begonias in between, or by the tall pink salvia which is so lovely with the *Sedum spectabile* as a groundwork. All in all this terrible winter, and the havoc among my roses and herbaceous perennials, have taught me a lot; as usual necessity being the key to success. I had lost all my fox-gloves, tritomas, *Anemone japonica*, all the finer sorts of hypericums, nearly all English and Spanish irises, the Spanish squill, and numberless small and pretty alpines and other dwarf plants. And so, to save what could be, I had to put all my energy on to raising a sufficient stock and variety of annuals, and make much use of half-hardy and sub-tropical plants from seeds and cuttings; my cannas, castor-oil plants, red and green-leaved Indian hemp — solanums, caladiums, and the three kinds of nicotiana, mixed with batches of gladiolus, montbretia, amaranthus, and choice cactus dahlias, made a huge tropical-looking bed of most decorative effect. All were put out the second half of May, some on a trampled foundation of manure to help rapid growth. The tall daturas made a lovely clump close by, sunk, and planted out in the grass, the base of the clump hid by polygonum and *Desmodium penduliflorum*. My sweet-peas, the earliest, were raised in the greenhouse, sown in February in pots and planted out in April; the second batch in a cold frame in pots and planted out late in May; the third batch sown in place and just a little thinned, watered and manured. They are still now in full beauty—September 20—but I never get them as tall or the blooms quite as large as the best English ones. Still they are good Eck-ford varieties, and I do not grumble, for, no doubt, I shall learn by and by. My greatest pleasure among all my annuals has been the thorough success of *Nemesia strumosa*, sown in three or four different little patches, in pots, in boxes, and out of doors; some pricked out into sandy soil and sunny places, some pricked off into boxes or large shallow pans, where they bloomed and seeded profusely. The biggest plants and profusest bloomers were those sown in place. They began to bloom the end of July, and are still lovely, and have yielded masses of cut blooms and good seed too, though it takes a lot of gathering.

'A paling, topped by a little red-brick roof, has been prettily overgrown with a lovely new (to me) creeper called *Solanum Wendlandi*, tender, and to be treated like *Plumbago capensis*. It has large clusters of lovely lavender or mauvy-blue flowers, is a strong grower and covers a large space in one summer. It was very pretty growing alongside of a mass of good snow-white everlasting pea, with *Plumbago capensis*, and a mass of different shades of heliotrope at the foot of it scenting the air at a distance. And now we are on the verge of death and decay to all this quickly conjured-up loveliness, and the long lottery begins again which always brings surprises, and will perhaps force us next spring to have even more recourse to annuals. The long and short of all my efforts is to conclude that gardening in this climate is not really worth the trouble, as it means five months of life and seven months of death, and a yearly renewed effort to produce and reproduce plants that grow like weeds in the blessed climate of your British Isles.'

My friend's sad grumble reminds me of two or three lines in one of Robert Lytton's letters : ' Friendship under the chill veil of absence is like a garden covered with snow. The roots, the germs, the bulbs all are there, but where are the flowers ? ' A garden is never without hope. What one thinks is dead springs up strong from the root, and what seems to have survived till the spring often proves to be dead.

September 19*th.*—To-day we drove to Kew. A lovely day here, but on arriving there we were met by a dark smoke fog from London. It is very sad how black Richmond and Kew have got lately, and the plants are suffering very much. One does not notice it in the early spring, but at this time of year the leaves are black and shrivelled, and the glories of autumn will not be for them. Oh, why cannot something be done to save us from this curse of smoke? The difference during the twenty years I have lived here is astonishing.

I was disappointed with a great deal of the planting at Kew; certainly large fat beds on green grass are very unsuggestive and commonplace. The spring garden was pretty in general effect, the autumn garden only interesting in detail. There were some plants pruned back to form interesting sub-tropical-looking foliage, which were far stronger and easier of cultivation than the real sub-tropical plants often used. They remain where they are planted and are merely fed, I suppose, and cut back in spring to one shoot. *Paulonia imperialis* filled one bed. The plant comes from Japan, and so might be injured sometimes by late frosts in spring. The leaves in this bed were huge, and would make a very fine effect in any large place where there is plenty of room, and when in flower it must be a very handsome plant. It does not do on a sandy soil, nor on a very cold damp soil. The stag's-horn sumach (*Rhus typhina*) was treated in the same way, leaving only one shoot to grow.

The *Nicotiana tomentosa alba* was quite a failure this wet year, and had not one flower, though magnificent foliage. I should say it had been put into rather too good soil and so had all gone to leaf. There were large beds of *Clematis Davidiana*, not effective. It is a beautiful thing, and two or three plants put together would make an interesting feature in any large garden, but neither form nor colour is good enough for a large bed. The white double *Rosa rugosa* made a splendid group on the grass. A great many of the shrubs were well pruned back even when not in regular beds. The bays are the hardier for this, and the *Rhus Cotinus*, if much pruned back, is a different-looking plant. *Rhus laciniata* is low growing and has lovely foliage. *Indigofera Gerardiana* is another plant well worth growing in a clump, and it also wants nothing but pruning back hard in early winter. The useful tall polygonums at Kew were not nearly so fine as my own. I do not know whether this is from the soil being colder and heavier than mine or from not doing what seems indispensable to their successful growth— thinning out their shoots in April to five or six. Properly treated, both *P. sachalinense* and *P. cuspidatum* are superbly decorative, beautiful plants, but I never see them what I call properly grown. I saw no montbretias planted in the grass, and yet they look most beautiful done in that way. Even in a wood, and given moisture, they seem to stand a good deal of shade. So do Japanese anemones, which look far better planted on the edge of a wood, or under some small fruit-trees, than in an ordinary border. They hate being disturbed, and look most lovely growing of their own accord in the shade ; a little top-dressing in spring helps them. They make a great show of seeding, but I believe only in one garden in Ireland has the seed ever really ripened. The rose 'Caroline Testout' was flowering very well; it is a good pink rose. All the others seemed over ; with roses everything depends on its being just their best day.

Perhaps what interested me most in this visit to Kew was a cool house where the creepers were admirably managed and so well pruned and grown ; they were most healthy and covered with bloom. The large conservatories of the rich are built far too high, and so the pruning is

not done, and the flowers, poor imprisoned things, stare
out at the sun through the glass at the top. Here the
creepers were all planted in the ground inside the conser-
vatory, so their growth was strong and their flowers
abundant. Blue plumbago, an abutilon—'Golden Gem,'
I think—and that shy flowerer, *Hidalgood Werokalsi*, was
right across the house and covered with its good red
orange blooms. Its foliage and growth are pretty and
refined as a creeper, but the flowers are very like a single
dahlia on a thinner stalk. There was an old plant of a
fuchsia called 'General Roberts,' that looked very well
trained against the roof and falling down. That lovely
thing, the *Lonicera sempervirens*, was flowering well. I
never can get it to do very satisfactorily here out of doors.
In the sun it gets too dry, and in the shade it does not
flower freely, and becomes blighted ; in this house it was
seen to perfection. On the shelves there were *Solanum
Melongenas*—egg plants—more curious than pretty, some
lovely pots of *Campanula isophylla* variety *Mayi*, also *C.
Loreyi*, both upstanding, and worth growing for variety
of colour in a greenhouse at this time of year. There
were many South African plants new to me, but the chief
interest of all was a large collection of South African
pelargoniums. Many of these I had never seen, except
in Andrews' book, 'The Botanist's Repository,' where a
great number of these plants are beautifully figured under
their old name of geraniums—which plants they resemble; but pelargoniums are entirely indigenous to
the southern hemisphere, while geraniums belong to this
hemisphere and are all hardy ; which, of course, pelargoniums are not in England, except quite in the South, and
even there they require some protection. I have amused
myself for the last few years collecting Cape pelargoniums,
but they are impossible to buy, and trying to get the roots
straight from the Cape seems rather hopeless, at least so

I have found. I have about thirty varieties, which I have collected with difficulty from different people, but none retain their bulbous roots. The prettiest and most curious is one figured in the ninth volume of Curtis' 'Botanical Magazine,' called now 'Moulton's Gem,' but by Curtis *Pelargonium echinatum*. I have also a rich dark red variety spotted in the same way. Rollinson's 'Unique' is another with a particularly attractive flower. Among the sweet-leaved varieties the best is the true old 'Prince of Orange.' It is not very easy to get now, and I would give cuttings to those who care to write and ask for them in August and September. This plant is rather tenderer than the generality of these sweet-scented pelargoniums, and does best planted in the ground in a rose-house or vine-house and well watered, when it grows into a large plant. It can then be cut nearly the whole year round.

I saw last year a gardening tool which I have had made and slightly improved upon. It is a navvy's crow-bar, to which is added a heavy knob of iron worked down to a fine point. It goes well into the ground by its own weight, and is most useful for planting small plants, on rockeries or in full beds, or for bulbs in grass. When the instrument is in the ground it can be shaken about to make quite a large hole. In the bottom of this hole is powdered some light soil which the weak roots easily penetrate. This diminishes the necessity of much watering in dry weather. An iron tip put to an ordinary dibber makes it more useful than when only made of wood.

I increasingly keep seed gathered from the best blooms after marking them carefully. In this

Y

way, I have no doubt, finer plants are secured than from the bought seeds.

Gerbera Jamesoni, a native of the Transvaal, is a handsome, glowing scarlet flower. I bought it two summers ago and was told it was hardy. I covered it with fern, but that was no use, and it died; so I was relieved to find they grow it at Kew in a cool greenhouse all the year round. I shall certainly buy it again, as it is well worth growing in that way. It is, to my mind, quite as interesting and as desirable to grow a variety of flowers under glass as out in the open, for those who have greenhouses at all. How few people, even of those with numberless greenhouses, grow the beautiful Cape heaths, of which there are such endless varieties! They are well figured and described in Andrews' 'Heathery,' the folio volumes of which I have not got, but it is a superb book; date about 1804. A few of these heaths are grown by dealers for the London market, where they are a constant bait to those who like flowers in their rooms. But, thus imprisoned, they quickly turn yellow and fade and die, as they are especially fresh-air-loving plants. And let any-one watch, as they walk or drive through London, how rare it is, except just in the summer, ever to see an open window. Paris is just the same. The house is supposed to be aired by the housemaid in the morning—which it often is not—and that is the amount of fresh air it gets all day. Growing plants are much to be encouraged in living-rooms, especially in those of children and invalids, as their healthiness or otherwise is a proof of whether the air is fresh or not.

I have always rather snorted at the modern large violets, because I cannot succeed with them, and because they are so different from the much-loved ones of my youth. But I must own that, when grown to perfection, in soil of the strength and moisture loved by the cabbage tribe, in full sun, the 'Princess of Wales' is a splendid variety, and has a sweet violet smell when first picked. I must try again, across the kitchen-garden, in soil as rich as I can make it, and then trust to the wet summers we are supposed to be going to have.

I think that, in large places, Michaelmas daisies, grown all together in big beds each side of a path, make a lovely graduated colour-mass, as Miss Jekyll recommends. But, in smaller gardens, I have an idea that they look best as individual specimen plants ; the earlier ones planted in half-shade and the later ones in full sun. *Aster ericoides* seems to have several varieties. I have a very pretty white one which flowers latest of all and looks well in water.

If the seed of good gladiolus is sown directly it is ripe, the plants will flower the second year.

At page 71 of 'Pot-Pourri from a Surrey Garden,' I mention that I hoped to get a white crown imperial. A kind lady offered to send me some. On their arrival, and when they grew the next year, they turned out to be white Martagon lilies. Messrs. Barr & Son have since brought to my notice that there is no white variety of the fritillaria except *F. Meleagris*, which is quite a low-growing one.

As an example of how the old English duplicate names for a flower often conveyed a strange contrast in their meaning—like 'Love-in-the-mist' and 'The Devil-in-the-bush'—I give the following interesting little anecdote which appeared not long ago in the 'Spectator.' I believe the flower alluded to is *Amaranthus caudatus*, often called 'Love-lies-bleeding.'

'In the current number of the "Cornhill Magazine," in an interesting paper entitled "On a Few Conversationalists," the writer tells an amusing story of Browning, and how he received certain flowers from a lady, who, on being pressed to give their English names, shyly confessed

they were called "bloody noses." I happened many
years ago to be staying in a country house when Browning
told this story in his inimitable way, and he ended with
the following lines, which I then and there committed to
memory, and which will, I think, interest your readers:—

> "I'll deck my love with posies,
> I'll cover her with roses,
>   Should she protest
>   I'll do my best
> To give her bloody noses." '

In the early days of what was called the æsthetic
movement sunflowers were much grown in gardens. The
gardener, however, found it a greedy feeder, and the
'chaff' against the 'Greenery-yallery Grosvenor Gallery'
school brought it into disfavour; but I still think the
branching kind grown singly is one of the very handsomest annuals we have, and useful too, as poultry are
fond of the seed and goats like the leaves. A newspaper
account the other day set me wondering whether in some
parts of England farmers could not, with advantage, grow
the variety named below, which, as is well known, yields
the seed; it is eaten as a dessert-nut all over Russia.

'The first year of the twentieth century closed with a
curious sale on the Baltic of a cargo of sunflower seeds,
which changed hands at 11l. 5s. per ton. Though a small
trade has been done in sunflower seed for close on two
hundred years, this transaction was the first in which a
whole cargo—three hundred tons from Odessa—was dealt
with. In Russia, where the cultivation of the sunflower
and the manufacture of oil from its seed is conducted on
a large scale, the *grandiflora* is the variety grown. This
species rises in a slender stalk five feet high, producing one
monster head, the average yield being as much as fifty
bushels of seed to the acre. So rich is it in oil that that
quantity of seed will yield fifty gallons of oil, while the

refuse of the seed, after this quantity of oil has been expressed, weighs 1,500 lbs. when made into cattle cakes. Few people in England who grow the sunflower for ornament have any idea of its usefulness. It is among neglected crops in which there is money, as is shown by the price paid a few days ago. Besides the seed, every other portion of the plant can be utilised. The leaves furnish an excellent fodder, while in Russia the stalks are prized as fuel, and their ashes, which contain 10 per cent. of potash, are readily sold to soapmakers. Naturally, in Russia the chief virtue of the sunflower lies in the oil contained in its seed. The oil is of a clear, pale yellow colour, almost inodorous and of an agreeable, mild taste, so that it is in great request as a table article. Why sunflowers are not cultivated on an extensive scale in England it is difficult to say. Poultry and cattle like the seed either in its natural state or crushed and made into cakes. No plant produces such fine honey and wax; when the flower is in bloom the bees abound in it.'

OCTOBER

Solomon's love of nature—An old letter—Zola and fresh air—Old
Harwich inn and curious specimen of *Clematis Vitalba*—Mesem-
bryanthemums for cliff gardens—An old monastery fruit-wall—
Three Pergolas—A long-wanted book on trees and shrubs— An
old Suffolk breviary—Stories—Wild flowers for garden culture—
Wellingtonias on a German hillside—Chrysanthemum culture—
Mr. Morley's gift to Cambridge.

October 1st sees me once more on my dearly loved
East Coast, with its splendid air, its open skies, its flat
distances, its boundless seas, and, for me, its kind friends.
This summer brought me back an old letter written in
my middle age to a young niece during one of my first
visits to Suffolk many years ago when out of health. At
that time I had worldly ambitions, though rather for
others than myself, and a simple, unartificial rural life
would have been impossible to me in spite of my strong
love of Nature. The only vocations which seem to combine creative work with the simplicity of an unworldly
life, and yet give scope to great ambition, are those of the
artist and the author, and as neither of these was for me
my love of Nature found its main vent in admiration of
those in whose lives it had played a dominant part. Let
those who have this love of Nature never crush it, for
did not the writer of old thus describe the wisdom of
Solomon ?—' For he was wiser than all men ; than Ethan
the Ezrahite, and Heman, and Chalcol, and Darda, the
sons of Mahol: and his fame was in all nations round

about. And he spake three thousand proverbs : and his songs were a thousand and five. And he spake of trees, from the cedar tree that is in Lebanon even unto the hyssop that springeth out of the wall : he spake also of beasts, and of fowl, and of creeping things, and of fishes.'—1 Kings iv. 31–33.

The finder of my letter writes: ' I have been all the morning in the attic, which is truly an Aladdin's palace of riches and surprises. We were choosing books to have in our London home, and out of a " Life of Benvenuto Cellini " fell this fat human document. We have read it, and I send it to you, knowing it will be a thrillingly interesting memory to you. We have so loved reading it, and I feel tempted to shake *all* the books in the hopes of finding more ! '

' Southwold, July 23, 1886.

' . . . About friendship, dear, you *can* judge, about love—don't think me a horrid old thing coming over you with that odious thing experience—you must allow me to say you *cannot* judge, either its nature or its power, because you have not tried it. It will come some day, and you will be the first to admit it is different, and that the grasp of the smiling boy is all powerful.

' I have come across here to-day one of those strange stories of life which are more moving than any novel. We went to see, M. and I, an old lady, a fisherman's widow. She was about seventy, strong and handsome, and weather-beaten, very rough at first, and then the intelligence, refinement, and talent came out in her talk, and you forgot the wild, even dirty exterior. As a girl, quite young, she had run away and married a handsome fisherman. Her family disowned her and cut her off. She was the daughter of a rich Liverpool citizen, brought up in every luxury, taught German, French, Italian, drawing, and natural history. She must have had a wild,

strong nature, for she has been very happy. She had
one son and she dearly loved her husband. She lived
his life and for years sailed about with him, studying the
stars and knowing every fish of the sea. Her brother,
S—— by name, was a Fellow of the Royal Society and
a great naturalist, and she has beautiful books, something
like my botanical ones, illustrated by him. She wanted
to see me because of the Earle name—a well-known one
in Liverpool. I wish I had known her sooner, it would
have interested me, though I have tried here to know no
one; I wanted to get away from humanity and its trials
and temptations. The old lady has six cats, to all of
which she is devoted. What is strange is that she should
have kept up her French and been so proud of her intellectual gifts, yet that she should be content to live in so
dirty a cottage and be so untidy in her *person*. She has
the artistic temperament, and that, I am afraid, left to
itself, does not care always for cleanliness. All about
her was eminently picturesque and eminently untidy.
She has for years written the letters of these villagers,
and they talk of her, not as one of themselves, but as " a
real lady." Is it not curious? As M. says, in another
age she would have been called a witch. Her husband
has been dead many years, but she talks with pleasure of
the happiest days of her life, sailing on the wild ocean
and coasting along the shores in his small open boat,
and knowing all about the curious live things that
came up in his nets. I shall probably never see her
again, but I shall not forget my afternoon with her. . . .
You must not think from what I said about Zola that
I like him. I have always hated him, and can seldom
read him; only with this book I was agreeably disappointed, as it has both power and truth. Some of
the things you say about truths of some kind being put
into a novel, many people would say of your friend " Tom

Jones." There are things in that not much more pleasing to me than the butcher's work you describe. Perhaps you have forgotten. However, I think in books, as in life, even *striving* at truth has a great charm for me, and though gazing at corruption may be sickening, I doubt if it is as bad for one as the most beautiful of whitened sepulchres. However, you and I often mean the same thing, only we express it differently. I certainly do not call a doctor a brute for publishing his experience of the most horrible diseases and operations. Those who are not interested need not read; to those who are, the beauty of the facts effaces entirely the ugliness of the detail. And so in the description of the human heart, if the disease is possible and true, it has a distinct interest for those who study human nature, though it may not be beautiful at all. Zola's book deals with that terrible question—the very narrow line between genius and madness, which is one of the saddest problems of poor suffering humanity.'

As I was copying this old letter into my chapter came the news of Zola's death from bad air. The pathetic account of so slight a cause having so big a result is almost allegorical in its significance; for convinced as I have always been that the motive of his work was a noble seeking after truth, the work itself was yet full of the miasmatic atmosphere which rises from the lowest strata of human nature, and I think there is no bathos in saying that if he had habitually slept with open windows, he would not only have lived longer, but his work would have been much more wholesome. The French newspapers call it ' that stupid death which sends the literature of all countries into mourning and is deplored by the whole world.'

October 9th.—Never till this year have I seen the old town of Harwich, my previous acquaintance having been

limited to its name, and the shed at the Great Eastern
Railway station on the way to the steamboats. In old
days Harwich was an important place ; the King's pack-
boats started from the town itself, and the early Georges
always sailed from here on their joyful visits to their
beloved Hanover. In the town there still exists a curious
old posting-inn called by the unusual sign of 'The Three
Cups,' and the room where Nelson slept is still shown.
The reason of my visit to the inn was to see one of the
strangest natural adaptations of luxuriant growth, of a
kind seldom seen except now and then with an old vine.
In the corner of the inn-yard is planted a *Clematis
Vitalba* (traveller's joy), which I should think may be
two hundred years old ; the stem is as thick as an old
apple tree, retaining of course its twisted, rope-like
character ; the long arms, viz., branches, of the plant
have been carried across the yard in all directions,
supported in the middle by a strong post. The plant
grows in a damp corner next the house ; it must have
been put there and encouraged by some plant-loving
landlord, and has been carefully trained and pruned ever
since. The effect is so rare and so charming that it
might be tried and carried out with advantage in many
places. It was such a splendid illustration of what I
consider one of the first rules of gardening—clearly to
show the hand of man, even to the extent of a certain
artificiality, and then Nature being allowed to assert her
sway. Thanks to pruning and care through a century or
more of growth, this magnificent specimen of a wild plant
still owes some of its charm to artifice. Given a sunny
position, an old wistaria might be trained in the way
described which is exactly how they are treated in
Japan.

Loudon says, ' The first clematis brought to England
was *Clematis Viticella* in 1569, Virgin's Bower. During

the reign of Queen Elizabeth the name of Virgin's Bower
might be intended to convey a compliment to that
sovereign, who, as is well known, liked to be called the
Virgin Queen.'

I never now visit seaside gardens except on the East
Coast, where the houses and gardens are close to the sea,
but it is interesting to observe there how well experience
teaches, for the clothing of cliffs under the spray of the
sea, which at one time seemed impossible, has now
gradually succeeded. My host told me the other day that
he had had the greatest difficulty in getting *Pinus austriaca* to grow. I immediately said, 'Oh, I am so glad !
I hate them.' He answered, 'You are wrong : their
branching habit and sturdy growth from their youth
make them endlessly useful to us as a protection. But,'
he added, ' you will be interested to hear, in confirmation
of what you have often said, that whereas we lost
hundreds of transplanted plants bought from nurserymen,
of our own seedlings we do not lose one in a hundred.'
I am sure this is sound gardening, and corresponds with
my experience with pears, already mentioned. Wherever
a plant is difficult to grow and adapt to any particular
soil, then grow it yourself from seed. It is a slow process,
but is best in the end.

The sand cliffs along this Suffolk shore, the feet of
which the sea is always licking, and at high tides gradually
undermining and carrying away, have to be most skilfully
preserved in both the gardens I know best. The fight
with the powerful element seems a continual excitement
to the owners, but to others it appears as a pathetic
struggle against an irresistible foe which is known to be
slowly devouring England both on the East and West
coasts.

One of the most interesting growths on these garden
terraces, clothing all the artificial rockwork in the most

perfect way, is due to various perennial and annual
mesembryanthemums, which bravely withstand the cold
and cutting winds. Nothing suits them so well as sand
and gravel, and they luxuriate and blaze in the sunshine
and pure air. The large lilac-flowered *M. Pomeridianum*
revels in the driest pocket and falls for yards like mer-
maid's hair over the hot bare surfaces.

All Cape and Japanese plants can be played with in
these gardens where the myrtle and fuchsia flower
abundantly every year. The Cape annuals, *Octotis
grandis* and the orange venidium, also do gloriously, but
even here the beautiful belladonna lily will not do
against an ordinary south wall. It must be against a
greenhouse wall where it gets a little heat from the
inside pipes, or it comes into bloom too late. In one of
these gardens where it is well established and never
interfered with, it blooms from August to October in a
right royal manner, with richer pink blooms and browner,
stronger stems than I have seen inland. All the genistas, from least to largest, do splendidly if cut back after
flowering.

I complained to one of my friend's clever gardeners
that my pears would go quickly in the middle. He said
he thought the cure for that was very early picking.
They have a tendency in hot soils to ripen first in the
middle. I long ago discovered how desirable it was to
pick medlars while still hard and let them ripen in the
fruit house.

In an old Suffolk rectory garden, once the property
of the monks, I heard of a fruit wall with this peculiarity :
small cupboard-like recesses were built into the wall
between the fruit-trees and cleanly finished with stone
slabs, and in these the peaches, nectarines, and apricots
were placed on being picked to ripen in the sun, and yet
not be exposed as on the tree to rain, flies, wasps, &c.

The entrance to the recess was closed temporarily after the fruit was in by a curtain of muslin or a piece of glass. I have never seen a fruit wall built on these lines, but I expect those old monks knew what they were about, and easily brought to their table each day such fruit as was ripe and ready.

My Suffolk friends have taken to constructing strong wind-resisting Pergolas. In one garden the piers are made of rough stones with strong iron girders sunk into cement at the top, and they run from north to south, which seems to me the best aspect, as in that way the roots on one side are in perfect shade, or in full sun according to the requirements of the creepers. The other Pergola runs east and west, and the columns are built of brick, while the top is made out of the curved beams of the sides of wrecked vessels thrown up on the beach. What a calm sylvan home for the poor storm-tossed beams to come to in their old age! Another Pergola I saw this year entirely made of young larch saplings was very pretty, though less substantial than the two I have mentioned. Alternating with three poles bound together and three across the top, there came a single pole and a single one across the top, a far prettier arrangement and more substantial looking than when the poles are all single. On a flat piece of ground it would be an improvement to sink the walk and have raised stonework on each side, slightly after the manner of the Amalfi Pergola referred to in April.

I think the small-flowered hardy clematises are not nearly enough grown on Pergolas, and they are so light in their growth that they hardly injure roses or anything else they climb on. I mean such kinds as *C. paniculata*, which is even later flowering than *C. Flammula* ; but in Jackman's catalogue they are all so well classified and described that there is no difficulty in ordering what is wanted. The new hybrid types raised from *C. coccinea* are very pretty,

but expensive. I am very fond of two herbaceous clematises called *C. erecta* and *C. maritima*. *C. Davidiana*
and *C. cærulea*, too, are beautiful plants well grown.
They do not like a dry place. None of these are climbers.
The hardy yellow *C. graveolens* does well in these sunny,
wind-swept gardens. It goes on flowering for months,
the blossoms lasting long after the early ones have
formed their pretty fluffy seed-tufts.

In gardens where enlargements are constantly being
made, it always surprises me to see how much repetition
goes on. The new beds are generally filled with plants
that are already flourishing in other parts of the garden.
I think, if I had a large garden, I would try to keep plants
apart, and a new bit of ground or a sheltered corner should
be entirely devoted to new kinds that were not in any
other part of the ground, or to plants which had failed
elsewhere. It is so rare in English gardens to find any
distinctive planting or originality, and I think it must be
partly due to idleness and laziness, for I can never take
up any gardening book without seeing thousands of things
which I still lack.

October 12th.—I see advertised to-day a book on
'Trees and Shrubs,' part of the 'Country Life Library,'
by Mr. Cook, editor of the 'Garden.' I am impatient to
see this book, as it is one I have been waiting for for years.
Up to now, there has been no really good one that I know
of, except the rare old 'Arboretum' of Loudon. I hope this
book may supply a very real want, for, as I have repeatedly
said, the ignorance about the pruning of trees and shrubs
is very great. It is one of the most important parts of
good gardening, whether practised in the tiny square of
ground at the back of a suburban villa containing only
two or three shrubs, or in the largest pleasure grounds of
the United Kingdom, and yet bad pruning does more
harm than none at all.

While correcting my proofs this book has appeared, and I am sure will be a disappointment to no one; it supplies a great want, and no garden library can afford to be without it. The photographs are many of them beautiful. Photography seems to suit the reproducing of trees and shrubs as well as it seems to me to do badly for roses either growing or as cut blooms. The book is suited to gardens and woods of all sizes, and touches on all the varieties of methods and growths for which various plants are suitable. The word 'English' of course stands for the British Isles. Towards the end of the book there are admirable annotated lists, divided into columns under the following heads :—Name—Country or Origin and Natural Order—Colour and Season—General Remarks. Another list gives hardy trees and shrubs for beauty of foliage and growth. With a careful study of this work no one can go wrong or leave out of his planting any desirable or beautiful shrub. The tender shrubs that only flourish in the south or west have a chapter to themselves. In spite, however, of its great merit and general usefulness, the book can in no sense compare with J. C. Loudon's wonderful 'Arboretum' which is almost a unique example of labour and industry in *garden* literature. While engaged on this work, which took him years and caused him to die heavily in debt, Loudon was in the habit of visiting all the large places in England, and on one occasion wrote to the Duke of Wellington asking for leave to inspect his beeches. The duke answered very stiffly in the third person to Bloomfield, Bishop of London, forwarding a pair of breeches to London house for the Bishop's inspection, having mis-read Mr. Loudon's wish to see the beautiful beeches at Strathfieldsaye, as a request from the Bishop of London to see the breeches he had worn at Waterloo !

Even in the last ten years, the marvellous increase in

the beauty of seaside gardens makes one realise the power
of man over plant life. This ought to be a great encouragement to everybody in all parts of the world. Give
plants more or less what they require, and you are sure
to be amply repaid.

Crabbe, who lived in Suffolk, and whose botanical
observations, as Anne Pratt said in a book I had as a
child, 'had led him to mark the inferiority of the vegetable
kingdom in the neighbourhood of the sea,' gives the following melancholy picture of the plants of a small town on
the coast :—

> Where thrift and lavender and lad's-love bloom,
> There fed, by food they love, to rankest size,
> Around the dwellings docks and wormwood rise.
> Here the strong mallow strikes her slimy root ;
> Here the dull nightshade hangs her deadly fruit ;
> On hills of dust, the henbane's faded green
> And pencill'd flower of sickly scent, is seen ;
> At the wall's base the fiery nettle springs,
> With fruit globose and fierce with poisoned stings ;
> Above (the growth of many a year) is spread
> The yellow level of the stonecrop's bed ;
> In every chink delights the fern to grow,
> With glossy leaf, and tawny bloom below :
> These with our seaweeds, rolling up and down,
> Form the contracted flora of our town.

This year an old local book has lately been published,
with notes by Lord Francis Hervey. It is called 'Suffolk
in the XVII. Century : the Breviary of Robert Bryce,
1618,' now published for the first time from a manuscript
in the British Museum. One sentence in his agricultural
descriptions naturally struck my eye. He says, ' As for
the goat hee is a stranger with us, hee likes not our fat
fertile soil, hee comes from our Western parts, where hee
delights in the hungry feed amongst the sharp rockes, and
steep mountains, he is very seldome with us unless some

for raritie and pleasure doe entertaine them, or for phi-
sick's several uses doe maintaine and breed them.'

I think perhaps few of the crowds of tourists who go
to Suffolk in summer know the two excellent little old
books descriptive of the East Coast, which Cassell & Co.
have republished in their 'National Library' for the
modest price of 3*d*. One is called 'Travels in England
during the reign of Queen Elizabeth,' by Paul Hentzner;
the other is a 'Tour through the Eastern Counties of
England,' 1722, by Daniel Defoe. Defoe describes how
in those days the poor turkeys were driven to London by
the road 'in droves from three hundred to a thousand.'
I think their legs must have grown harder and their fat
less in the process. Of Ipswich he says what might be
said to-day, 'An airy, clean, and well-governed town, a
very agreeable and improving company almost of every
kind. A wonderful plenty of all manner of provisions,
whether flesh or fish, and very good of the kind.' He
mentions the decline in the shipping trade even in his
day. He only casually notices Harwich, from which place
he sent round his horses and took a boat up the river
Orwell to Ipswich. He mentions that the late Dutch wars
injured the coal trade. The modern grumble is that the
harbour is silting up, that there is no money for dredging,
and that the fish are killed by the sewage of Ipswich, and
that the river is being choked by a luxuriant weed that
grows in the sewage. The future man, who invents something that solves the sewage question, wet or dry, will
indeed be a benefactor to human kind.

I came across an old book last year which I have tried
to get everywhere and have failed. It is called 'Theory
and Practice of Landscape Gardening,' by H. Repton,
London, 1805. The idea of the illustrations and teaching of
this book could be adopted by anyone wishing to improve
a large and woody place. What the author did was this :

z

He drew a faithful sketch of the landscape as it was, with every tree and shrub and undulation marked. You lift up this drawing, and he shows you how he proposes to alter it and how it will look when done. In some cases he merely opens out the view by cutting down the trees, in others he throws up the ground on one side and lowers it on another, so showing up a valley which had been hidden before. In another case he dams up a stream and turns two or three swampy fields into a large lake. The whole book, though old-fashioned and hardly to our modern taste, is full of suggestive ideas. It is a pity it is so scarce. The non-illustrated edition is useless, and one must get the first edition of 1805.

While staying away from home the other day, a kind fellow-guest wrote me out the two following stories, which made us laugh, and which I at least had never heard before.

At a rent-audit dinner, the squire noticed that a new tenant of his, sitting in the place of honour on his right hand, was taking nothing to drink, so he said, ' Well, Johnson, this won't do, you are drinking nothing,' &c. Johnson replied, ' No, squire ; I never drinks nothing with my meals.' ' How's that ? ' said the squire, ' are you a teetotaller, or suffering from rheumatism or anything, and acting under doctor's orders ? ' ' No, squire, t'aint that. It's this way : if you take a bucket full of watter you can't get no taters into it, but if you puts the taters in *fust* it's wonnerful what a lot of watter you can get in afterwards.'

A philanthropic old lady in Exeter, very keen on the drink question, got hold of a very bibulous old sailor whom everyone had given up as a bad job. He had lost a leg and one eye, and used to do odd jobs about the market place. He told the old lady that, if he could once get a fair start on his own account, he would try to reform, many of the jobs he now did being paid for in drink. The old lady, after much thought, purchased for him a tray to hang round his neck with a broad strap and a supply of nice gingerbread, and she taught him the following sentence to repeat at intervals :

> ' Will any good kind Christian
> Buy some fine spicey gingerbread
> Off a poor afflicted old man ? '

When he had sold a shillingsworth, he congratulated himself on his strength of abstinence, and thought he would treat resolution to just one half-pint. This, needless to say, led to two or three more, and when he resumed his station on the pavement his cry became a little mixed, and in a loud voice he appealed to passers-by with :

> ' Will any poor afflicted Christian
> Buy some good kind gingerbread
> Off a fine spicey old man ? '

Trade became very good, and he again treated resolution, with the result that his cry became :

> ' Will any fine spicey Christian
> Buy some poor afflicted gingerbread
> Off a good kind old man ? '

In return, I told him the following : ' I am told that in the Bankruptcy Court the bankrupt is always asked by the Judge if he can give any reason for his failure. A young man who was being thus examined promptly answered, " Oh, yes, quite easily ; fast women and slow horses." I did not know him, but I heard with regret that this poor witty young fellow died in the war in South Africa.'

The fashion of everyone discussing health—his own

or others'—is so common a one now that the taste sometimes assumes extraordinary developments, and I have
been told that, two or three years ago in Paris, cinematographs of hospital operations were the fashionable diversion
at evening parties. A check was given to this kind of
entertainment by one of the guests fainting on recognising
a friend on the operating table. How history repeats
itself! In Ten Brink's latest book on the French Revolution, he gives a graphic account of Paris dinner-
tables decorated with toy models of the newly invented
guillotine, a doll filled with red scent representing the
victim, and ladies dipping their handkerchiefs in the
sham blood which spurted out as the head fell into
the basket.

To wind up with something more human, I give the
following letter from Major A. C. Hamilton, 6th Dragoon
Guards, to his mother, who kindly sent it to the press :
' Barberton, September 26.—I have a little moral story
which you can publish in a paper if you like with names.
I dined with Van de Post, whom we took here. He was a
Free State commandant and Speaker of the Volksraad. He
told me after the engagement at Ramah he found a soldier
who had been out all night bleeding to death and nearly
dead. His doctor said, " It is useless to do anything for
him," and suggested amputating both legs. Van de Post
looked in his pocket and found a letter from his mother
to this effect : " Dear George,—I am so anxious about
you in this terrible war, but I hope you will be always
merciful to the wounded and respect women, and cause
as little pain to others as you can, for Christ's sake.—Your
Mother." He was so touched he wired for his carriage
and the best doctor, sent the man to his own farm, which
was near, and he quite recovered. His name was George
Cowan, Mounted Infantry. It would be nice if his mother
knew this.'

In the 'Spectator' of October 18, 1902, there was a review of 'Ballads of the Boer War; selected from the Haversack of Serjeant J. Smith, by "Coldstreamer."' The article was so sympathetic and appreciative that I instantly sent for the book. Although I agree with the 'Spectator's' praise, I must say I think that many of the poems are better than those selected by the reviewer. The one which shows a real imaginative power of a certain kind deals with a very difficult subject, I should have thought—'The Queen's Chocolate.' It is too long to quote entirely, but these verses, taken without the beginning and the end, explain themselves :—

> I never 'ad no truck with gals ;
>   Soft, stuck-up things, they seems to me
> An' though I'd h'often see my pals
>   A-settin' of 'em on their knee,
> It ain't a thing H'I ever done,
> Cos why ?   I didn't see the fun.

> Them gals was nice enough, no doubt,
>   I ain't a-contradictin' it ;
> I see'd young fellers walking out
>   When I was polishin' my kit,
> An' each 'is bit o' chintz 'ad got ;—
> Well, they was welcome to the lot !

> And, h'as I'm talking, H'I can say,—
>   Tho' p'raps it sounds a funny thing,—
> No woman to this blooming day
>   'As give me e'er a brooch or ring ;
> No trunkets, nor the like o' that,
> Nor yet no ribbon from 'er 'at.

> An' h'only one, as I recall,
>   In all these weary months o' war,
> 'As sent me h'anythink at all,—
>   (An' she wont never send no more ;)—
> Ah ! what was that, an' 'oo was she ?
> I'll tell you if you'll 'ark to me.

The Queen was driving h'out one day,
　　And, 'appening to pass a shop,
She calls the drive of 'er shay,
　　' 'Ere John,' sez she, ' 'ere coachman, stop !
You wait a while outside for me,
H'I wants to buy some sweets,' sez she.

The shopman, knowin' who she were,
　　'Urries respeckful to the door ;
' I've thought,' sez she, a-smiling there,
　　' As 'ow my soldiers at the war
Would like a suck o' chocolit,
So I come in to h'order it ! '

' Just send a million boxes round,'
　　Sez she, an' quick she writes a cheque —
' I'll pay at once, ten thousand pound,
　　You'll find the signature correck ;
That's it, Victo-ri-a, Har. I.,
Take care, the h'ink is 'ardly dry.'

Then h'out she goes, an' drives away,
　　Without the slightest sort o' fuss ;
The boxes they comes round next day,
　　An' h'off she sends 'em out to h'us ;
An' that was 'ow I got a bit
O' Queen Victoria's chocolit.

The h'only present, fust an' last,
　　As any woman sent to me
In h'all them weary years as passed
　　Since first we sailed acrost the sea ;
I've never 'ad no gifts before,
An' so I prizes it the more.

An' 'ere's my box, as good as new ;
　　I 'aven't touched the chocolit,
Nor yet I ain't a-goin' to do,—
　　'Cos why ? Because I values it.
You'd like to buy it, eh ? Good Lor
Wot sort o' cove d'you take me for ?

Suppose a gal, some New Year's Day,
  Sent off a box o' sweets to *you*;
Would *you* go off an' sell it, eh?
  Is that the sort o' thing *you*'d do?
Wot would 'er Gracious think of it,
If I should sell 'er chocolit?

' She'd never know!' sez you? May-be!
  (Gawd rest 'er soul!) Per'aps you're right.
But still I likes to think as she
  Is watching 'ow 'er soldiers fight,
An' smiling somewheres in the sky
A-seeing 'ow 'er soldiers die!

But, h'if she knows or h'if she don't,
  This blooming chocolit is mine;
D'you 'ear? An' part with it I won't,—
  So there!—for all you talk so fine.
I wouldn't sell it now, you swab,
For fifty, let alone ten bob!

Two pounds, sez you? You'll make it three?
  Well you're a gentleman, H'I'm sure!
Don't push your blooming coins on me!
  You thinks to tempt me 'cos I'm poor?
I may be so; h'it ain't denied;
But still I 'as my proper pride.

No use a-h'arsking me to sell,
  I'd feel a villain if I did
(So you an' yours can go to 'ell!)
  I wouldn't, not for twenty quid.
Take back your money, h'every bit!
I'm richer,—with my chocolit!

October 20*th*.—Cultivating to perfection some of the
wild flowers of our own country is, I think, a delightful
thing to do, at least I do it with several plants, such as
the blue *Geranium armenum,* one of the handsomest of
our wild flowers and worthy of both a good place and a
bad in every garden. The difference in situation will
make a fortnight's interval in its flowering, for it grows
both in shade and sun. The wild yellow toadflax

(*Linaria Cymbalaria*) grows poorly in our hedges about here, but it always has a place in my garden and flourishes exceedingly from July to November. I grow, too, both the willow herbs (*Epilobium angustifolium.*) The pink one is rather a weed, but the white variety (called French because it is not a wild English plant) is more restrained in its growth, and groups charmingly with the blue geranium.

The orange hawkweed and the little pale yellow one I consider both worthy of the garden. I have not many varieties of ferns as the dryness is unsuited to them, but of the Welsh Polypody (*Polypodium cambricum*) I take some care as its best time is late autumn when all others have turned yellow.

I have derived immense pleasure from the branching larkspurs, the seed of which I brought back from Florence two or three years ago. I noticed there that it grew all over the place like a weed, as my œnotheras do here. The seed I brought home has grown most satisfactorily, and I always save my own seed now. It is certainly slightly different from any that I can buy in England, which seems more cultivated and inclined to be double, causing it to grow a little less freely. In a number of the 'Spectator' some time ago, I saw a letter, to me very interesting, by Mr. H. B. Cotterill, from Switzerland. I venture to think that those of my readers who did not notice it in the 'Spectator' will be glad to see it: 'Sir,—In an edition of Milton's "Lycidas," lately published by Messrs. Blackie, I gave what I consider to be rather strong grounds for believing the "flower inscribed with woe" (*i.e.*, the ancient ὑάκινθος) to be the larkspur (*Delphinium Ajacis*). Correspondents inform me that they have examined larkspurs growing in English gardens and have been disappointed by not discovering the "A.I.A.I." marks. Perhaps you will kindly allow me to suggest that possibly the larkspur examined by them was some cultivated variety of the *Consolida*, which grows wild in these parts, and on which I myself have also failed to find the marks in question. Or possibly there may be English varieties of the *Ajacis* which do not possess these marks. I can only say that the larkspur commonly to be found in Swiss gardens is the *Delphinium Ajacis* (a native of Southern Europe, said to be found sometimes "escaped" in England), and that I have never failed to discover the "A.I.A.I." or "I.A.I." on its petal. I am, Sir, &c., H. B. Cotterill, Clarens, Switzerland.'

On reading this letter, I rushed out into my garden to see whether I could find the markings on my flowers. On the petals of the blue variety, the descendant of the Italian seed, I found the white letters 'A.I.A.I.'—the cry of lamentation in the pagan world—quite plainly marked. In the cultivated, or double kinds, raised from seed bought in England, all trace of them had disappeared. I also grow the red valerian. Its old cottage name was 'Pretty Betty,' and Chaucer calls it 'Swete-wall.' It grows in all sorts of places about the garden. It is one of those kinds of plants that like dryness and poor soil, but as much sun and air as they can get. It is, I believe, quite rare to find it wild in England. I never tire of varieties of wallflowers (*Cheiranthus*), and this year Mr. Thompson, of Ipswich—from whom, as I said before, I get all my uncommon and non-nurseryman's seeds—had a long list of these plants in his catalogue. One which he called 'Dresden' is very like the wild wallflowers found on old buildings. It began to bloom here early in October and so gives a cheerful promise of spring before winter begins. Anne Pratt, in one of her early books, says it was regarded by the Troubadours as the emblem of faithfulness in adversity because it smiles upon the ruin. In Saintine's 'Picciola,' that character-making book which I so loved in my childhood, the real heroine is a wild wallflower. It is a pleasure to me to find that, in my old age, the books that I loved most between fourteen and eighteen I enjoy now with a renewed freshness. Second childhood I suppose. However, ' Picciola '—the ' new book ' when I got it in 1851— has become, I believe, a schoolroom classic for girls learning French. How much worldly wisdom is taught from that description of a political prisoner's life ! In my enthusiastic youth I thought that imprisonment for a righteous cause—for liberty, for the good of the people— would not have been hard to bear, especially if consoled by one fragile little plant. In those days how little I knew what imprisonment meant ! A man who has him-self experienced it, for what he considered a righteous cause, refers to it thus : ' Liberty ! Who that has not himself been once imprisoned can appreciate what this means ? Who that has not had to look forward with aching heart and longing soul for days, months, years, to the time when he would once again find himself unfettered, free to talk to his fellow-man—at liberty to exercise the rights of his nature's manhood, undeterred by prison rules or the threat of warders' reports, can realise what that heaven-born word implies ? I was liberated once— unexpectedly set free, after seven and a half years of close imprisonment, and I am almost inclined to say that the punishment involved in a penal servitude of that duration would be worth enduring again to enjoy the wild, ecstatic, soul-filling happiness of the first day of freedom. . . . Everything which meets the gaze of the liberated prisoner, every thought of the present and the future assumes a brighter hue and wears a more blissful meaning from the terrible recollection of the felon degradation, the narrow cell, the stinted sunlight, the loathsome daily task, the brutal warder, and the weary, heart-longing expectancy for the hour of deliverance.'

This quotation is from a book published by Mr. Michael Davitt, called 'Leaves from a Prison Diary,' when he returned to the world after his long confinement.

There is another well-known testimony to the sufferings of prison life in a volume of poems written by Mr. Wilfred Blunt called 'In Vinculis.' He says in his preface that these sixteen sonnets were written when actually in prison, and he adds : 'They record an episode in the writer's life to which, in spite of many austerities and some real suffering, he cannot look back otherwise than with affection. Imprisonment is a reality of discipline most useful to the modern soul, lapped as it is in physical sloth and self-indulgence. Like a sickness or a spiritual retreat, it purifies and ennobles ; and the soul emerges from it stronger and more self-contained. Alas, that these influences should so soon lose their power! And yet, fall as we may from the higher level, they do not wholly perish, but remain for us a wholesome recollection and a standard of all that we can imagine best for this life and another.'

To return to my plants. Playing experiments with the sowing of flower-seeds is not entirely without danger. This year, instead of ordering the handsome, often-grown *Helichrysum bracteatum incurvum*, I thought I would order *H. Gnaphalium*, in spite of the warning of '*fœtidum*' added to its name in the catalogue. Words fail me to describe how the horrible smell of this plant haunted the garden for at least six weeks. It was like the most evil of he-goats. Once I picked some, and the garden gloves had to be burnt, as they scented the whole hall. My other experiment was *H. setosum*, described as 'new and from the Transvaal' ; described, too, as a half-hardy perennial instead of an annual. It is rather a nice little everlasting with a pretty growth.

The pretty maidenhair tree—its Japanese name is *Ginkgo*—grows well in any strong soil in the south of England. In Northamptonshire, I saw it grown like an old pear-tree on a west wall and, as it had been a good deal pruned back, the leaves were large and handsome. They turn such a beautiful, clear yellow in autumn, that it seemed to me a plant to be recommended for wall covering.

Saxifraga Fortunei is a plant seldom seen. It is perfectly hardy, but, as it does not flower till the end of October or November, the delicate plumes of blossom are injured by cold rains or frost. The way I grow it is in pots sunk outside in shade all the summer and well watered. As the buds form they are brought into the room or greenhouse. I think few orchids are as pretty, or repay one so well for so little care.

The beautiful Californian *Delphinium cardinale* I find very difficult to grow. I have not before named as a stove flower for November and December the pink plumbago. Like the half-hardy blue one, it lives well in water if floated.

October 23rd.—I have not much increased my little collection of orchids, though I have propagated those I had. At this time of year the cypripediums are invaluable. Stuck into a small Japanese wedge with a few red leaves of the azalea, they form a perfect autumn bouquet, and when all Nature is fading away their persistency and unfadingness have a peculiar charm. I can add to my list *Lycaste gigantea*: it is in full flower now, and is very ornamental in its pot with three long ribbed leaves gracefully curved, and at their foot a bunch of seven or eight unusually shaped pale yellow green flowers. This forms a naturally growing Japanese combination. I repeat what I said before, everybody who has a small stove ought to insist on growing a certain number of these easier orchids.

October 27th.—In a letter to-day from my German friend there are two gardening paragraphs, one a great example of how we sow and our descendants reap, and the other a practical experience which may be useful to anyone. 'I must tell you how I went to visit a friend half-way between Darmstadt and Heidelberg in the Oden-wald. She took me a drive into her hills and woods and showed me a whole hillside (north-east) very sheltered and entirely planted with wellingtonias, *Abies macro-carpa*, and enormous Thuyas and Thuya Globosas, simply lovely and quite unique, I should say, in this part of Europe. The wellingtonias that are so hideous as single specimens on a lawn were grand in a big mass, and so healthy, not one yellow twig—so picturesque, like huge bluish feathers. And think of the variety it made to have that hillside of blue in the midst of the black Tannen-baums and the golden oaks and beeches in their autumn robes! It was all planted sixty or seventy years ago by my friend's father-in-law, Count Berkheim, who was a Frenchman, the Berkheims being the oldest feudal family of Alsace, and though this one branch is now German, they considered themselves French, and talked French in the last generation.'

Her next subject is chrysanthemums. She says: 'Another thing I have quite made up my mind about is that the summer and autumn culture of amateur-grown chrysanthemums, as I have seen it practised in nineteen out of twenty gardens in England, is not so good as ours. Now—October 25th—I have not one yellow, diseased, or missing leaf on any single one of my 150 plants. I am sure it is entirely due to the plan of sinking the pots well and deep into a heavy wet soil, and especially giving them only half as much stimulant all the summer, and hardly any artificial manure at all; plenty of hoof-parings and bone-dust in the soil at each re-potting, and always a little very-much-diluted liquid manure from a sink-tank where there is plenty of cabbage water. I must tell you that the director of the celebrated Palmen Garten, the German Kew, came the other day and said he had never seen finer plants than ours anywhere, and only wished his were a patch on them. I only tell you this because I have always felt sorry to see chrysanthemums diseased, and the plants so straggling and empty of leaves in most English private gardens, for, after all, from October till the new year they are the staple material for one's indoor decoration, are they not?'

There is a good deal of truth in my friend's remarks, and there has been much chrysanthemum disease about in England in the last four or five years. I have tried sinking the pots, even in this light soil, with great benefit. The wet summer has also helped; but I am quite sure that the tendency to overfeed cattle, plants, and land is one of the dangers in England at this moment.

The cultivation of chrysanthemums is a bother, but at the same time we cannot do without them, and I suppose as long as gardens last we all shall feel more or less what is charmingly told in the following poem from the 'Westminster Gazette':—

CHRYSANTHEMUMS

We spring from the earth at the winter's birth
  When the ground is bare,
And we reign supreme, like a passing dream,
  Till the world is fair.

We brave the gale, and the rain and hail,
  Afraid of none!
While our petals blaze like the golden rays
  Of the setting sun.

From bronze we shade, and to pink we fade,
  As the rosy morn;
And our lips are kissed by the dewy mist
  Of the early dawn.

To mauve we turn and to crimson burn—
  Then fiercer glow.
But our souls delight in our petals white
  As the driven snow.

Some of the critics of my other books twitted me with
my admiration of Mr. John Morley, but that admiration
has been even heightened by his consistent and dignified
attitude in political life during the last three years. The
few people whom I have seen during the present week
had, I found, entirely missed the letter in which he
announced his magnificent gift to Cambridge; and if this
can happen within sixteen miles of London, I feel it may
the more easily occur to those abroad, or ill, or temporarily
debarred from seeing the daily papers. These may be
grateful to me for giving the full text of the letter here,
for is it possible to express generosity more graciously
and interestingly than in these words, which, apart from
all else, are a model of noble English prose?

57 Elm Park Gardens : October 20, 1902.

MY DEAR DUKE OF DEVONSHIRE,—You may have
heard some months ago that what I hope will be known
as the Acton Library passed, by the signal regard of a
friend, into my hands.

For some time I played with the fancy of retaining
it for my own use and delectation. But I am not covetous
of splendid possessions; life is very short; and such a
collection is fitter for a public and undying institution
than for any private individual. After due inquiry and
deliberation, and with the possible reservation of an inconsiderable portion of quite secondary importance, I

have decided respectfully to ask the University of Cambridge, in which you hold the high office of Chancellor, to do me the favour of accepting this gift from me.

The library has none of the treasures that are the glory of Chatsworth. Nor is it one of those noble and miscellaneous accumulations that have been gathered by the chances of time and taste in colleges and other places of old foundation. It was collected by Lord Acton to be the material for a history of Liberty, the emancipation of Conscience from power, and the gradual substitution of Freedom for Force in the government of men. That guiding object gives to these sixty or seventy thousand volumes a unity that I would fain preserve by placing them where they can be kept intact and in some degree apart. I am led to believe that at Cambridge this desire of mine could be complied with. There is no other condition that I wish to impose.

In this way, I believe, Cambridge will have the most appropriate monument of a man whom, though she thrice refused him as a learner, she afterwards welcomed as a teacher—one of the most remarkable men of our time, extraordinary in his acquisitions, extraordinary in the depth and compass of his mind. The books will, in the opinion of scholars more competent to judge than I, be a valuable instrument of knowledge ; but that is hardly all. The very sight of this vast and ordered array in all departments, tongues, and times, of the history of civilised governments, the growth of faiths and institutions, the fluctuating movements of human thought, all the struggles of churches and creeds, the diverse types of great civil and ecclesiastical governors, the diverse ideals of States—all this will be to the ardent scholar a powerful stimulus to thought. And it was Acton himself who said that the gifts of historical thinking are better than historical learning. His books are sure to inspire both, for, multitudinous though they be, they concentrate the cardinal problems of modern history.

I need not say that it will be a lasting pride and privilege to me that my name should, even for a transitory moment, be associated in the mind of the University with the establishment of the Acton Library within the precincts of a home so famous.

Believe me to remain, yours sincerely,

JOHN MORLEY.

October 28*th.*—Weekly, now, the troops are coming home, and one wonders how they will settle down to the tame life here after all they have seen, and felt, and done in South Africa. Not the least regrettable influence of war is that it is such a bad preparation for peace. Even with young officers, the same thing applies. One has to remember that the very natures which do best for war—wild, reckless, gallant fellows—whose parents were glad and proud that they should volunteer to go out, are the very ones to be a considerable puzzle, both to themselves and to their parents, on their return. The world has grown so wise and practical lately, one is apt to forget that the old-fashioned parents still exist who glory in imperialism, blood and thunder, &c., and who may be singularly blind with regard to the consequences of war as an effect on character. I can quite imagine a young gentleman, who joined the Volunteers or Yeomanry to go to South Africa, being a sore trial, on his return, to the said type of father who would probably scold and stop supplies. If, by misfortune, the mother were of another old-fashioned type, who with tears in her eyes would pelt him with texts, both parents might easily send him flying down-hill *via* music-halls, racecourses, and pawnshops.

The ethics of war trouble the hearts of so many mothers that I must speak of a great book which has

A A

only just come to my knowledge, and which remains unknown to many people of the West in spite of the fact that it is one of the bibles of the world. It can hardly be without interest to a fighting nation that one of the most enlightening and inspiring books of the world's literature should have been given to a soldier as he stood irresolute on the battlefield, palsied in heart by feeling, 'Better in this world to eat even the beggar's bread, without slaying . . . than by slaughtering . . . to enjoy on this earth alone blood-stained pleasures, lusted after by those desiring possessions.'

To him the Teacher says : 'Thy business is with the action only, never with its fruits ; so let not the fruit of action be thy motive, nor be thou to inaction attached. . . . Surrendering all actions to Me, with thy thoughts (resting) on the Supreme Self, from hope and egoism freed, and of mental fever cured, engage in battle. . . .

'Taking as equal pleasure and pain, gain and loss, victory and defeat, gird thee for the battle ; thus thou shalt not incur sin. . . . Therefore, without attachment, constantly perform action which is duty, for, performing action without attachment, man verily reacheth the supreme.'

There are two excellent translations of the 'Bhagavad Gita,' one by Annie Besant, to be had for 6d. in paper covers, one by Mohini Chatterji, a large volume full of most helpful notes, at 10s. 6d.

After the publication of my second book, I received kind letters and even presents from unknown American and Canadian readers. One was an especially pretty little drawing of *Anemone thalictroides*, and some books and poems gave me great pleasure. I tried to answer and send thanks for everything as it came, but my personal sorrow and the black shadow of the war may have prevented me at the time from remembering all   If there was any omission, I feel sure that in the circumstances it will have been forgiven, and not have been put down to ingratitude. If I could write verses, I should like to have written these which appeared some time ago in the ' Westminster Gazette,' for they express what I so often feel—that however far we are from our friends' ideal of us, the higher they think of us the more they help us to become something better than we are :—

VALE !

I am not fair,
    But you have thought me so,
    And with a crown I go
More rich than Beauty's wear.

I am not brave,
    But fear has made me so,
    And dread lest I forego
The honour that you gave.

I am not wise,
    But you loved wisdom so
    That what I did not know
I learned it in your eyes.

I am not true,
    But you have trusted so
    That faithfully I go,
Lest I be false to you.

If Heaven I win,
    I can no virtue show
    But that you loved me so.
Will they not let me in ?   E. C.

THE JOURNAL OF A TOUR IN THE NORTH OF EUROPE IN 1825–26

(REPRINTED FROM THE ' CORNHILL MAGAZINE ')

I FOUND the other day, when looking through some old family papers, a journal written by my father-in-law during a tour through Denmark, Sweden, and Russia in 1825–26.

Charles Earle was born in the last days of the preceding century, and was consequently in the flower of his youth at the time that he writes this journal. He seems to have been a young man of a simple, genial nature, and in these pages, written only for his own amusement, he continually expresses his gratitude for all the kindness he received. He was apparently quite unruffled by the discomforts and dangers he went through by land and water. He narrates several times with the utmost simplicity his hairbreadth escapes from carriage accidents. In these days an occasional railway accident startles us because of the numbers it affects, but the risks that our luxurious parents experienced when travelling in their solitary grandeur were far greater than those incurred by any individual in modern travelling. Charles Earle had the good fortune to be in St. Petersburg at the time of the death of the Emperor Alexander I. and the accession of the handsome and interesting Nicholas I., who began his reign by weakening the power of Turkey and helping on the independence of the Greeks, but closed it with his ambitions crushed by the alliance of France and England with the Turks, and died leaving his peace-loving successor Alexander II. to sign the treaty disastrous to Russia which closed the campaign in the Crimea. I have had immensely to curtail the journal, only selecting those passages which seem of some general interest. The prophecy at the end, in the light of after-events, is distinctly remarkable.

He begins his journal at Hamburg on July 31, 1825, with the following words: 'After having trodden for two years in the beaten track of European travellers, and after having visited most of the countries south of the Baltic, my account of which was stolen from me at Warsaw, I have determined, in spite of this misfortune, to begin another journal of the tour I am on the eve of making through the North of Europe.'

He leaves Hamburg without regret, and apparently finds Lübeck much more interesting. He writes several pages in the ordinary handbook style, of which one remark may be noticed :

'I was particularly amused by a picture by Holbein, called "Death's Dance," which occupies three walls of a chapel in one of the principal churches,' &c., &c. Living before the critical age, he simply accepts the attribution of this picture. Nowadays we know that the 'Dance of Death' was one of the favourite subjects of northern mediæval painters ; and the best critics state that there is no authentic record of any 'Death's Dance' by Holbein, and throw considerable doubt even on the celebrated engravings.

He left Lübeck on a steamer for Copenhagen with his Italian travelling companion, Signor Rossi, and says: 'On arriving at Copenhagen, I paid my respects to Professor Oïrsted, to whom Dr. Young had given me a letter

of introduction, and who procured admission for me to the
fine cabinet of minerals.'

He constantly mentions visiting museums, and in this
interest for science there is probably a touch of romance,
as he was intimately acquainted with Dr. Thomas Young,
whose wife was the elder sister of the lady whom Charles
Earle already knew and afterwards married. Thomas
Young, M.D., was a scientific man of great distinction.
Among other things, I read that he discovered the law of
the interference of light, though I know not what that
means. He also evolved the process of investigation by
which the received interpretation of hieroglyphics has
been arrived at. It was he who deciphered the Rosetta
Stone in the British Museum, and I believe he is recognised as one of the most ingenious and original philoso-
phers of his time.

'They show the tomb of Hamlet in a garden not far
from the town. A simple stone marks the spot where he
was supposed to have been buried. On arriving at
Elsinor we paid a visit to the castle of Cronenburg, a
fortress commanding the Sound. It is a Gothic structure, and from its lofty tower there is a view of which
travellers are wont to speak in raptures. In order, however, to enjoy the prospect, the permission of the Governor
is necessary ; no easy matter to obtain, as experience
proved. After passing along corridors innumerable and
traversing dirty and deserted apartments, we at length
found the animal in his lair in the inmost recesses of the
castle. I had seldom seen, and it would be difficult to
describe, anything in the shape of a civilised being so
filthy as the said Governor. His uniform, covered with
the accumulated grease of years and in tatters, contrasted
with the numerous orders that glittered on his breast.
' Nein, mein Herr," was the only answer we could obtain
to our humble request to be allowed to mount the tower

to enjoy the view of the Sound. He assured us the responsibility was so great that he dared not incur it. At length, however, he yielded to our earnest entreaties ; but a dirty, slovenly sentry, to whom he gave the order to admit us, declared that he had refused. In vain we begged he would return and bring the permission we had been promised ; so our errand was after all a fruitless one, and we descended in no good humour from this impregnable fortress.' In the midst of drastic reducing of the journal I keep this little anecdote of the visit to Elsinor as characteristic of the want of civilisation at that time.

'*August 22nd.*—On the way to Carlstadt we had had one of our numerous carriage accidents, so were obliged to remain a few days in the town for some necessary repairs. There being so few resources within the walls, I was glad to accept the invitation of some brother sportsmen to accompany them to a neighbouring island, which they said abounded with game. After passing a wretched night in a peasant's cottage, walking twelve hours the following day, and narrowly escaping from drowning, I returned without firing off my gun or exchanging a word with my friends, who could speak nothing but Swedish. The result of the chase was one hare, which, not being able to escape from the island, was hunted down by the dogs.'

My father-in-law remained throughout his life the keenest of sportsmen. He shot woodcocks in the snow a few days before his death at the age of eighty-two.

'We spent two or three days at Stockholm. The streets are for the most part narrow, badly paved and very dirty. No capital I have seen contains so few good houses as Stockholm.'

It is rather amusing to note that a modern tourist, writing to me while taking this same northern journey, says, on beholding the quays of St. Petersburg, 'I have never seen so fine a bit of town ; the *only thing at all approaching it is Stockholm.*' So great a difference has grown up in the last fifty years.

'Whilst at Stockholm I made the acquaintance of a Mr. Lloyd, better known as "the Bear-killer." He had just returned from an unsuccessful expedition against the Russian bears, who, living under a despotic government, are secure from the attacks of the stranger, since no one can penetrate into the interior of the country without a particular kind of passport. Mr. Lloyd was desirous of obtaining a general permission to go where he liked, as in Sweden ; but his ideas and those of the Russian police on this subject did not exactly harmonise. Whilst at Stockholm he received a letter from the author of a tour in Sweden and Lapland, requesting him to pick up as many anecdotes as he could for the second edition of his work. He showed me his characteristically laconic answer : "Dear B.,—Your work already contains so many lies that to add to them is not the wish of yours truly."

'*October 6th.*—We set sail in a packet for the shores of Russia. Our voyage was a prosperous one upon the whole, and we sailed with rapidity to Bomarsund and that cluster of islands, the navigation of which is often attended with so much danger. These islands are barren and for the most part uninhabited. A pole with a red flag was the pilot from the hidden rocks, which are so numerous as to render sailing through the islands by night impossible. We accordingly passed our second night under the lee of the shore, and the following morning ran into Abo in good time. Being the bearer of despatches for our Ambassador at St. Petersburg, I found no difficulty in getting the carriage through the Custom House. Abo is a dirty, uninviting place, only famous for the treaty signed there between Russia and Sweden in 1743. There is no country on the Continent where the posts are so well served as in Finland, and, provided you have a coachman who understands putting on the harness, you never experience a moment's delay. It was our intention to have reached Björsby the same night, but at a wretched village two posts from that town the rain descended in torrents. Notwithstanding this, we determined to carry our intentions into effect; but scarcely had we left the post when the peasant ran us off the road, with the wheels of the carriage in such a position against a rock that the least movement from within or without would infallibly have overturned it. We fortunately succeeded in taking the horses off, and, being guided by a light glimmering in a window, we waded through mud and water until we arrived at the door of a Finland hut, the interior of which baffles all description : those, in fact, who have had no experience of these unfrequented countries would never believe that Christians could live in so swinelike a manner.'

Apparently things have not much improved, as my modern tourist before referred to gives the following description of a hut he arrived at while on a bear-shooting expedition in Finland :—' I am writing now in the peasant's hut. It consists of one room, of wood, and contains the family (ten in number), selves (two), and drivers (two). In one corner is the huge stove, in another twelve chickens in a cage. There is a table and a bench. I don't mention the other animals, though, I am sorry to say, I have already made their acquaintance, and am bitten from head to foot. There is a baby in a cradle that creaks horribly (the cradle, not the baby) and is curiously constructed. The baby is put in a basket hung on the end of a long sapling which is attached to the rafters. Then somebody crosses his legs and kicks it up with his foot, and the spring in the sapling rocks the baby to sleep. I hope there is not a cock among the chickens in the corner! . . . We turned in at 8.30 last night, and as there was no room on the floor I spread my shouba on the bench, and with my tobacco-pouch as pillow was soon asleep. Of course nobody undressed, and everybody snored wonderfully; and there *was* a cock in the corner! '

The journal continues : ' The Finlander's habitation consists of three houses ; one for the summer, one for the winter, and the third for a kitchen. These houses are joined together; they are of wood and resemble a Swedish cottage. On entering the one in question we found a large family assembled in the room, a part of whom were in bed, and the others preparing for a wedding that was to take place the next day. We in vain entreated them to let us have one of the blackcock then roasting for our supper. The good lady of the house, however, promised us something to eat, showed us into the adjoining room, and lighted a fire. The supper then arrived, but, in spite of a *faim de loup,* I could not touch it. The Finland peasant, though small in stature, is well made, and, being constantly employed in fishing or hunting, is more active than his neighbour the Russian. Their language is peculiar to the country, though many of them speak Swedish. They use the Gothic character in writing ; the sound of the language is indescribably harsh. The Finns are still governed by Swedish laws ; there are no nobles. In seasons of plenty they eat five times a day; yet, from the severity of their climate and the sterility of their soil, there are no people so often reduced to want. So great, indeed, is the scarcity during long and rigorous winters that they are often obliged to mix the bark of the fir-tree with their meal. In spite of all these disadvantages they are remarkable for their longevity.

The circumstance of the necessaries of life being only
procured by great and constant labour, which con-
duces so to health, may be the cause of their attaining so
great an age. The costume of the women is singular
enough ; they rejoice in immense earrings, usually made
of glass, as also their necklaces, though the latter are
often composed of pieces of money strung together. The
men adopt the dress of the Swedes. Whilst the marriage
feast was preparing I was in vain endeavouring to sleep
on the table, awaiting the first ray of light, which we
agreed should be the signal for our departure. I had not
the good fortune to see the bride during our short stay
under her mother's hospitable roof, but before we started
we heard a bustle which ushered in a day of no small
importance. A large party was already assembled, and
their clean and smiling faces proclaimed the occasion no
common one. The bride in Finland is obliged to present
each guest with three yards of cloth and a pair of stockings ; tho guoct immediately gives their value in money,
which belongs to the bride.

'*October* 16*th.*—As we drew near to St. Petersburg our
desire to see some part of the celebrated city was pro-
portionately great ; but, although we took advantage of
every rising ground, there was nothing to be seen on all
sides but a plain of boundless extent, little cultivated,
thinly inhabited, and offering a prospect of which there
are few examples in the vicinity of a large capital. But
at length the river with its interminable quays burst on
our view. On reaching the middle of the Neva I ordered
the postilion to stop, that we might contemplate this magnificent city. Nothing I have as yet seen can compare
with the *coup d'œil* this spot presents. Every object that
the eye ranges over is grand ; the noble river confined in
its bed of granite, the quays lined with sumptuous palaces
and extending further than the eye can reach, the

Admiralty with its gilt steeple—in fact, everything we beheld was calculated to make a deep impression on the mind.'

Want of space prevents my giving his further descriptions of St. Petersburg, the sights he saw and the interesting people he met. I must pass on to his account of the revolt which broke out at the accession of Nicholas I.

'On December 9 the melancholy tidings of the death of the Emperor reached the capital—the news of his illness and of his death arrived almost simultaneously. Everyone appeared panic-stricken at this distressing intelligence, for the Emperor was equally beloved by all classes of the community. His fortitude in adversity and moderation in prosperity have justly procured for him the admiration of the present age, and I only hope his example may be followed by the monarchs of succeeding ones.

'The following morning the troops took the oath of allegiance to the Grand Duke Constantine. From this time until the courier returned from Warsaw on the 25th the public mind was a good deal agitated on the subject of the successor to the Imperial throne, for it was known that Constantine had formally announced his intention of abdicating in favour of Nicholas his brother. On the morning of the 25th the solemn renunciation of the Grand Duke and on the 26th the manifesto of the new Emperor were published, and the Imperial Council, the Senators, and the Holy Synod took the oath of allegiance to Nicholas. At about 11.30 the Colonels of the Horse and Chevalier Guards, of the Préobajensky Guards, and of six other regiments announced that their respective soldiers had taken the oath. From the other regiments nothing was heard, but this was attributed to the great distance of their barracks. At twelve o'clock intelligence was brought to the palace that several officers of the Horse Artillery, having manifested some opposition, had been arrested.

'It was at this moment that I passed through the Place d'Isaac on my way to the English Library. There was a vast concourse of people in the square, and the shouting was vehement and uninterrupted. Upon asking my *laquais de place* what they were calling out, he told me that it was in favour of the new Emperor. Thinking this very natural, I went on towards the Library. I had not been seated many minutes when the wife of the librarian rushed in, half-dead with alarm, and told me that the soldiers were fighting in the streets, and that the police had ordered the doors of all public places to be closed. On leaving the house I met the librarian, who informed me that General Count Miladorovitch had been shot. 'I immediately got into my sledge, and drove towards the scene of action, but found all the avenues to the square guarded. Not being allowed to pass, I was proceeding to my lodgings by a circuitous route when I met Count Doernberg, who was in search of Lord Strangford. He got into my sledge, and we went together to the Hôtel de Londres, where we found Lord Strangford, who had seen Miladorovitch fall. We all went back together to the square. It was then about one o'clock, and the Moscow Regiment had formed a square after having refused to swear the oath of allegiance. Upon seeing such a concourse of people assembled, the Emperor went out unattended and penetrated into the midst of the mob, by whom he was received with enthusiasm. The presence of the military now became indispensable, and the Emperor, ordering a battalion of the Préobajensky Regiment to march to the square, put himself at the head of it and marched towards the revolted troops. He was determined, however, not to have recourse to violent measures until conciliation had failed. Shortly after this several regiments made their appearance, as it became necessary to oppose force to force. The mob began to be very riotous, and, breaking down the palings opposite to the Church of St. Isaac, armed themselves with bludgeons, and, surrounding the Colonel of the Horse Guards, threatened the life of that officer, whose forbearance was beyond all praise. It was quite evident that in the present temper both of the revolted troops and of the people all conciliatory measures were useless, and the Chevalier Guards were ordered to charge. They were received by a volley from the rebel square. A horse was struck close to the spot where I was standing with Prince Schwartzenberg; the ball entered the poor animal's shoulder, and it was with much difficulty that the soldier dragged him from the ranks. Everyone who was present must bear testimony to the noble bearing of the Emperor upon this trying and melancholy occasion. Neither the tumultuous attacks of the mob nor the obstinate resistance of the revolted troops could shake the resolution he had formed of not using violent measures till all others had proved useless. It was not, in fact, until the Metropolitan had harangued the soldiers of the Moscow Regiment, and until these had fired several rounds at the troops in the square, that he could make up his mind to employ a force which would bring matters to a speedy conclusion. Such a measure was rendered the more indispensable as the daylight was fast disappearing. Accordingly the cannon were brought up and charged with grape. In the hope of intimidating the rebels the first shots were fired over their heads, as the marks on the wall of the Senate House prove. Seeing this measure had not the effect of dispersing them, the next discharge was levelled at the square formed by the revolted troops. The distance was not more than 100 yards, so that the effect of ten guns firing grape on a compact mass may readily be imagined. A charge of cavalry immediately followed; the mob and revolted troops were dispersed in every direction and pursued and sabred by the Chevalier and Horse Guards. It will never be known how many fell on the occasion of this revolt, for the bodies were immediately put under the ice in the Neva. The son of the man from whom I hire my carriage fell a victim to his curiosity, and many others narrowly escaped a similar fate. The troops that had been engaged during the day bivouacked in all the principal streets and squares. The most profound tranquillity reigned throughout the city that night; not a sound was heard save a sentry's challenge and the soldiers talking round their fires. The Cossacks, as they were seated round the fires, with their long lances and remarkable costume, their horses tethered behind them, presented a novel and interesting sight. The following morning the troops which had revolted were drawn up before the Admiralty, and the colours taken from them the previous day were restored to them by the Emperor. This was an affecting ceremony. A temporary altar had been erected in the snow in the middle of the square. Mass was performed by the Metropolitan; and the soldiers, prostrate before their Sovereign, seemed at once to implore his forgiveness and to atone for their misconduct by a sincere repentance.'

The following account of the sledge journey to Moscow is amusing :—

'*January* 21*st*, 1826.—We left St. Petersburg for Moscow, placing our carriage on a sledge drawn by six horses abreast. The cold would have been by no means insupportable had it not been for the high wind, which was so violent at times as to make us fear the overturn of our carriage. The unevenness of the road soon proved to me how erroneous were the ideas I had formed of sledge-travelling in Russia. The bridges too, except on the main road, being only temporary and constructed for sledges with one horse, were not wide or strong enough to support us. Our peasant coachman continually left the main road and took us across country to save a few versts, so next night we were upset into a snowdrift, miles from help of any kind, and with no moon or stars to show us where we were. Our coachman threw himself down in the snow and burst into tears, and from this position nothing could induce him to stir. We could do nothing but await the return of day. The cold was intense, the roof of the carriage covered with ice, and our voices even appeared to freeze. Making therefore paths on the snow, we began to walk briskly up and down to keep up the circulation in our benumbed limbs. At length our wretched coachman began to listen to the voice of self-preservation, though he had been deaf to that of reason, and, rising from his bed of snow, mounted one of the horses and rode off for assistance. Hour after hour passed, and no signs of his return. At last at daybreak we heard bells, which proved to be our friend returning with two or three miller's horses, whose efforts, combined with our own, succeeded in extricating us.

'O ye inhabitants of civilised countries, ye dwellers amongst hills and vales and verdant fields, think with compassion upon those travelling over this dreary waste of snow, where there is not a sign of anything animate or inanimate except the animals dragging and those driving the carriage. To arrive at a *relais*, which often consisted of a solitary hut of wood, was an event that almost gladdened the heart, so dreary and desolate was everything around us. All the days we spent on the journey, not having an appetite for tallow and other Russian delicacies, we were well-nigh starved as well as frozen.

' At last we saw, to our inexpressible delight, the gilded and painted domes and spires of the ancient capital of the Czars. But nothing announced the approach to a great city; not a carriage, nor cart, nor the sign of anything living, relieved the dull monotony of the snow-covered plain till the Russian sentry at the gate put out his barbarian hand to receive the passport of civilisation. Inside the barrier there was an almost equal degree of desolation. A street, the end of which no human vision could reach, and which we, blinded as we were with the snow, could hardly see across, led to the more frequented part of the town. We passed many palaces, which seemed to have parks as pleasure grounds; and contiguous to these habitations of the great were to be seen the miserable abodes of the wretched serfs. Never did I behold such apparent luxury in contact with such misery.'

He stayed in Moscow nearly six months, but states that this part of his journal had again been lost!

He merely says :

'The hospitality of the Muscovites was boundless, but confined to small family reunions. The day before my departure the Princess Zénéïde Volkousky (La Corinne du Nord) provided me with many letters of introduction, much valuable information, and a number of maps of the Crimea, &c., she had herself made for me.

'*June* 17th.—At midnight we started for Odessa, and after traversing several of the long and melancholy streets of this vast city we arrived at the stone bridge from which the Kremlin is seen to the greatest advantage. I know not whether it was because I beheld it for the last time, but I had never before been so struck with the singularity and magnificence of its gilded towers and antique domes. There is a mixture of European and Asiatic architecture in all the public edifices of Moscow which certainly gives it an interest that no city I have yet seen possesses. The perfect stillness that reigned all around us at the moment we took the last long look at the Kremlin, in the uncertain light that the moon cast upon every object,

B B

produced a sensation it is difficult to describe. We passed through Tula and Batourin, then to Kiew, formerly the capital of Russia. Nothing can exceed its filth and wretchedness, but it is of great historical interest.

'Some days after leaving Kiew we came to the steppes. The weather being dry and the road good, we travelled with rapidity, this being the only circumstance which alleviated the monotony of the journey. You see the sun rise and sink below the horizon as when at sea.'

In spite of the immense increase of luxury in travelling nowadays, the modern tourist still complains bitterly of the discomforts attendant on this same journey. My correspondent writes in the train on his way from Moscow to Sebastopol: 'I left Moscow on Sunday morning. They arrange to stop the train for food, so at five o'clock I got out at Tula—"the Russian Sheffield," says Murray, "and famous for its hardware." To test the hardware I ate a cake, and certainly it was hard—harder even than the Dover bun—though perhaps not so strong in another sense as Russian eggs are. The tea is so weak and the eggs so strong that I sent for the manager and advised a better distribution of force, but he did not understand. . . . I have now been fifty-three hours without sleep or wash, travelling over boundless steppes which I am glad to have seen, and hope never to see again!'

The journal continues at Odessa :—

'*July* 20*th*.—On our arrival we drove to the Club Hotel, the appearance of which was certainly not inviting. Every article of furniture was covered with dust, which so filled the atmosphere as to render objects on the opposite side of the street almost invisible. The streets of the city are broad and uniform, and the houses for the

most part of one storey. It is evidently a town in its
infancy; in fact, the ground on which it now stands was
a barren steppe thirty years ago.'

He gives a long and interesting account of his stay in
the Crimea, where he was the guest of Prince Woronzow.
How little he could have anticipated when visiting Sebastopol and Balaclava, which he minutely describes, that
he would one day have two sons fighting the Russians on
that dreary plateau!

The journal ends at Brody, a town inhabited chiefly
by Jews, where he was detained through difficulties with
his passport, and where he put on paper the following
reflections suggested during his stay in Russia :—

'The menacing attitude assumed by Russia towards
her neighbour Turkey, the revolution now going on at
Constantinople on account of the suppression of the
Janissaries, and the known tendency of Muscovite policy,
will make this country, probably at no very distant period,
the battlefield of Europe. Every step which Russia
makes towards the Dardanelles must be considered as the
advance of barbarism against civilisation. Whether this
inroad of the Northern hordes will take place in the lifetime of anyone now living, who can tell? But come it
must, sooner or later. The Russians even now talk of
Constantinople as theirs, and sometimes pronounce the
words Malta and Gibraltar. What they aim at is
universal dominion in Europe and the annihilation of our
power in the East. The Imperial Court is increasing its
influence every day in Germany by the matrimonial
alliances it is contracting with the different reigning
families in that country, and at the same time it is con-
ciliating the leading diplomatists of Europe by munificent
presents. The boundary of the Empire will soon be
removed from the Pruth to the Danube ; and with the
mouth of that river in the power of Russia, Austria is

annihilated and the commerce of Germany intercepted, if not ruined. Some great European convulsion is not by any means an impossibility, seeing how restless and dissatisfied France is under her present government. And Russia, separated as well by her institutions as by her geographical position from the contact of liberal opinions, will be ready to pounce, as it were, upon Europe, torn and distracted by internal dissensions. We have only to observe the preparations Russia is everywhere making, to be sure what her designs are. I saw fifteen sail of the line in the harbour of Sebastopol. What enemy can they fear in an inland sea like the Euxine? Is it not manifestly assembled to attack Turkey, just now enfeebled by the revolt of the Janissaries and impoverished by a defective administration of her affairs? The Russians are in no hurry to arm Sebastopol, because they have no one to fear; but English engineers are now employed there, and I make no doubt that long before any necessity arises that place will be rendered impregnable. If we turn from the South to the North, what do we see going on at Kronstadt and other fortified towns on the Gulf of Finland? What a fleet is at Kronstadt! I forget how many sail of the line, but quite disproportionate to the exigencies of the State in time of peace. That the blow is meditated no one can doubt; all that is wanting is a pretext, a favourable opportunity. And when the vast resources of the country are organised, then the question is: Who is to oppose this inroad of the Russian hordes, this second and perhaps final irruption of the Huns? It is clear Turkey can oppose no effective resistance; her tottering dominions are already falling to pieces under the weight of her own intestine disorders. Prussia will not—she is so bound to Russia by ties of consanguinity and diplomatic cajolery—and indeed, if she were willing, she *could* not stem the torrent. Austria,

with her heterogeneous and discontented population, has
too much to do at home to send an army to the banks of
the Pruth.   The Greeks are the co-religionists of Russia.
What, then, remains?   England and France.   But a
cordial alliance cannot be reckoned upon between two
countries who have drawn the sword against each other
for twenty years.   As for the Scandinavian Powers, they
are prostrate at the feet of the great Autocrat.   I really,
therefore, see nothing that can prevent Russia from walking across Europe whenever she is so disposed.

'These somewhat melancholy forebodings were made
whilst waiting for my passport, and no doubt were in a
great measure inspired by the wretched place in which I
then was!   The filth, the degraded aspect of this town,
with its loathsome miserable population, no words can
describe, and great indeed was my joy on seeing my
courier arrive from Lemberg with the passport which
was to emacipate us from this Jewish city.   The rest of
our journey was through the fertile plains of Moravia,
and we made a triumphal entry into Vienna on September 6th.   Those only who have been for some time
in contact with barbarians can duly appreciate the happiness of once more finding themselves amongst a civilised
people'

THE LAST LETTERS OF CAPTAIN SYDNEY EARLE, COLDSTREAM GUARDS

IT is impossible for me to judge whether these simple letters of one of the best of sons will be of the smallest interest to those who did not know him. A friend offered to publish them in a magazine shortly after his death. But then I felt I could not stand giving to the public what had been written in confidence to his family. Now it is different. It was he who encouraged me, when under the shadow of another sorrow, to write my second book, and it is indeed to his memory, and with absolute confidence of his approval, that I dedicate my third. In all great sorrows one longs to realise that, in spite of external appearances, 'Time's effacing fingers' are not in any way really dimming the memory of what is most dear. I am in no sense a worthy mother of soldiers. I gave him very grudgingly to the army, and he knew quite well all it meant to me when his earnest desire to go to South Africa was granted, and he was ordered to join one of the earliest batches of Staff officers who were sent out. My own view of the war at that time was that it had been brought about by a mistaken policy, and was not a just war, and that he was not without sympathy for my attitude of mind is shown by the letters he wrote on board ship. I think he knew quite well as we walked round the garden for the last time that we should never meet again. I gave him to die for his country not willingly at all, and I publish these characteristic letters, because it is a pleasure to myself to see them in print. It is to me the same feeling as the preservation of any remembrance, the crystallising, and putting as it were into a glass case, of 'a footprint in the sands of time.' As with so many in the early days of the war, he had the strongest feeling of his coming doom, and when he bade good-bye to his soldier-servant at Orange River camp, he said he would never see him again. Three of the letters came, of course, after the fatal much-dreaded telegram. His last words, written two days before he was killed, ' I wish it were all over,' expressed a desire that peace might come, not for him by his death but in a very different sense. The following little notice of his short career appeared in the ' Pall Mall Gazette ' of December 1, 1899 :—

'A POPULAR GUARDS OFFICER.—Captain Sydney Earle, of the Coldstreams, killed at the Modder River, was, besides being one of the best liked and most unaffected officers in the Household Brigade, one of the most promising of the younger generation of officers in the British army. After a brilliant career at the Staff College, he was appointed Deputy-Assistant Adjutant-General for instruction in the Home District. In this position he conceived and initiated an experimental course of instruction for Volunteer officers in applied tactics, field engineering, and military sketching, and himself conducted the first series of classes at Wellington Barracks in the autumn of last year. So successful was his scheme that the authorities decided to make it a permanent institution in the Home District.

' The handful of Volunteer officers who spent so useful and agreeable an eighteen days last year, not only at Wellington Barracks but at the Guards' field-days on Wimbledon Common, at open-air exercises on the Surrey hills, and in studying demolition work at Sandhurst under his guidance, must share the deep regret with which his brother officers in the Brigade of the Guards have heard of his death on the field of battle. No more genuine man or valuable officer served in the brigade.

'Captain Sydney Earle was the eldest son of Charles William Earle of Cobham. He was born in 1865, and entered the army when twenty. His father was killed two years ago in a bicycle accident.'

His name was the only one among the junior officers of the rank of Captain mentioned in the 'Times' account at the end of 1899 of those who had fallen in the war. Will anyone who lived through them ever forget those two dark months, the November and December of that year?

I offer my sincere sympathy to all those who have suffered as I have done, to those who gave more than one son, and to those who suffered even more from their loss being of an only son, and perhaps an only child, for they cannot realise the one compensation that can come on the death of a child which is that it brightens the eyes of love for those that are left.

STARS

Stars in the North !—world-fragments, that through space
Æon on æon ran their darkling race,
Strike fiery-white against Earth's airy wall,
And luminous in dissolution fall.

Stars in the South !—dim souls, that could not shine,
While life's dull orbit did their course confine,
Now, devious hurled on war's opposing breath,
Flash in a brief magnificence of Death.

J. RHOADES.

No. 1

Monday, 9/10/'99 : 'Braemar Castle' off Gib.

DEAR MOTHER,—Never can more auspicious start have been made ! Everything has gone well, fine weather, perfect ship, and good company. We started soon after five, a little after Lionel left, with cheers from a large crowd of people. More cheers from a big battleship, the 'Australia,' I think, lying in Southampton Water. My cabin companion is called Captain B—— of the Army Service Corps, and he seems a nice fellow. We have got a good roomy cabin on the main deck, port side, just about under the saloon and bridge. There are about six or seven men I knew before on board, and, of course I have already made acquaintance with others. They are almost all Army Service Corps, doctors, Army Ordnance Corps, and special service (like me). There are just a few going out to join their respective regiments. Just before starting I received a wire from General T—— wishing me luck and sending messages to Major S—— and to Miss A——, a nurse from the London Hospital. She seems a nice woman, but has been rather poorly from sea-sickness.

There are six or seven army nurses on board ; the only other woman is a Mrs. B——, wife of an A.S.C. officer ; she, I believe, is a native of the Cape (not black). I sit at dinner and all meals at the Captain's table. I put my own name down for it. There are about eighteen at it, headed by Col. G—— and Prince C—— V——. I sit between Major S——, a special service officer, and the ship's second officer, a good fellow. The food is excellent, but I am dieting myself strictly. There has not been much in the way of combined amusement started yet, but I believe a newspaper has been started, as I have been asked to contribute, and I hope to get a game of cricket this afternoon. It is feeling a little muggy and warm now, but I have not begun to put on summer clothes. I feel aches about the back and legs, as I always do at first on board ship. We have formed a class for early morning ' Sandow' exercises in pyjamas on deck. I attended this morning. I have got my name down for a very good bath hour—viz., 7.30 ; this is *the* most fashionable and sought after time. Don't know what has happened to Davison [his soldier servant], haven't seen him for about forty-eight hours. There is nothing for him to do, the steward looks after me and my clothes. This ship, though new and splendid, is a slow one, as she is intended for freight ; she only goes thirteen knots an hour, so we shall not get to Cape Town till about Saturday fortnight—*i.e.*, the 27th or 28th. I can gather no ideas of what we are going to do. Prince C—— V——, though he knows a good many things, hasn't the faintest notion. We have most of us been given secret books compiled in the Intelligence division on the Boers and the Transvaal. We have just been attending a lecture on inoculation for enteric by one of the doctors on board. I am not going to be done myself. People rarely get it over thirty, and the fact of having had typhoid is an additional safeguard.

Tuesday.—Getting hotter, but I have not begun summer clothes or khaki, though I expect to do so to-morrow. I have got pains, as I often do on board ship; shoulders, loins, and legs, liver, and rheumatism I suppose; it will be all right probably to-morrow. Mind you let me know comments on 'Pot-Pourri' No. 2, especially any unfavourable ones. I like to hear the other side. I have written a heap of letters to different people; they are taken back to London and posted, which accounts for the penny stamp doing. I see the penny stamp is to return to its original red after this year—a good thing too. It will be a long time before you hear from me again, and then it will only be a tiresome ship letter with no news. I shall wire from the Cape any important news. The name of place, say 'Kimberley,' would mean I have been ordered to Kimberley, the purpose I will try to put in after. Thus 'police,' 'defence'—to prepare defences; 'Irregulars,' to recruit and drill an irregular force. Those who wish to land to-morrow at Las Palmas have been directed to send in their names to-day, but nothing has as yet been said about uniform. Good-bye, dear mother, let me know all you do.

Your affectionate son,
SYDNEY EARLE.

No. 2

Monday, October 16, 1899 : 'Braemar Castle.'

DEAR MOTHER,—I find no mail goes home till three or four days after our arrival at Capetown, but still I will begin this letter early and jot down things as we go along. When we arrived at Las Palmas we found one of the Artillery ships beginning with a Z was in there, though she left Liverpool about a week before us. F—— W—— (R.A.) who was with me at S. C. and who married a Liddell, came with another officer alongside of us in a boat. I gave

him a shout, and he told us they had had no news since
the 2nd, and had come for newspapers; he dashed on
board and collected a few, and then off again, like a
shot, which was accounted for by the fact of their ship
being on the point of starting; in fact, she was off before
they left our ship. We heard that seven horses had died
of pneumonia, that they could not get enough water at
Canaries, and were going to put in at St. Vincent, in the
Cape Verde Islands. I hope they will twist somebody's
tail over that business; it is time that the idea that anyone can swindle the Government in time of war should
be exploded. It was a bitter blow to us to hear that
there was no news at Las Palmas; however, before leaving, we got an answer to a wire which the gunners had
sent home to Intelligence Department asking for 3*l.* worth
of information—it was to the effect that Boers had sent
an ultimatum, and that mobilisation had been ordered;
the gunners had gone before it arrived, but we were
very grateful to them. We have not had much incident.
A man died on Saturday night of pneumonia and was
buried on Sunday morning. I have never seen a funeral
at sea before. We have got eight trained nurses on board,
very nice women indeed; they only heard the man was ill
on Saturday. On the Superintendent applying to give
assistance, she was met with a flat refusal from the
Principal Medical Officer. She insisted on seeing the
man, and saw he was very ill; but they still refused to
allow her to nurse him, and he was left to the tender
mercies of the hospital orderlies. I must say that the
army doctors make me boil with rage; they undoubtedly
killed a man of the Scots Guards this year, by keeping
him with pneumonia in a tent with a daily range of temperature of about forty degrees, and in this case on board
ship, if they had put their miserable jealousies on one
side and called in the help of a good nurse, I have little

doubt the poor man's life would have been saved, or at any rate his death would have been made pleasanter. Another Crimean repetition! I have no doubt they would have squashed Florence Nightingale if they could. We had a good view of Cape Verde while passing. I have seen it before, but I don't like it any better, and have no intention of purchasing a villa there. I have forgotten to say that I had intended to go on shore at Las Palmas with S—— (see below), but on seeing the place all glare, sand, and dust, we decided that we should be better on board. Those who did go on shore seem only to have exchanged a very good luncheon for a very bad one, and to have spent about a sovereign to achieve the result. It seems to me, as far as I know of them, that we have never had an expedition with such good commanders or such an efficient Staff. Things ought to go smoothly and well when the army corps gets out, provided the contractors have done their work as regards transports and provisions. The sending out of Army Service Corps and Army Ordnance Corps to make preparations is very sound and wise, though rumour says that Lord L—— was much opposed to it. I view with a certain amount of apprehension the present condition of affairs. If the Indian troops have arrived, Sir G. W—— ought to be able to hold his ground in Natal, especially as they have withdrawn with great wisdom from the dangerous New-castle district. I am not sure that they would not have been wiser to withdraw still further (I hear this has not been done on account of the coalfields near Glencoe). They have no superiority in artillery. This is, I think, a pity, but I suppose it can't be helped. With regard to the scattered forces round the Northern and Western boundaries of the Republic, I look upon them as dis-playing a strategy of so mean an order as to point to political interference. There is no doubt that whenever

politicians interfere with military disposition they court
and invite disaster. Probably in this case, people
clamoured for protection, saying that they expected to
have their throats cut ; of course this is not likely to
occur, as, however barbarous the Boer may be, he could
hardly kill defenceless people ; the real reason of their
clamour is that they want their property protected.
I should encourage them to remove the property by
denying them protection and by giving compensation to a
moderate extent for property damaged that could not be
removed, such as mine machinery. My forces I should
concentrate sufficiently far from the frontier as to render
an attack in overwhelming force improbable. I fully
expect to hear that the different detachments at Tuli,
Bulawayo, Mafeking, Kimberley, &c., have been mopped
up : if they are not, the Boers have only themselves to
blame ; we have played their game as much as the Khalifa
did Kitchener's at Omdurman. I also expect a pretty
general Dutch rising in the Colony, which also would
have been less likely to happen if our total available
forces had been concentrated somewhere near the line
Colesberg-Burghersdorf. Major S——, whom I mentioned
above, is one of the special service officers. He is a Staff
College officer and was on the same staff with me at last
year's (Wareham) manœuvres as D.A.A.G. He is a cousin
of Sir Leicester S——, who was a distinguished general.
Before coming out, he was Major at the depôt of his
regiment, the 21st (R. Scots Fusiliers) at Ayr. He brought
out a draft years ago to his regiment in Natal. The day
they landed was the day the Boers cut up the British
regiment at Bronkerspruit. He was hurried up with his
draft to the front, and was present at Laing's Nek, where
he was employed to take ranges for the Artillery, who
could not take their own, as they were not at home with
the new range-finder and were really garrison gunners,

not field ditto. He was in the camp the day of Majuba, and, after the armistice, went up to Pretoria, where he remained for some time with his regiment—very interesting man to be with. I hope we may keep together, but of course we have no idea of what we are going to do. We have had several tremendous tropical storms of rain, just as if all the fire brigades in the world were playing on us. To-morrow we cross the line, and then we hope to meet the head wind which will make things cooler and dryer. Your thermometer has been a comfort. I sleep with it close to my bunk; the heat has not been excessive, about 89° on the bridge is the most I have heard of. I have had no difficulty in sleeping soundly. The greatest misfortune I have had at present is with my pyjamas. I asked Davison to get a beautiful pair of wide floppy ones washed, and they have come back like acrobatic tights. I am reading Ball's 'Story of the Heavens' out of the ship's library—excellent! I also read a good bit of 'Pot-Pourri' No. 2, and find it also excellent. I see you state that nothing grows under beech; I saw beautiful shrubs with a sweet white flower (very well known, only I can't remember the name) growing under the beeches at Amesbury, the Antrobus' place. I find that in the 'Globe' I bought just before starting, there was a criticism very favourable and complimentary, but nothing really in it; still I was glad to see just one notice. I forgot to make any arrangements about having newspapers sent out. Something in time to catch the weekly mail which would give a summary of the week's news would be very acceptable. If there is a big war, a file of some daily paper would also be an interesting thing to keep for me to look at on return.

Thursday, 19/10/'99.—We are now well over the line and have met the trade winds; it has never been oppressively hot, and now it is perfect as regards temperature. I think the Government treat the men on board ship with considerable meanness ; almost all the men on board are skilled men of some kind or another, and as such receive ' Corps Pay,' the lowest rate of which is 6d. a day ; this they lose from the moment they step on board. Of course the idea is that their skill is not required on board ship ; but still they don't go on board of their own choice, and they feel that their loss of pay is a grievance, and I agree. They have no beer, and I believe no liquor ration on board. Now, moderation is a great virtue, but enforced total abstinence for no definite reason seems to me to be unwise. While actually on a campaign it may not only be advisable, but necessary, but it should be universal ; if Pte. T. A. is compelled to be a teetotaller, so should the G.O.C. and all other officers. For future reference I am going to jot down my ideas on the conduct of a war against the Boers. A campaign of this kind necessarily forms an exception to several of the best known rules of strategy, as the advantage of rapidity of action is greatly diminished from our point of view in consequence of the English becoming stronger and the Boers weaker (owing to the consumption of supplies, the diminution of wealth, sickness of men and horses, desertion and evaporation of enthusiasm). Political reasons—i.e., the protection of both Cape Colony and Natal—oblige us to commit the gravest of strategical errors—viz., a separation of force without any possibility of mutual support. This misfortune should, it appears to me, be reduced to a minimum at the outset of the campaign by massing each of the dividing forces at such a distance from the border as to render an attack upon either a difficult and hazardous operation ; in fact, we want to gain time. We undoubtedly must expect loss of prestige by leaving places undefended and exposed to the mercy of the Boers, but that is better than the destruction or capture of one or more small garrisons. The next move depends upon whether the Dutch inhabitants of Cape Colony rise; if they do, no forward movement should be made against the Boers, until the country has been properly policed by the soldiers; in fact, let them (the Boers) stew a bit. The police duties must be carried out with rigour, and should consist of total disarmament and of partial confiscation of property of disloyal colonists. Arms should be given to loyal colonists, and armed local police forces raised. If there is no trouble of this kind, or if it has been put down, I should devote my attention to the establishment of an advanced base somewhere on the line Colesberg-Burghersdorf. The railway bridge over the Orange River will probably have been destroyed, but the passage of the river, and ultimately the repair of the Bethulie bridge, will probably present no very great difficulties, though it should not be attempted till a considerable force has arrived at the front (especially artillery); the general advance on Bloemfontein should not be begun till the whole of the army corps and line of communication troops have arrived at the front. From this point the advance must be made in conformity with the rules of European campaigns—cavalry screen, advanced guards, &c. If the enemy is met with, he probably would not be in great strength, as it is difficult to see how the Transvaal would be able to take many men or guns away from the Transvaal border. But whether he is strong or weak, he must be gently 'levered' out, a considerable display of artillery in his front, and turning movements of cavalry and mounted infantry round his flank. There may be opportunities of getting him on the run, when horse artillery and cavalry can act with boldness, but they must be carefully chosen; the one thing to avoid is running up against a selected position to be shot, as the Boer himself puts it, like buck in the open. Now the Boer, owing to being mounted, has great tactical mobility; but in order to keep the field for any length of time, he must have his waggons, which serve him as tents, kitchens, and store-houses; he has therefore very little strategical mobility, and it ought therefore to be easy either to separate him from his waggons, or at least to drive him and his waggons back on Bloemfontein. In either case it is difficult to see how he is to avoid a disaster, his inferiority, especially in artillery, is so evident. I am rather of opinion that one *serious* reverse to the Boers would be sufficient to stop the whole war. If not, the whole proceeding must be begun again, this time with Bloemfontein as the advanced base. Our line of communications will by this time have assumed gigantic proportions, and any advance will of necessity be a slow matter. However, as I have already pointed out, that is attended by little disadvantage except expense. We have, I believe, eleven battalions to guard our communications, and these, with all available local forces, will have their work well cut out for them. Great strictness must be enforced, and if the line is tampered with, the people of the district should not only be fined to make good the damage, but be compelled to work at it with their own hands. The further advance must be conducted on similar lines, except that immediately after crossing the Vaal River steps must be taken to remove any troops still blocking the line from Natal. Pressure upon their line of communication with Pretoria, &c., will, if they are in large force, probably effect this; if their force is small, it may be necessary to detach a force from the army corps to co-operate with the Natal force. After our forces have got within supporting distance of each other, further actions must depend upon what the enemy do. If they, as they most probably will, fall back upon the Pretoria forts, I look upon the campaign as at an end, probably without a further engagement ; if they retreat in some other direction, they at best can only hold out a certain time ; their best course is to take up some intermediate and, if possible, flanking position, to bar our advance on Johannesburg and Pretoria. If we avoid the frontal infantry attack, we still ought to have little difficulty. These are my views before seeing the opening moves ; they.will of course be modified when I see what happens. Colonel R——, our senior special service officer, tells me that he anticipates great difficulties from the fact that the Boers take women with them in the waggons to cook and look after the teams, &c. Our business will naturally be to go for the waggons and try to shell them, capture them and destroy them in all possible ways ; there is no doubt that the women will shoot and will probably get killed, and of course a howl will be raised about fighting against women. It is a serious business, as people will not accept the only rational answer, which is that the women should remain at home, where they would be safe.

Friday.—Three days ago we espied the ' Mexican,' the mail ship which left a day after us, lying across our bows in the far distance flying the signal ' unmanageable' ; great excitement on board—we thought she had broken down, and that we should have to tow her to St. Helena. However, when we got within a mile of her she steamed off on her true course, and we have never seen her again. The probable explanation is that she had got some small matter wrong with her machinery, and had stopped to put it right. Another excitement— we yesterday passed the homeward-bound mail-boat ' Dunvegan Castle,' but she was two and a half days late. Why ? we can't tell—a possible explanation is that, as the interior mail trains have stopped running, there were delays in getting the letters from all the seaport towns.

Perhaps they delayed her in case she might have been wanted to bring troops from Natal to Cape, or *vice versa.* We expect to find Capetown crowded with refugees, so much so that it is doubtful whether we can get in at an hotel. S—— and I propose trying the new hotel, ' The Gardens,' beyond Government House ; we may of course be sent on at once.

Sunday.—Weather turned quite chilly and cool. There is rather a heavy swell, though no breeze to speak of. We have gone back to serge clothes, having been in khaki all the rest of the time. I wish now that I had brought more clothes of all sorts ; it is much better to have too much than too little, and no questions were asked as to the amount of baggage we had. I am also afraid that I left my notebooks, that I looked out carefully, behind ! However, I think I can replace them somehow. The captain and officers of the ship have invited us all to supper on Wednesday next— very civil of them. I hope the company will bear the expense. I have recently read the ' Convention of '84, London '—have you ? I never thought we had much right in this affair, and now I am convinced we have absolutely none. I am not surprised we would not consent to arbitration ; no impartial person could possibly give it for us. If Lord Derby did not intend to resign all rights of any kind over the Transvaal, he was jolly well taken in by old Kruger and Co. ; but one would have thought that anyone could have seen to having what they wanted put down on paper—a clerk in a bank could have done that.

Wednesday 25th.—We expect to be in to-morrow afternoon. Weather improved, but quite cool, and last night a tremendous thunderstorm, which is most unusual in these latitudes at this time of year. I read ' Sybil,' by Dizzy, out of the ship's library the other day, and on turning over ' Pot-Pourri' I see you mention it as giving an account of the times. I thought it very interesting, but of course the whole tone is in the fashion of the last generation. The swells talked and thought in a way hardly conceivable at present. I had no idea Dizzy had such liberal ideas. Why didn't he do a bit more ? For instance, he is always jeering at foreign Missions ; while home heathen or miserables exist, I should like to see them made a criminal offence. The excitement about news is of course intense. The possibilities are so wide I wonder whether the telegraph will have broken down. I don't think Kruger knows his business if it hasn't. It is just possible we may be in time to catch the in-going mail if she is late, so if this letter ends abruptly and without signature, you will know that I have just sealed it down and chucked it in in time to catch post.

Saturday 28th.—Arrived yesterday afternoon and heard the exciting news of first three battles. No one had any orders for us, but to-day I went to Headquarters and have received orders to go to De Aar, where the advance base is expected to be established. There are a good few troops there, and I think I have been lucky. I shall be working under Major H——, who was at the S.C. with me—a great comfort to have a friend. Du C——, who is Staff officer, has been more than kind. I have got a Horse Artillery reservist as groom. J—— D—— has also been most attentive and helpful. Please thank Uncle Henry for his letters. I had an interview with the bank manager, and he was charming, and said he would do all to help me. I have not seen Lady E—— C——, she is at Rhodes's place. I am going to leave your book for her if possible. I have bought some provisions, which may be useful or not, but still it is a safeguard to have them. I have no idea what my work is going to be ; probably preparing for the advance of the army corps ;

they ought to be here in about a fortnight. S—— is going to East London, not nearly such a good post. Everyone is very jolly and cheerful here, and pleased with our having held our own. I have only seen the General in the distance, not to speak to. I must close now to make my final preparations, and get down to the station. I am dining with Du C——. Much love to all. Rather disappointed at not getting any letter by the 'Mexican.' I suppose they were posted too late. I have sent a wire to let you know my whereabouts; you had better still address Standard Bank.

Your affectionate son,

SYDNEY EARLE.

No. 3

(From Cape. My only letters are to you.)

Sunday, Oct. 29, '99 : De Aar.

DEAR MOTHER,—On arriving here I found your comforting little note, and the extract from the 'Globe' which, as you will see by my other letter, I had already read. I was also given the telegram about Max, which was indeed good news, though I am afraid it will greatly increase your worries. However, by the time he gets out and up to the front the serious business will be over. Alas! I hadn't a moment at Capetown to get the photo done. It was all rush from beginning to end. I gave 'Pot-Pourri,' No. 2 to J—— D—— to give to Lady E—— C——. I received a wire on my way up—it was about twenty-nine hours' journey—to say that my saloon was going to be put in a siding, so that I could go on sleeping peacefully; but I found that that was impossible, so I bundled out at about 1.30 A.M., and got a good night's rest in an empty carriage. This morning, soon after I got up, I met my new chief, Major

H——, Camp Commandant, who met me in a most friendly manner. He told me I was to sleep in the office, which is a large, airy, well-found waiting-room. Nothing could be more comfortable, and it has far exceeded my wildest hopes. Meals—very fair—we get from station contractor, in company with a large number of newspaper correspondents. One of them, P—— L——, I know, a friend of Lord B——'s. The Lochs will tell you all about him; he is always civil to me. Immediately after breakfast I met Col. C—— B——, who commands the regiment here—the Lochs can tell you all about him—and to us came Major Henry Earle, who is the second in command; he was very civil, and it certainly is very refreshing meeting friends at every turn. After I had settled down a bit, I rode round the place with Major H—— on Col. B——'s pony. Everything is in perfect order, and in an advanced condition. We first inspected the spot selected for General W——'s camp and mess, who is expected soon; quite a nice garden with shady trees, and water laid on. A railway official has prepared for him a bath-room and kitchen, as he had the necessary range, &c., to put in a house he was building. We next went on past the little hospital and the medical camp (chiefly Volunteers) to the remount officer, and, practically, bought a couple of horses. I am going to try them with a saddle on to-morrow. They are poor-looking beasts; but I dare say they are willing—one of them is about thirteen years old, but has good strong legs. I think that one will be my first charger. I had the good fortune to pick up a Horse Artillery reservist as groom in Capetown. This will be a great comfort, as he can ride my second horse and carry kit. I have, of course, hardly got to know what my work will be; but it will consist chiefly of making arrangements for the large number of men and horses that will use this point as their advanced base when the army corps lands.

Monday, Oct. 30th.—In the afternoon I walked round
the defences with Major H—— and an intelligent young
sapper called W——; he has done excellent work all
through, and has been particularly clever in making a gun
carriage out of iron rails, chiefly with his own hands.

After Kimberley was invested a Hotchkiss gun and
ammunition came through here, but no carriage. However,
that is all put right by W——; he has tried his gun, and
bar the first shot, which missed the hill aimed at altogether,
and is supposed to have fallen on a Dutchman's farm
about five miles away (he has asked us to discontinue
this practice), he has done very well with it. Except for
the fact that the defences are much over extended (about
three and a half miles for one battalion), they are very
satisfactory, and I think it would require a very large
force to do us any harm. We have small look-out posts
all day long. Scouts ride out just before dawn, and all
night the forts are held by strong picquets. Some of the
forts are very neatly made, and without artillery I don't
think the Boers could ever turn us out. I dined in the
evening with F—— of the A.S.C. at a little mess of Army
Service Corps, R.E., and A. Ordnance. I got up in the
middle of the night to receive Col. G——, the cavalry
A.A.G., who was passing through on his way to Capetown. Several correspondents also arrived—Knight, of
the 'Morning Post,' and Bennet Stanford, of the 'Daily
Mail.' There was also a very suspicious Irish-American
who came out in the 'Mexican.' He is supposed to have
married Kruger's daughter; he refused to drink the
Queen's health; he stated he was going to Aliwal North
to buy horses, and had 1,000 rounds of ammunition and a
Mauser gun. This morning we have wired on to have
him stopped, if he cannot show a pass or give a good
account of himself. A wire was received to-day from
the Postmaster-General asking who gave military people

authority to act as censors to telegrams. No one did, but it is hardly likely we should allow all sorts of rumours and real reports of important moves to fly about all over the country. Heaps of telegrams come to me, but we hardly stop any, unless they contain anything which may disturb the existing quiet as regards local farmers, who are pretty friendly on the whole—*i.e.*, they are willing to sell us anything. This morning I went to try my two quads ; one is so yellow that I propose calling them Mustard and Cress, which ought to please a vegetarian mother. I think I shall take them both. After breakfast I went out to decide upon camping grounds for cavalry brigade and infantry division ; they had been mapped out before, but in a confused manner ; they are, I think, now square. There is of course plenty of room, but we want all the camps near water, which is also abundant, but we have had to wire for more pipes, &c. We have had absolutely no news of any kind lately, but that is good news ; every hour we get stronger and the enemy weaker, and the Dutch colonists less likely to rise. This afternoon I am going for a ride with Major H—— (with whom I am getting on admirably, and who is evidently a very good man) and Col. B—— to consider a portion of the defences, and to visit two Dutch farmers in the neighbourhood.

9.40 P.M., *Monday.*—Mail closes at 10 P.M., so I have just left Station Restaurant, where I have been dining with H—— and B——, to close this. The day has been hot ; the railway engineer tells me that the thermometer showed 94° in the shade at his house, but your thermometer shows maximum 86°, and is now at 75°. The dust was bad, with a strong wind ; my evening bath was heavenly. The dryness of the air is rather trying, and the regimental buglers can hardly play, and one's nails get brittle at the end ; but it makes one feel fit and well.

The difficulty is to avoid drinking. I got down to a very small ration on board ship : half-cup of tea at breakfast, half-glass water and lime-juice at luncheon, and one glass (half-pint) very weak whisky-and-water at dinner, but I think that is too little here ; so I am increasing my evening allowance—it encourages perspiration, which is a relief in the dryness. We had about six drops of rain in the afternoon. I am thoroughly well and happy, and consider myself quite out-of-the-way fortunate, as I think the most interesting part of the show will come my way ; but I am reconciled to the thought that I may be left behind here, in which case, however good my work, I shall get little credit ; it is only the advertisers who can hope for much, but I expect to have the satisfaction of knowing that I have made their way easier and smoother. Much love.

Your affectionate son,
SYDNEY EARLE.

No. 4

2/11/'99 : De Aar.

DEAR MOTHER,—Your dear letters received ; most cheering, as I was a bit depressed at news of C——'s force in Natal. They are making the mistakes I expected, detaching small forces to be mopped up in detail. From the meagre information to hand, it is hard to see what the authorities wanted in Natal, unless it was to score a victory before being superseded, which is a cause that often makes mistakes. We hear the Boers are crossing the Orange River at Norval's Pont and Bethulie, but that has been expected for a long time ; and General B—— has ordered troops at Naauwpoort to concentrate here ; I felt sure that he would not have had all these detached posts to tempt the enemy. Kimberley and Mafeking seem to be holding out well, and no doubt we shall do the same. We have been strengthened to-day by a battery and a half of artillery, among whom is F—— W——, whom I saw at Las Palmas. I am very busy here; got up at 3.30 this morning to receive gunners, and have been at it ever since (5 P.M.). The Naauwpoort garrison will probably be arriving all night. I think it is evident that we have underrated numerical strength of Boers, and that we shall have a harder nut to crack than even I expected. We shall have a pretty strong force when the garrison from Naauwpoort arrives; we want a good many, as we have a lot of valuable stores and a large extent of ground to watch. I have paid for my two ponies, and have got them down in the camp, not far from my office, but I have been too busy to ride. The weather two days ago became suddenly cold. Your thermometer was down to 49° in my room, and it must have been nearly freezing outside at night. The flies, which are a nuisance, manage to revive by day. I have got Davison out of the camp of the Yorkshire Regiment, which he hated on account of the dust, and he lives now near the station, in the guard-room, much more convenient for me. We are expecting General F—— W—— to come up here, or even General B——, but we know nothing definite. H—— gets newspapers sent out, which is a comfort, as I read them. I wish I had made arrangements to have some sent out. I have just heard a rumour that wires to Ladysmith have been cut, so we sha'n't get any news from them for some time. It is difficult to understand how Boers are able to keep the field so long; their organisation must be very good.

L—— has taken a photograph of me on my pony ' Mustard,' which I shall perhaps be able to send home to you. I shall number my letters, and send them off to you as I write, as the mail service might break down. I am going to write to Max now, so that he shall get something on arrival.

Your affectionate son,

SYDNEY EARLE.

Let me hear all about book.

No. 5

Sunday, November 5/'99 : De Aar.

DEAR MOTHER,—Things are still very busy here indeed, but we don't get much news. On Friday the troops at Naauwpoort were brought here with all their stores for the purpose of concentration; it was very well and smoothly carried out; the railway people worked wonders; but, of course, it has created a bad effect in the country, as it is practically an evacuation and change of plan. Major Rimington's Guides form part of this force; they are an irregular force of colonials, and do excellent work as scouts and spies; they have been more or less in touch with the Boers all the time; they seem to think that the Free State Boers don't mean to come down far this way; on the other hand, there are rumours that about 4,000 of these have crossed at Norval's Pont, and are coming this way, but they are still sixty miles off. In an open country like this we ought to get three or four hours' warning of their approach, and so be ready for them. I don't think they can possibly have any artillery, and we have got a good bit. General W—— arrived yesterday to take over the command. I hope this will be a permanent arrangement for some little time, as we have been suffering from changes in command as senior men arrive.

The weather is unpleasantly dry and windy, cracks one's lips, but we do not suffer from heat or cold—about 54–77° in my room. I am sitting now (7 A.M.) facing the

bright sun, on the platform, very comfortable and warm.
On Friday we had a most terrible dust storm, covered
everything about two inches thick and gave one a thick
layer of it between shirt and skin. The same evening we
had a thunderstorm with a little rain. We get dreadful
dust every afternoon. Martial law was proclaimed for the
district on the 3rd, though we only heard about it yesterday. We stuck up a notice about registration of people
and closing of bars. Anyone found selling drink to
soldiers or natives is to be locked up and his goods destroyed publicly. This has given me a lot of work getting
out the passes which I shall have to continue to-day and
this evening. There were rumours last night again that
the rail had been cut between this and Capetown, which has
delayed all the trains. It was, of course, not true, as the
wires were working, but still I expect every moment that
it will be done ; the people are evidently very disloyal.
The proclamation of martial law may make a difference
if they carry it out strictly. Major E—— seems a very
nice fellow, and I should think a good soldier. He
certainly gets a lot of work out of his men, and his praises
are always being sung by the station-master. The railway officials and telegraph are beyond all praise. Mr.
C——, traffic manager ; Mr. C——, station-master ; and
Mr. F——, chief telegraphist, are all A1. A farrier
corporal-major of the 1st Life Guards is standing about
two yards off, which seems home-like ; we have already
had a talk. We have large working parties out every day
under the Engineers. They are making excellent works,
but our whole line is much too extended, though it is difficult
to know how to avoid it, owing to the nature of the ground
and the extent to protect. We have got enormous stores
here, thousands of animals, so that it would be a tempting place for the enemy to make a dash for ; but I fancy
he would get more than he bargained for. We are going

to have some experimental alarms just to see how the disposition works out. The General and his Staff (Majors B—— and C——) are forming a small mess, and H—— and I are going to join them; but we shall probably not get as good food as at the Station Restaurant, which would be quite good but for the flies. We think it probable that the first troops to arrive will come up here, unless they are wanted in Natal, or unless they decide to concentrate them further back, say at Beaufort West. They should begin arriving on Monday or Tuesday at Capetown, and I suppose Max will come about the middle of the week. It would be nice if he came up here. I am now writing at noon in my office and bedroom at the station; there was a momentary lull in the business in which I had time to write a few words. We have just had news—only railway news, which is, as a rule, very inaccurate—that there has been a victory in Natal. Let us hope it is true. It is now about 6.30, and I am still at work and likely to be for some time to come, though I shall get some food about seven o'clock. I am getting very little exercise at present, which is tiresome.

Very much love to you all. I may possibly get off another letter to-morrow.

Your affectionate son,

SYDNEY EARLE.

The Boers are evidently coming into Cape Colony, and the line to Norval's Pont has been ordered to be destroyed by us.

No. 6

Monday : De Aar.

DEAR MOTHER,—Just a line to say all well. It looks to me as if Boers were not coming our way after all, they seem to me to be just wandering through their own friends' country doing nothing. We have arrested one or

two about here ; one, a ganger on the line, was found with
a very dirty Mauser which he said he had borrowed to
shoot rock rabbits.   I have had a letter to-day from
my old Colonel, who is on the Staff at Capetown.
L——, of the ' Times,' tells me he is trying to get a
rider into Mafeking, so I propose trying to get a letter
in for N—— C——.   We had a rehearsal to-day of
what to do in case of attack.   Our guides, Rimington's
Horse, appear to be excellent scouts, and we shall probably
get good warning of any attack.   I have been in De Aar
now just over a week, and I feel as if I had never been out
of the country or town.   I hope I sha'n't be here for ever.
The station and telegraph people still work wonders.

Love to all,
Your affectionate son,
SYDNEY EARLE.

No. 7

9/11/'99 : De Aar.

DEAR MOTHER,—Things going on well ; rumours of
enemy being near are getting less frequent, which causes
disappointment in camp ; people are getting jealous of
those in Natal.   We have heard to-day that the mail is
in, but she is very late : two days we have heard nothing
of arrival of troops.   I have written and wired to Max to let
me know of his arrival.   We think it almost certain now
that the Guards' Brigade will come up here—jolly lucky
for me to be with my friends.   I have had a line from
V—— C—— to-day.   She seems to like your book, and is
going to appropriate it, though I only meant to lend her
my copy.   As you wrote my name in it, however, you must
do it in another.   I have forgotten whether I have
mentioned what luck I have had in my clerk—S. Major
S——, of the B.S.A. Police.   He was on his way to join his

corps when he found he couldn't get on, and was commandeered here; he is invaluable and such a nice man. He was one of the raiders, but does not seem bloodthirsty; he says that most of them were under the impression that the raid had something to do with the natives. We have got a bad dust storm to-day, otherwise the weather has been perfect; in fact, I have nothing to complain of in any way as regards comfort, except an all-consuming thirst which comes from the dust particles in the air. I had the misfortune to crush my fingers this morning. I was tying my pony to the iron railway bridge here, when he jerked his head up and caught my hand between rein and iron railing. I am afraid I shall lose some nails. I have had it bandaged by doctor to keep it clean, and it is now quite comfortable. We sent a portion of our artillery away to-day. I believe they contemplate a reconnaissance towards Belmont to put a shell or two into the Boers. My idea is that all the Boers near us are merely doing a sort of promenade round the country to visit their friends on the side of boundary. We caught three beauties to-day. A father wired to his two sons, students at Stellenbosch, to come home at once, and, from letters and information, evidently for purposes of joining Boers. The farm lies north of here, but at the next station south, the lads got frightened about passing through De Aar and sent for their father, who came a long way round in a cart to fetch them. They are now all in gaol, and it will go pretty hard with them I expect, though they are no worse than all the rest. I don't think the preparations generally in South Africa are in such a forward state as was imagined. I expect the war will last at least a year. If they had at once pushed preparations, war might have been saved, though only to lead to another crisis. Longing for your next, very well and happy. Turner forgot to put in my regimental bridle with my saddlery; that will take the shine off my entry into Pretoria.

Your affectionate son,

SYDNEY EARLE.

No. 8

11/11/'99 : De Aar.

DEAR MOTHER,—I heard yesterday of the death of poor C—— K—— F—— ; so sad for his wife—he was a real good fellow, and his loss is a great one to all. The news was only telegraphed by correspondents to-day. We knew that a reconnaissance was going to take place to disturb enemy and keep him on the move a bit: it appears to have been successful in attaining its objects. The proportion of officers hit appears extraordinary—four officers and two men. The reason is that in a reconnaissance an officer constantly goes with a small party of men to some place to get a good view. The men of course lie down while the officer is observing, and then, when finished the party retires with the officer last. It is most difficult to tell here the difference between an officer and a man in khaki at anything over 400 yards, except from the way he moves about. There is some talk of giving company officers guns to complete the resemblance, but I don't think that will be done. It appears now that all the first arrivals have been diverted to Natal, and I suppose Max and the Guards' Brigade will be among them—a great disappointment to me, as I had hoped to see them here, but possibly a good bit of luck for them, as apparently there is little doing on this side. We are expecting Lord M—— to-night, only there seems to be some doubt as to his movements, as we had a wire to say on his arrival ' Tell him to go back again.' The wire must have been sent just about the time he was starting on a thirty-six hours' journey with many

D D

stops on the way, so it is perfectly inexplicable. There doesn't seem to be quite so much business doing here now for me, though mules and stores and railway material of all sorts are coming, and we are holding camp ready for arrival of troops. I now hear that Guards are probably coming up here : if so we shall have a good time. There is a rumour that another poor fellow has died at Orange River ; his wound was supposed to be only a flesh one, and he was reported going on well ; it may prove quite untrue. The weather has been good—lovely bright days (except for sand storms) and clear cold nights, nearly freezing, I should think. The flies bother me a great deal. I tried to remember fly trap out of 'Pot-Pourri,' bread and jam on the top of a jar and soapy water underneath, but it didn't do much good.

Your affectionate son,
SYDNEY EARLE.

Love to all. I hope to see Douglas to-night.

No. 9

Sunday, 12/11/'99 : De Aar.

DEAR MOTHER,—Things are mighty quiet to-day, very little doing. Lord M—— arrived at about 1.30 A.M. this morning. I waited up for him, but they were all asleep when the train came in. I saw a clerk, however, and he told me that Douglas had been kept behind, and was to join some irregular force. I am afraid he will be disappointed, but it may turn up trumps for him. Three pieces of baggage of his are here. I have wired to him for instructions, but can get no answer. It was a disappointment not to see him, but everything was forgotten in the joy of getting your letters—you did not say what Max was coming out as, though you say something about police, so I gather that you mean assistant

provost marshal to a division. I call the notices in the
paper most complimentary—much more intelligent praise
than last time ; the Leeds paper was not among them,
though. Lord M—— did not stop long; he went to
Church Parade, at which we had ' Child she-bear ' hymn,
and then went round the camp and on to Orange River
at about 12.30. He seemed in excellent spirits and
form; his two A.D.C.s I know. Most of troops now
here are moving to-morrow to Orange River, and others
go on Tuesday; we shall be left with three guns,
which have been mounted on the neighbouring kopjes,
half company mounted infantry to act as scouts,
and probably two battalions of infantry when they
arrive. We are expecting almost at once a portion of the
1st Battalion Highland Light Infantry, who were with
me at manœuvres this year, a very nice lot of fellows.
I shall be glad to meet them again. As regards my own
future, I really can't make out what I shall do; it would
really be very bad luck to stick here all the time, but I
don't see how I can hope to get on far for some time.
Please tell Lionel that I am most grateful for his letter
and for forwarding my things, which have not yet reached,
but I have no doubt that they will do so soon. He
has acted quite rightly with regard to signing of cheques.
I had a letter from Messrs. H—— explaining the matter,
but I sha'n't answer them—it is not worth while after such
a lapse of time. Major Rimington's Guides are an excellent
body of men, half-soldiers, half-detectives ; they ought to
get great praise for the work they are doing: hardly a
mouse stirs in the country without their knowing all about
it. The rather vague rumours of large bodies of Boers
moving about in various directions towards Colesberg and
Hanover Road, Philipstown and Petrusville, have almost
entirely stopped, but we don't know here where the
enemy have gone to, as we don't see reports except in our

neighbourhood.—I don't like this forward move on Kimberley; it again smacks of politics and playing up to the B.P. Detached forces of this kind are unsound in principle, and as far as the ultimate result of the campaign is concerned, it would be much better to let Kimberley fall than to risk the entanglement of a whole division. M—— is the man to make it a success, if anyone can. It makes matters much more interesting knowing so much of what is going on as I do. Every wire is brought to the office, and I act, usually, as censor, and very strict we are ; we let none of the usual reporters' bosh through. They meet a man they have never seen before, who tells them of something he thinks has happened 100 miles off, and they wire it off as gospel, or rather they would do so if it wasn't marked ' (Stop. S. Earle),' which means that the message does not go on except by post to the P.-M.-General. Am going to bed for a bit. Have only had about three hours' sleep the last two nights. Much love.

Your affectionate son,
SYDNEY EARLE.

No. 10

13/11/'99 : De Aar.

DEAR MOTHER,—Just a line before mail goes. Nothing to record except that General W—— (who nearly turned Gladstone out for Mid-Lothian) arrived here. Also the 11th Co. R.E., with whom I lived at manœuvres this year. Quite nice to see old friends. We are expecting 2nd Bn. Coldstream Guards to arrive to-morrow morning. Davison, who has nearly shaved all his hair off except a sort of thatch on the top, quite beamed when I told him. We are worried off our heads about returns, figures, &c. ; they are never correct, but we get them somewhere near. The remount officer finds he has got thirty more mules than he bought, though he has lost a lot through straying. We are to have half the Highland Brigade here, including, I think, my friends of the Highland Lt. Inf. who were at manœuvres. We had good news from Mafeking to-day; they seem to be holding their own well. We tried our Hotchkiss gun to-day on its new mounting; it shot off well, but it only fires an armour-piercing shell for torpedo boats. It would frighten more than hurt. Mail just off. Good-bye; write often. Correspondents here have been 'Times,' 'Morning Post,' 'Daily Mail,' 'Sun.' Collect any notes about De Aar.

Your affectionate son,

SYDNEY EARLE.

Douglas Loch wired to have his luggage sent.

(On a little extra sheet was written.)

In some of our seized correspondence we sometimes come across some amusing yarns. One story told in a letter as a fact is that the British soldiers have had to be tied together, two and two, all the way up from Capetown. They are so terrified of meeting the Boers that without this precaution they would all escape and run away on the way up !—the absolute unconcern and indifference on the part of the real Mr. T. Atkins is most marked. I don't think he either knows or cares whether there is a Boer in the country. He doesn't like being without a canteen; on the other hand, I think he prefers doing fatigue and out-posts to drill and barrack life.

No. 11

15/11/'99 : De Aar.

DEAR MOTHER,—Things are fairly humming now. Troops are being pushed up as quickly as possible to Orange River with a view to an advance on Kimberley— rather a wild scheme, in my opinion. I hope they are not under-rating their enemy; they can't have a very good opinion of him, as, of course, there has been nothing except promenades and the sieges of Kimberley and Mafeking—open towns defended almost entirely by irregulars, and yet they have been unable to take them, or at any rate they have stood out for a long time. But to advance against the enemy with such an exposed line of communication as we have got would, in ordinary circumstances, be the height of folly. There is a large force of the enemy only about fifty miles from here, and he could get in and bag us all with stores if he was determined; but, instead, he remains at Colesberg and sends in the traders of that town to Naauwpoort to ask the railway authorities to hand over any foodstuffs that there might be there; the latter wired for instructions, and were told—'Certainly not; send foodstuff south.' A most amusing way of conducting a raid. Our garrison has been considerably weakened now, but it will become stronger again later, and we shall have troops in trains sleeping here every night, though they would hardly be much use, as they would not be prepared for a sudden attack. Of course we have got scouts on the look-out at a considerable distance out. The 2nd Battalion Coldstream Guards passed through yesterday, and a portion of them slept in their train here; they were looking very fit and well, and Davison and I thoroughly enjoyed seeing our pals, and a lot of reservists greeted me in one way or another. Unfortunately, their first train took us by surprise. Capetown had not notified their coming, so I had nothing ready for them for breakfast; however, they soon got some coffee, and they were all right. The second half I did 'slap-up,' tea and biscuits, provided by Colonel B——, oranges, bread and a pint of beer on sale; dinner ready for the officers. The Scots Guards arrived to-night, and an ammunition column. The New South Wales Lancers are here. We fairly worried them ; we had no warning of their arrival, and they had orders to get out here from Capetown. As soon as they 'were out,' and had got their camp in order, a wire from Orange River said, ' Send them on ' ; back they bundled, horses, baggage, and everything, into the train. It was then discovered that they had no transport, and were deficient in many equipment things. Of course they were not expected to form part of the force, so we wired for instructions, and now they are finally out here, and very glad we are to have them ; from what I saw of them at manœuvres I should say they will make admirable scouts. They took their worrying too, well, considering how annoying that sort of thing is. ' D—— that Staff ! ' Things are now working smoothly at Capetown.

The line of communication station staff is not in working order yet ; I have no doubt it will improve. I hope to send you a poem culled from a local paper on Martial Law at De Aar to-morrow, if I can get a copy. I see little chance of my moving from here for some time ; it will be rather dull to spend many months at De Aar. Love to all.

Your affectionate son,

SYDNEY EARLE.

P.S.—I haven't written to a soul except you ; you must make my apologies for me. I am very busy. I started 4.15 this morning and did not get a wash till 1 P.M., and have been at it till now, 7 P.M., and shall start again at 8.30.

No. 12

16/11/'99 : De Aar.

DEAR MOTHER,—I got a wire early this morning to say I was to go at once to Orange River to join mounted infantry of Methuen's force, in what capacity I

do not know. I am glad to get up to the front, but I am
sorry that I am going to take up a job of which I know
nothing, as it is a bad time to learn. However, one can't
tell what luck one may have, but I don't think I am
suited for the work. I am too heavy physically. The
troops I shall be with will be probably the Northumberland Fusiliers, but I don't know yet. I am leaving
almost the whole of my kit here in charge of a man
named McI——, a storekeeper; I dare say he will
look after it all right. It is possible that I shall be
separated from Davison; that will be an awful blow,
both for him and for me, but it can't be helped. I
expect my letters will get few and far between now, as
the opportunities for writing will almost cease.

Your affectionate son,
SYDNEY EARLE.

No. 13

18/11/'99 : Orange River.

' DEAR MOTHER,—I have arrived here and joined the
Mounted Infantry of the Loyal N. Lancashire Regiment
(the 47th Foot), and I now command a section in it. The
captain of the company is, I believe, junior to me, but I
have to do subaltern's work—*i.e.*, take my turn of orderly
officer; it is the very last job in the world that I expected.
However, the job seems interesting, and I will do my
best to do the thing well; the reason I and some other
officers have been taken is that there were so many
casualties when poor C—— K—— F—— was killed.

They say we are to start on Monday on our march to
Kimberley; but I have my doubts as to whether we shall
go there at all, as it has been so advertised that we are
going there, it has been officially announced, and the war
correspondents allowed to wire it, that it looks to me as if
our real 'point' was to be somewhere else, say Norval's

Pont or Bloemfontein. We are to take nothing with us but blankets and a waterproof sheet and one or two things on our horses or in our pockets. I am going to take my groom Gasten on with me, but Davison I must leave behind. We shall live entirely on the ration, tinned meat and biscuit, and I hope some lime-juice. We shall probably have great difficulties about water. I am sitting writing now in W—— B——'s tent, which I am sharing. It is very hot; your little thermometer, which I shall have to leave behind, now reads 88°. I shall, I am afraid, lose almost all my kit, unless Davison can manage to get it on somewhere. In consequence of my sudden move I have not managed to get hold of my mail letters for this last week—a dreadful deprivation. I have not heard a word of Max; they have evidently got confused over him and me already. I was put in orders to-day as belonging to the Grenadiers; and Colonel C—— told me that he was under the impression that Max was up here, and he wanted to 'commandeer' him to make him serve in his battalion in the place of poor A—— T——. Two of Max's letters have just arrived here, re-directed to me; rather distressing. I am not even taking any writing things with me now, except a notebook and stylo. Much love. Your affectionate son,

SYDNEY EARLE.

No. 14

19/11/'99 : Orange River.

DEAR MOTHER,—I have got the mail that I thought had gone wrong. . . .

I hear Max is in Capetown now, not Natal, as I thought, but I have had no wire from him, though he might have got my address from the bank, even if my wires missed him. They are going to try to get him up here to fill T——'s place. I have been trying to get moved from the mounted infantry to one of the battalions of the Brigade, but without success. We are moving across the river to-morrow, to encamp on the other side, near where the Guards are. What we are going to do then we don't know, though it has been carefully given out that we are going to Kimberley so carefully and so openly that I have my doubts. I should not be surprised myself if we made a sudden turn off to the right and tried to get down to Norval's Pont. This would relieve matters, I think, at Kimberley, and we would effect a junction with troops coming up *viâ* Naauwpoort. They are even loading trains for Kimberley, but that may be another blind, as they could send them *viâ* De Aar and Norval's Pont, where the railway is re-opened ; anyhow, we expect exciting times the next few days.

I shall be glad to get away from the discomforts of this camp to the discomforts of the open field. We are taking no baggage at all, and shall probably be unable to wash or even take off our clothes for a fortnight or more. I shall have no stationery with me except a field note-book, so you mustn't expect letters.

Your affectionate son,
SYDNEY EARLE.

No. 15

Tuesday 22nd: Near Belmont, on sort
of outpost duty, watching enemy.

DEAR MOTHER,—We marched the day before yesterday across the Orange River, and bivouacked on the north side, not far from the Guards' camp, where I got a wash in Col. S——'s tent, a mug of beer and a little soup and some bread and tinned meat, so I didn't do badly. I had a wire from Max to-day to say he was coming up to join the 3rd Batt. We were ordered to start off at 12.30 A.M., but that was afterwards altered to 2.30 A.M. My section consists of about twenty men, all old soldiers, and dreadful ruffians in private life I should think, but not bad fellows to deal with, though totally without discipline. The non-commissioned officers are also worse than useless ; they do nothing but sleep. We went as portion of the baggage escort, and we marched without incident to Witteput. I saw a good many friends on the road and heard —— use some dreadful language on having his eye nearly put out by a native mule driver. The march was short and easy, but the transport seemed quite up to its work. There were a farm and stores at Witteput, and we managed to get a tin of herrings and some bread, so we did well. We took our horses out to graze, and just as we were bringing them in there was an alarm that some of the enemy were advancing. We saddled up and dashed off, then heard that Lord M—— was missing. However, he was discovered quietly sitting at the foot of a kopje watching events. The alarm was false ; the enemy had fired five shells at the 9th Lancers, and killed a horse, I believe. In the evening I borrowed a bucket from the New South Wales Lancers and had a good wash. My men know nothing about looking after themselves, they are like idle children. I am teaching them a few useful habits. After a good night I was roused at 2.30 A.M. and at 3.30, having had a little coffee and bread, the whole of the M.I. trotted off in the moonlight towards Belmont, the place where they had the skirmish the other day. On getting to Belmont Farm we dismounted, and lined the kopjes, and except for small changes we are here still, though it must be about 9 o'clock. My watch stopped yesterday—a great misfortune. I suppose the dust has got into the works, or I have banged it about somewhere. Soon after we arrived here we saw numerous small parties of the enemy, I should think about 300 in all, coming out from behind one hill and going behind another, without much aim apparently, except to graze their ponies ; the men are almost all out of sight, but I can see their ponies with my glasses about three miles away. We expect to be reinforced to-day at 6 P.M., so we have a good long wait before us, and mighty little prospect of any food, though we have a day's ration on our horses, which are about 400 yards away. We heard a few shots fired on the outpost line—probably someone got the jumps. I expected to be excited the first time I saw the enemy, but it seems as natural as anything now. The 2nd Staff (Methuen's) came out with us this morning. I waved to Douglas from my humble position. My captain seems nice, but he is not the man to keep these rough diamonds in order. He is too serious ; they only get sulky. My brother section commander is called L—— of the Bombay Army. It seems a queer turn of the wheel that I am commanding 20 men of the N. Lancashire Regiment, while Max will command 120 Grenadiers with two officers. In addition to having to live my time away from friends, my subordinate position will quite prevent my getting any professional advancement; it is *bad* luck. We have got E—— as Staff Officer to the M.I., and though he has been very civil to me, he probably hates the Guards as much as his father does, which accounts for my being put with this company. I have just jumped up in a great hurry on seeing a nasty looking snake making for me. He was apparently as much frightened as I was, as he has nipped under a stone. I have no stick to kill him with.

About 4.30 *p.m.*—My picquet was withdrawn about 11 A.M. to the farmhouse, which belongs to an Englishman named Thomas, and where we joined the reserve. I got some milk, and later a little bread and jam from the farmer. All of a sudden the Boers opened fire on us with

a gun, not a very big one—a six-pounder I should think—
and from a long way off they made quite good practice,
and during the afternoon they have put about fifteen
shells into the small amphitheatre surrounding the farm
pond and spring.   Taking horses to water is the signal to
shoot.   I expected to feel anxious, but instead, I only felt
curiosity and excitement.   Of course we all ducked under
big stone walls on seeing the smoke.   Then came a pause
before hearing the report, then the whistle of the shell,
and then the explosion of the shell.   Some did not explode
at all ; one went thump into the rocky ground about
twenty yards from me without exploding.   Another
missed the farmhouse just over, and exploded in the hillside beyond, about two feet from a goat who was taking a
snooze, but without hurting him.   The ostriches seemed
alarmed—it is a great game here for the soldiers to pursue
the birds and snatch out their feathers, which they then
put inside their helmets to keep their heads cool, and
as a present for their girls on return.   We are expecting
the advanced guard of the force this evening at 6 P.M.
We shall probably have a battle to-morrow, which we
ought to win easily, if we don't throw masses of infantry
against the Boers before thoroughly shaking them with
artillery.   We are travelling as light as we possibly can—
nothing but blankets and a waterproof in cart, great-coat,
&c., on horse.   We draw rations with the men ; good bread
or biscuit ; the tinned meat might be better, some is good
and palatable, other tough, stringy, and nasty to look
at—a question of economy ; but I am sure soldiers would
sooner be spared dinners and extravagances on their
return if they might have the best possible on service.
One should remember the initial value of the article is so
little compared with the value of the article delivered
here.

Am giving this to groom to post if possible, 2.30 A.M.

Thursday.—We expect battle to-day.   We protect advance of Guards' Brigade.

Your affectionate son,
SYDNEY EARLE.

No. 16

26/11/'99 : Near Graspan.

DEAR MOTHER,—I hope you got my fifteenth letter, dated from Belmont; it was entrusted to a wounded sergeant of the Coldstreams, who promised to post it. We have had an anxious three days. Thursday we fought the battle of Belmont, and though we turned the Boers out of an almost impregnable position, we lost heavily in so doing. My company was operating on the right flank of the Guards and artillery, but we were unable to do much; we began by drawing the fire of the enemy to enable the artillery to have a target. We remained under fire until the position was taken, and then we dashed round the south side of the great kopje and tried to catch the enemy trekking away; but it was impossible to distinguish friends from enemy; in fact, the Grenadiers fired at a portion of one company. We got a few waggons and a small laager of the enemy, we then tried to get still further round, but they got away so cleverly and so quickly, covered by the fire of a few men, that we never did any good. We then went away still further from the rest of the force (which in my opinion is wrong, as in these great plains it is most important not to lose all cohesion), and coming on to the top of a rise, a gun opened fire on us at a very short range, supported by rifle fire; we stopped there for some time, but did no good. If we had been able to get up some guns, it might have been all right, but still the detaching of small forces is what so often leads to disaster. We then returned slowly and got some water and Boer bread (an excellent sort of rusk) from this laager.   We destroyed about 15,000 rounds of their ammunition ; they have plenty more, though.   I saw Max, who, I hear, did splendidly on the afternoon, and heard about losses of Brigade.   I saw C—— and B—— with their wounds dressed walking about quite cheerfully. Max gave me your letter with father's ideas about war, in which I quite concur ; it seems the most ridiculous thing that we should be going through every sort of horror in order to shoot at total strangers who have never done us much harm.   They are splendid fighters too, from what I can see, and they make full use of the wonderful country in which every two miles there are perfect positions which fulfil all the requirements that a most ingenious engineer can contrive—a first-class position in Europe may fulfil five or six conditions, leaving four or five others unfulfilled ; here every one is fulfilled, absolutely ready made, except an abundant supply of good water.

The day after the battle our company started long before daybreak to support the armoured train, which was going to repair the line.   We sent out scouts in front and flanks.   I took charge of a small rearguard and the left flank scouts ; all went well till we got nearly here.   I was then sent off with a few men to a small farm about three miles to the right, where we thought we saw Boers and waggons, but on arriving we found only black farmers with oxen.   I purchased a drink of milk all round from them, and was just going to water my horse at a pond when I saw in the far distance a big gun fire at the armoured train, which proceeded to retire rapidly.   I at once set off to support it, and on arrival found that an officer and two of our scouts had been fired on at close range, and were missing.

We have since found that the officer and one man

were killed; the other is lying here dangerously wounded, shot through the back of the head. It was no good attempting to repair the line farther, so all retired to Belmont, where we stopped the night alone, all the rest of the army marching out north. Yesterday we did not start till 5 A.M., as our horses and men were very tired. I borrowed a basin from the farmer, whose house had been turned into a hospital, took it into his vegetable garden, which had of course been completely stripped, and had a fine wash, which was only spoilt by the near neighbourhood of a decaying horse—this was the evening of the armoured-train day. Next day we started off in pursuit of our friends, arrived just in time for the opening of what I suppose will be called the engagement of Graspan. We had no orders, but we had an idea that the 9th Lancers were round on the right flank (east), so we rode round, being shelled at two or three places without result, though the aim was very good. We found the 9th snug under a kopje. We could not tell how the fight was going, but our artillery fire, which had been going on for a couple of hours, stopped suddenly, and we saw Boers on three sides of us. Colonel B—— G—— was just going to retire when we perceived that the Boers themselves were retiring, so we went on further round, very well concealed, till we got to a place where we could see them leaving the kopje with their waggons. There is no doubt that the Lancers ought to have charged home here, supported by us, instead of which we dismounted and fired at them. I am afraid we frightened them more than we hurt them. Anyhow, there were hundreds of them going away, trekking across the plain; a perfectly orderly rabble, no hurry or confusion; it is always so, for every man works for himself in this open country. We got on our horses again and pursued, but got no closer; we gave them a parting shot or two, but they were still round us on two sides. The Lancers then retired, and we were told to hold a kopje to cover their retreat; we had a ticklish time, but got out of it all right, one man shot through the cheek and neck. They shelled us going back, but only got a horse. I don't know yet whether we are doing much good; we are camping on enemy's position, but that is only a barren honour, specially as the water is infamous, same for animals and men, and all the colour of coffee. I hope it won't give disease. I am trying to drink it only boiled, but the dryness makes one dreadfully thirsty. I am suffering, like most of us, from bad cracked lips, and have no grease to soothe them. To-day we have a rest day, but I expect we shall go out this afternoon. Much love.

Your affectionate son,
SYDNEY EARLE.

I wish it was all over.

INDEX

A——, Miss, 377
Abbott, Archbishop, 101
Abo, 360–361
Acacia, yellow, 160, 169
Ackland, Dr. W. H., 75
Acorns, eatable, 164
Acton, Lord, 351–353
Adenophora Potanini, 218
Adulteration of food, 32, 100
Advertisements, teaching of, 52
Agaves (aloes), 163–164, 170
Agneau-de-lait, 94, 96
Agnew, Lady, 107
Aix, 238
Albumen, daily allowance of, 85–86, 111; digestion of, 105 (*see also under* 'Proteid')
Alcohol, hereditary effect of excessive drinking of, 51; Plasmon as a substitute for, 55; Maeterlinck on, 73–74; effects of, 256
Alison, Mr. George, 220
Aliwal North, 392
Alkaline salts in potatoes, 71
— treatment, 44
Allen, Grant, 253–255
Allinson, Dr., 15; 'Medical Essays,' 30; on diet for uric-acid diseases, 45
Allman & Son, 34
Allopathy, 30
Almond cream, preparation of, 69–70
Almonds, prussic acid in, 62, 98-9; in diet, 69, 71, 130;

varieties of, 98; oil of, 99; amount of proteid in, 111
Aloes, 163–164, 170
Alternanthera, 208
Amalfi, 198, 333
Amaranthus caudatus, 323
American food reformers, 112
Amesbury, 383
Amomum, 272–273
Ampelopsis Veitchii, 271
Amygdalin in almonds, 99
Amygdalus Davidiana, 152
Anæmia, diet for, 45, 07
Analysis of food, 104; on the Continent, 100
Andover, 121
Andromedas, 233
Anemones, 165, 172, 354
Angelo, Michael, 173
'Animal Life,' 272
Anthropoid apes, teeth of, 28; diet of, 29
Antrobus family, 383
Appendicitis, 126
Appetite, importance of, 67
Apple and nut diet, 98
— dumplings, 128; pudding, baked, 129; meringue, 287
Apple-trees, 212; prunings of, for goats, 103; grown from pips, 260
April, notes for, 175–202
Aquarium, 98; at Naples, 185
Arabis alpina flore-pleno, 220
Army Service Corps in South Africa, 381

Arnold, Dr., 75
Art, pre-Raphaelite school of, 58 ; modern, 205
Artichoke, leaves of, for goats, 103, 227 ; Globe, 226–227, 260 ; how to serve, 285
Asparagus, 4–5, 17, 260 ; xanthin in, 48
Asphodels, 197–198
Asters, 265
' Astolat Press,' 100–102
Atlee, Mr., 295
Atwater, Professor, 27
Auckland, 250
August, notes for, 266–304
' Aurora Leigh ' quoted, 202
' Australia ' battleship, 377
Austria, and Russia, 371–373
Ayr, 382
Azaleas, 216, 231–233, 348

B——, Captain and Mrs., 377–378
B——, Colonel C., 391, 393, 406
B——, General, 394–395
B——, Lord, 391
B——, Major, 398
B——, W., 409
Baby food, 297
Backhouse, Messrs., 232
Bagnoles-de-l'Orne, 63
Baiæ, 179, 184
Ball, Mr., 250
Ball's ' Story of the Heavens,' 383
' Ballads of the Boer War,' 341
Bamboos, 170, 233, 235
Bananas, 115
Bankruptcy Court, story from the, 339
Barberton, 340
Barcelona nuts, 130
Barley and raisin pudding, 129
— bread, 126
— porridge, 112
— water, 70, 113, 289 ; receipt for, 128
Barr & Sons, 251, 323
Barr, Mr. Peter, 210, 249–251
Batourin, town 370

Battle Creek Sanatorium, 112
Bax & Sons, 126, 295
Bay-trees, 162, 236–237
Beaconsfield, Lord, ' Sybil,' 388
Beale, Dr. Lionel, 11
Beans, 4, 17, 57, 261 ; uric acid in, 18, 48 ; ancients abstained from use of, 49 ; used by the Greeks in balloting, 49 ; as food for the poor, 109 ; from Madeira, 144 ; preserved, 301
Beaufort West, 398
Beaumont, Sir William, 208
Bee-keeping, 124
Beech-trees, 211, 236, 383 ; at Strathfieldsaye, 335
Beer, 32, 50, 90, 121
Beer-money for servants, 145
Beetroot salad, 288
Bell's ' Life and Light ' series, 26
Bellet, 161
Belmont, 400, 410–411, 416 ; battle of, 414–415
Benham, Dr., 75
Berkheim, Count, 349
Berlin Conference on Tuberculosis, 88, 158
Bertschinger, Dr., 100
Besant, Mrs. Annie, 354
Bethulie Bridge, 385, 394
' Bhagavad Gita,' 354
' Bibelots ' series, 101
Bible on flesh foods, 48
Bignonias, 231
Bills, payment of, 138–139
Biscuits, use of, in travelling, 66 ; making, 252, 293, 300
Björsby, 361
Black, Dr. George, 75
Blacker, Dr., 75
Blackthorn, 153
Blatchford, Robert, ' Merrie England,' 249
Blight, 247
Bloemfontein, 385–386, 409
' Bloody Noses,' 324
Blunt, Mr. Wilfred, ' In Vinculis,' 347
Boer War, 89, 194, 227–228 ; incident in the, 340 ; ' Ballads

of the,' 341 ; Captain Sydney Earle's letters on the, 374-417
Bohn's Classical Library, 183
Boix, Dr., 27
Bomarsund, 360
Books, see under Cookery, Diet, Education, Gardening, Health, Hygiene, Poems, &c.
Booth, Charles, 139
Boracic acid in mucus colitis, 68
Boron in bread, 65
Borrow, George, quoted, 147
Botanical names and the ignorant, 170
— gardens, 229-230
Botticelli, painter, 174
Bouchard, Dr., 27
Boulouris, 161
Bourget's 'Homme d'Affaire,' 165
Bouvardias, 273
'Braemar Castle,' 377-379
Brams Hill, visit to, 274-275, 306, 309
Bran tea, 113, 294
Brassica oleracea, 252
Brazil nut soup, 130
Bread, Graham brown, 4 ; gypsum in, 65 ; receipt for, 296 ; amount of proteid in, 111 ; unfermented, 126 ; flour for making, 294-295 ; on bought, 295, 300 ; Boer, 415
— and fruit diet, 120
Breakfast, hour for, 4 ; omission of, 15-16, 27
Briar root for pipes, 161-162
Brighton, 279
Brindisi, 176
Brink's ' French Revolution,' 340
British Dental Association, 297
British Museum, 358
Broadbent, Mr. Albert, 26, 297
Broadbent, Sir William, on consumption, 96
Broadlands Home, 77
Broadmoor, 72
Broccoli, 252, 262
Brody, town, 371
Bronkerspruit, 382
Brooke, Mr. Stopford, 213-214
Brouardel, M., 158

Brown, T. E., ' Letters ' quoted, 3
Browning, Robert, 323-324
Browning, Mrs., 202
Bruccio, curd made from goats' milk, 96
Bruce, Hon. Mrs. F. J., 26
Bruges, nursery gardens at, 230
Brussels, 230
Bryce, Robert, ' Suffolk Breviary,' 336
Bulawayo, 382
Bull, sex vigour in the, 55
Bullant, architect, 167-168
Burgundy wine, 67
' Buried Temple, The,' quoted, 72-74
Burne-Jones, painter, 54
Bushey art students, 124
Butter, 88, 90 ; as an aid to digestion, 71 ; nut, 112 ; cocoa-nut, 117

C——, Colonel, 409
C——, General, 394
C——, Major, 398
C——, Mr., traffic superintendent, 397
C——, Mr., stationmaster, 397
C——, E., ' Vale,' poem, 355
C——, Lady E., 389-390, 399
C——, N., 399
C—— V——, Prince, 378
Cabbages, 252, 260 ; how to serve, 286 ; on cooking, 287
Cain, sculptor, 166
Cakes, how to keep fresh, 285 ; foreign, 293
Calendulas, 213
Calomel in eczema, 46
Cambridge University, 351-352
Camellias, 235
Camembert cheese, 90
Campanulas, 217-218, 320
Campions, 232
Canary Islands, 162
Cancer, cause of, 47
Cannas, 267
Cannell, Mr., 262-263, 267
Cannes, 162 ; visit to, 169-172
Cantani, Dr., 78, 80, 84, 86-87

Cape montbretias, 235 ; bulbs, 310 ; pelargoniums, 320-321 ; heaths, 322 ; plants, 332
Cape Colony, 398 ; Dutch rising in, 382, 385
Capetown, 249, 378–379, 388 *seq.*
Capetown Cold Storage Co., 250
Cape Verde Islands, 380–381
Caprera, Island of, 200
Capri, 196–197
Capucine convent, 198–200
Caracciolo, Admiral, 186
Carbolic acid in mucus colitis, 68, 188–189
Carisbrooke, 303
Carlsbad waters, 44
Carlstadt, 359
Carmichael, Dr. D. George, 75
Carnations, 230, 236, 263–264
Carnivora, 28 ; sex vigour in the, 55
Carnivorous habit in man, 25, 29, 72
Carrots, 261
Carter, Mr. Albert, 298
Carvalho, Miolan, 166
Casein, 78 ; in separated milk, 90
Cassell's National Library, 337
Cassia corymbosa, 263
Cataloguing a garden, mode of, 253, 262
Catalpas, 248
Cats' tongues (biscuits), 293
Cauliflowers, 252, 261
Ceanothus, 212
Cedars, 236, 327
Celeriac, 261
Celery, 261
Cellini, Life of Benvenuto, 327
Cenotaphs, 193
Centaureas, 232
Ceylon and the tea industry, 32, 51
' Chambers's Encyclopædia ' quoted, 99, 164
Champagne in dyspepsia, 61, 70
Charwoman and Cabinet Minister, 109
Chatsworth, 352
Chatterji, Mohini, 354
Chaucer quoted, 345

Cheese, 5, 78, 84–85, 88 ; melted, 7 ; curd, 8, 129 ; as a substitute for meat, 67, 69 ; in diet, 86–87 ; Camembert, 90 ; goats', 96 ; amount of proteid in, 111 ; for dinner in various forms, 113 ; and macaroni patties, 127 ; Gruyère, 127 ; cream, 252 ; salad, 285 ; boiled, 289
Cherry and raspberry jam, 285
Chestnuts, leaves for goats, 103 ; purchasing of, 115–116 ; how to stew, 128 ; as a sweet, 289
Chicken farming, 124, 150
Chicory, 261
Chilblains, cure for, 304
Children, management of, 30 ; diet for, 50, 52
Chinese method of preparing tea, 51 ; payment of doctors in health, 62 ; flowering shrub, 152 ; Guelder roses, 265
Chocolate in glasses, cold, 294
— toast, 285
Chou braisé, 286
— à la crème, 286
Choucroute, preserved, 301
Christmas roses, 153, 221
Chrysanthemums, how to grow, 349–350 ; poem on, 350–351
Cicero's essays, ' Old Age ' and ' Friendship,' 184
Cimicifugas, 264–265
Cinematographs of operations, 340
Cineraria maritima, 271
Claude, artist, 206
Clematis, 222, 318, 330–331 ; 'as a Garden Flower,' 251–252 ; at Harwich, 330 ; for pergolas, 333–334
Clergy and the diet question, 104
Cleveland, U.S.A., 284
Clodd, Mr., 255
Cobham, 294, 376
Cocoa, 17 ; xanthin in, 48 ; substitute for, 113
Cocoanut butter, 117
Cod-liver oil, 90
Coffee, 4, 17, 42, 50 ; uric acid in, 18, 48 ; condemnation of,

31, 141; goat's milk with, 93; substitute for, 113

Colds, open-air treatment of, 159

Coldstream Guards, 404, 406, 414

Coleridge's ' Table Talk ' quoted, 180–181

Colesberg, 403, 406

Colesberg-Burghersdorf railway, 382, 385

Colitis, 68, 71, 126

Comfrey for goats, 103

Congo plants, 230

Connington Hall, 307

Constantinople, 371

Constipation the proof of indigestion, 52; in vegetarian diet, 81–83, 87

Consumption, high feeding in, 41; treatment of in sanatoria, 83, 96, 178; Berlin Conference on, 88; Sir William Broadbent on, 96; suggestion for the King's new hospital for, 96; exemption of goats and asses from, 96; letter on the infectiousness of 156–158; National Association for the Prevention of, 157–158

Convention of London, 1884, 388

Cookery Books, 283; (see also Health books); Kellogg, Dr., ' Science in the Kitchen,' 5, 32; Fernie, Dr., ' Kitchen Physic,' 75; ' Dainty Dishes,' 287; Hilda's ' Diary of a Cape Housekeeper,' 284;—' Where Is It of Receipts,' 284; instructions for elementary schools on cookery, 105; Liverpool School of Cookery book, 284; Ross, Janet, ' Leaves from our Tuscan Kitchen,' 284

Cooking reduced to once a week, 116

Cooks, 140–141, 144

Cope, Sir Walter, 241

Copenhagen, 357

Copper stewpans lined with silver, 302

Cork-trees, 164–165

' Cornhill Magazine,' articles from the, 107 seq., 133–151, 323, 356–373

Corns, cure for, 303

Coronation decorations, 240–241

Correvon, M., quoted, 153

Corrosive sublimate as an antidote for microbes, 188

Corsica, 162; goats in, 96

Cottar farmers, art students as, 124–125

Cotterill, Mr. H. B., letter on larkspurs, 344–345

Cotton, Sir Robert, 307

Cotton-seed oil for frying, 116

' Country Life,' 103; ' Country Life Library,' 245, 334

Country life and slum-dwellers, 123; desire for, 124; advantages of, 147; expenses of, 148 seq.

Cowan, George, 340

Cows, artichoke leaves as food for, 103

Cowslips, 155, 206

Crabbe, poet, 336

Cranberries, 144

Crawford, Marlon, ' The Rulers of the South,' 49

Cream, no proteid in, 90

Cream cheeses, 252

Cremation, 200–201; Garibaldi on, 200

Crespi, Dr., 75

Crimea, 369, 371

Crimean war, 194, 228, 357

Crocuses, 155

Cronberg, 311

Cronenburg, Castle of, 358

Crops, rotation of, 260–262

Cruelty to animals in Italy, 177

Cucumbers, preserved, 301

Cud-chewers, teeth of, 29

Curd cheese, 8, 129

Currie, Sir P., 208

Curry, to make a dry, 290

Curtis, Miss Adela, 26

Curtis, A. C. and the ' Astolat Press,' 100–102

Curtis & Davison, booksellers, 32

Cyanthus lobatus, 310
Cypress-trees, 172, 239–240, 248
Cypripediums, 348

D——, J., 389–390, 406, 408
Dabbs, Dr. G. W. R., 303
Daffodils, 152, 155, 213 ; ' Lecture on,' 249–251
Dahlias, 267–268
' Daily Mail,' 77 ; war correspondent, 392, 405
' Dairy Farm in Miniature, A,' 103
Dàl Bhàt (Indian breakfast dish), 290
Dalmatia, 162
Dante quoted, 46
Danube, river, 371
Dardanelles, 371
Darmstadt, 349
Date pudding, 127
Dates, albumen in, 111
Datura cornigera, 266–267
David, statue of, 173
Davison, Mr., 378, 383, 395, 404, 409
Davitt, Michael, quoted, 346–347
De Aar, 389–410
' De l'Horlogiographie,' 168
De l'Orme, architect, 167–168
Deane, Dr. R. Edmund, 75
Defoe, Daniel, ' Tour through the Eastern Counties,' 337
Delphiniums, 348
Denmark, tour in, 356 *seq.*
Depopulation of rural districts, 108, 123 *seq.*
Depression caused by indigestion, 120
— on change of diet, 45–46
Derby, Lord, 388
' Devil in the bush, the,' 323
Devonshire, Duke of, Mr. Morley's letter to, 351–353
Dewey, Dr. E. H., on diet, 8 ; ' A New Era for Women,' 15, 27 ; ' True Science of Living,' 32
Dictamnus fraxinella, 226
Diet of the author, 4–7
— Dr. Haig's, 6, 8, 17, 41–42, 62,

75 ; Salisbury, 6, 46, 50 ; vegetarian, 6, 8–11, 25, 43–44, 60, 67 ; meat, 6, 10, 16, 18, 28, 46, 48, 50 ; moderation in, 8, 30 ; change of, 11, 53, 88, 121, 255, 300 ; of anthropoid apes, 29 ; of man, fruits and seeds, 29 ; Dr. Allinson's, 30 ; in illness, 34, 62–64, 66–68 ; difficulties of change of, 46–7, 59, 60–61, 67 ; wrong, the cause of disease, 40 ; of a doctor, 42 ; in uric-acid diseases, 45–46 ; depression on change of, 45–46 ; serious study of, 46, 51–52, 68 ; in old age, 47, 56 ; for children, 50–53 ; in schools, 54–55 ; the cure for obesity, 58 ; in indigestion, 61–62, 67 ; while travelling, 64–66, 204 ; dry, 69 ; the cure for skin-diseases, 71–72 ; Maeterlinck on, 72–74 ; homes for instruction in, 76 *seq.* ; a friend's criticisms on, and Dr. Haig's reply, 78–88 ; habit in, 87–88 ; the question of quantity in, 78 *seq.*, 84–87 ; of nuts and apples, 98 ; universal interest in the question of, 104 ; at three shillings a week, 109 *seq.* ; cheapest form of, 120 ; plea for a simpler form of, 120 *seq.*, 204 ; the cure for corns, 303
Diet, books on (*see also under* Health books) : Sir Henry Thompson's ' Diet in Relation to Age and Activity,' 13–14 ; Dr. Haig's leaflets on diet, 24 *n.*, 45, 48–49, 57 ; — ' Diet and Food in Relation to Strength and Power of Endurance,' 25, 41, 86 ; Eustace Miles's ' Muscle, Brain, and Diet,' 25 ; — ' Failures of Vegetarianism,' 25 ; — ' Better Food for Boys,' 26, 53–54 ; ' Forty Vegetarian Dinners,' 26 ; ' Science in the Daily Meal,' 27 ; ' Fruits, Nuts, and Vegetables : their Uses as Food and Medicine,' 27 ; Mrs.

Anna Kingsford's 'The Perfect Way in Diet,' 28; T. R. Allinson's 'Book F,' 30; A. W. Duncan's 'Chemistry of Food,' 32; A. W. Duncan's 'Foods and their Comparative Values,' 32; Hutchinson's 'Dietetics,' 32; Smith's 'Fruits and Farinacea,' 32; Albert Carter's 'Natural Food,' 298

Dieting, doctors on, 6, 11; weight in, 58, 61, 78 seq., 85 seq.

Dietists, dinner to, 20; and hotel-keepers, 64, 66

Digestion, 79, 87; mental irritation bad for, 60; aids to, 71

Digestive organs, in man, 29, 60, 79, 81; of the cud-chewers, 29

'Digit of the Moon: A Hindoo Love Story,' 37–39

Dinner, the hour for, 56

Dinners and teas to the poor, 35

Diosma capitata and *ericoides*, 211

Doctors, imaginary conversation between two, 42–43; and the treatment of vegetarians, 52–53, 61–62; list of those who advocate a non-flesh diet, 75; and vegetarianism, 83; and their bills, 138

Doernberg, Count, 365

Dorbay, architect, 168

Dorking, 295

Douglas, Mr., 263

Dover, 252

Drill Hall exhibitions, 152–154, 207, 210

Drinking, hereditary effect of excessive, 50–51; Maeterlinck on, 73–74; at meals, mode of, 112; stories, 338–339

Drunkenness, 48

Dry diet, 69

Du C——, 389–390

Du Cerceau, architect, 168

Dublin, 204, 230

'Dunvegan Castle,' 387

Dyspepsia, diet in, 61, 67; acid, 71; produced by milk, 86;

and tea drinking, 141–142 (see also under Indigestion)

Dyspeptics, diet for, 5; long life of, 6; meal hours for, 57; diet for, with reduced cooking, 118

E——, Major, 397

E——, Staff officer, 412

Earle, Charles, Journal of, 356–373

—— Charles William, 376

—— Major Henry, 391

—— Lionel, 377, 403

—— Max, 390, 396–402, 409–415

—— Captain Sydney, 253; last letters of, 374–417

Early rising, 142

Eating, love of, 59–60

Economy in housekeeping, 142

Eczema, 46; diet in, 71; the cause of a tragedy, 71–72

Edinburgh, 273

Education, Books on: 'An Ideal School, or Looking Forward,' by Preston Search, 35; 'Education and Empire,' by R. B. Haldane, 36

—— defects of modern, 149

Edward VII., King, new sanatorium for consumption, 96; and the diet question, 104; interest in the Society for the Prevention of Tuberculosis, 158; illness of, 241, 243

Egg-plants, 320

Eggs, uric acid in, 16, 78, 85

'Eighteen hundred a year,' 133–151

Elementary Schools, Instructions on Cookery for, 105

Eliot, George, 137; 'Romola,' 191

Elizabeth, Queen, 307, 331

Ellacombe, Canon, 247

Elsinor, 358–359

Elysée, Palace, 166–167

Emulsin in almonds, 99

'Encyclopædia Britannica' quoted, 161

Endive, 261 ; French mode of cooking, 287
Enteric, inoculation for, 378
Epilepsy, diet for, 45
Epimediums, 155, 235
Ericas (heaths), various, 207–210, 224
Erskine, of Linlathen, Thomas, 243
Esher, Lord, 274
Eton and the diet question, 52
Eucalyptus-trees, 169
Eucomis, 310
Euonymus, 195
Euphorbias, 232, 270
Evergreens in London, 195
Everlastings, 374
Eversley, 274
Exercise in indigestion, 61
— for invalids, 62, 86
Exeter, 338
'Exile's Mother, An,' poem, 215
Eye, training of the, 149–150

F——, of the Army Service Corps, 392
F——, Mr., chief telegraphist, 397
F——, C. K., 401, 408
'Family Budgets,' articles on, 107 seq., 133–151
Farming, various kinds of, 124–125 ; goat, 89–106
Fat, reduced by diet, 57–58, 61 ; digestion of, 105 ; in diet, 110 ; in goat's milk, 93, 119
February, Notes for, 152–159
Ferments, 99
Ferns, 195
Fiesole, Mino da, 173
Fig-trees, outdoor culture of, 268
Fig pudding, 291
Figs, albumen in, 111 ; stewed, 289
Finland, 361–363
Fir-trees, 211 ; Scotch, 275
Fish in rheumatism, 5
— salt, the cause of leprosy, 47
— uric acid in, 16, 47–49
Flesh foods, Bible on, 48 ; uric acid in, 48 (see also under Meat)

'Fleurs et Montagnes,' poems, 153
Flint, Dr., 75
Florence, 191, 284, 344 ; visit to, 172–173 ; letter on, 173–174
Flour, various qualities of, 116, 294–296
Flourens, M., on man's frugivorous nature, 29
Flower-growing, 124 ; in London, difficulties of, 194
Flowers, in table decorations, 144–145 ; floating arrangement of, 221–222 ; how to dry, 302 ; wild, 343–344
Fly-trap, 402
Folco, Sina, 174
Food, adulteration of, 32, 100 wrong, the cause of disease, 40 ; the study of, 51–52 ; chemistry of, 53 ; on board ship, 65–66 ; Maeterlinck on, 73 ; instruction in elementary schools on, 105 ; wholesome, for the poor, 107–32 ; Reformers, 108, 112 (see also under Diet)
Forget-me-nots, 207, 232
Formic acid in almonds, 99
Forsythia suspensa, 153
Fotheringhay village, 306–307
France, visit to the South of, 160 seq. ; 'Renaissance of Art in,' 167 ; and Russia, 372–373
Francoa ramosa, 268
Frankfort, 171
Fréjus, 166
French Exhibition of 1900, 280
— hours for meals, 57
— plums, albumen in, 111
Froude on Mary Queen of Scots, 307–308
Frozen provisions, 65
Frugivorous habit of man, 28–29, 72
Fruit, albumen in, 111 ; puddings, 113 ; and bread diet, 120 ; farming, 124 ; to bottle, 288–289
Fruit and seeds, man's proper diet, 29, 43

Fruitarians, 54; on board ship, 65–66
Fruit-trees for London, 195
— how to grow, 259–260
Fruit-wall, an old, 332–333
Fry, Mr. Roger, 206
Frying, oil for, 116
Fuchsias, 320
Furnishing, 138, 149

G——, Colonel, 378, 392
G——, Colonel B., 416
Gambling, 134
'Garden, The,' 207, 245, 334
Garden City Association, 109
Gardening, wild, 198, 208, 233–235, 282; recent books on, 203; in winter, 204; letters from Germany on, 217–219, 301–302, 310–315, 349; new tool for, 321
— Books on: Mr. Robinson's 'English Flower Garden,' 155, 198, 208–209, 219; Moggridge's 'Flora of Mentone,' &c., 161, 165; 'Riviera Notes,' 171; Jekyll and Morley's 'Roses for English Gardens,' 246; Nicholson's 'Dictionary of Gardening,' 219, 253, 273; Gerarde's 'Herbal,' 225; 'Century Book of Gardening,' 237, 252–253; Miss G. Jekyll's 'Home and Garden,' 244; — 'Wall and Water Gardens,' 244–245; — 'Lilies for English Gardens,' 245–246; 'Country Life Library,' 245, 334; Herbert Cousins' 'Chemistry of the Garden,' 246–247, 314; Violet Biddel's 'Small Gardens and How to Make the Most of Them,' 247; Forbes Watson's 'Flowers and Gardens,' 247–248; Edward Jesse's 'Favourite Haunts and Rural Studies in the Vicinity of Windsor and Eton,' 248–249; Richard Jefferies' 'Life of the Fields,' 249; Peter Barr's 'The

Daffodil' and 'Lilies of the World,' 249–251; Moore and Jackson, 'The Clematis as a Garden Flower,' 251; Mrs. Loudon's 'Lady's Country Companion,' 252; Edward Step's 'The Romance of Wild Flowers,' 253; Rider Haggard's 'Farmer's Year-Book for 1898,' 255–256; Vilmorin's 'Vegetable Garden,' 258; J. C. Loudon's 'Arboretum et Fruticetum Britannicum,' 271, 334–335; Andrews' 'The Botanist's Repository,' 320; Curtis' 'Botanical Magazine,' 321; Andrews' 'Heathery,' 322; Cook's 'Trees and Shrubs,' 334–335; H. Repton's 'Theory and Practice of Landscape Gardening,' 337–338
Gardens, advantages of, 150; ideal position for, 231; shelter for, 234; lily, 245–246, 249–250; mode of cataloguing, 253, 262; kitchen, 258–262, 283; rotation of crops in, 260–262; wall and water, 244–245; seaside, 331 seq.
Garibaldi, Guiseppe, 174; and the Naples Museum, 178; autobiography of, 186; residence of, 200
Garlic, wild, 232
Garner, Professor, 272
Garrod, Dr., 5
Gassendi on man's frugivorous nature, 29
Gasten, Mr., groom, 409
Gastro-intestinal inflammation, 70, 126
Gates, Mr., 91; on dry treatment of sewage, 92; his goat farm, 97, 102–104
Gauntlet, V. V., & Co., 235
Gaura Lindheimeri, 240
Gem Supply Company, 35
Genistas, 264, 332
Genny, Dr., 75
Genoa, 161–162, 201–202; 'Flora of,' 165

Geraniums, 219, 320; wild, 343
Gerbera Jamesoni, 322
German waters, 44, 46
Germany, treatment of tuberculosis in, 41; analysis of foods in, 100; letters on gardening from, 217-219, 301-302, 310-315, 349; fruit-preserving in, 301-302; and Russia, 371-372
Gibbon's 'Roman Empire,' 179-180; Coleridge on, 180-181; Frederic Harrison on, 181-183
Gibraltar, 371, 377; goats in, 96
Gingerbread, receipt for the best, 292
Girls, training and education of, 35-39
Gladiolus, 210, 323
Gladstone, W. E., 404
Glasnevin botanical gardens, 229-230
Glencoe, 381
'Globe,' 383, 390
Gnocchi, receipt for, 286
Goats, 89-106, 347; sex vigour in, 55; Mr. Bryan Hook's 'Milch Goats and their Management,' 91 *seq.*; Toggenburg, 91, 96, 102 *seq.*; fat globules in milk of, 93, 119; food for, 94, 103-104; Mr. Pegler's book on, 94; their yield in milk, 95, 102; prejudice against, 95-96; cheese, 96; their exemption from tuberculosis, 96; 'A Dairy Farm in Miniature,' 103; milk in London, price of, 119; in Suffolk in the 17th Century, 336-337
'Golden Treasury' series, 184
Goode & Co., 302
Gorse, 224, 264
Gourds, 262
'Gourmands, Des,' 33
Gourmet boiler, 127
Gout, 34, 63; 'La Physiologie du,' 32; water for drinking in, 35; diet for, 45-46, 61, 71, 84; Plasmon in, 55; milk and

cheese as the cause of, 78, 86-87
Gramnivorous animals, sex vigour in, 55
'Grape nuts,' 8
Graspan, 414, 416
Gray's 'Elegy,' missing stanza of, 248-249
Great Barton Mills, 295
Great Bookham, 263
Greek civilisation, high level of, 193
Green & Nephew's glass bowls, 221
Green sauce, 286-287
'Greenery - Yallery Grosvenor Gallery,' 324
Greenhouses, how to keep dry, 230
Grenadier Guards in South Africa, 409, 412, 414
Grouse à la crème, 291
Gruel from Graham flour, 296
Guards' Brigade in South Africa, 399-402, 410, 414
Guelder roses, Chinese, 265
Guildford, goat farm at, 91; a day in, 97, 100-102
Gums in underfeeding, 8, 61
Gypsum in bread, 65

H——, Major, 389-395, 398
H——, Messrs., 403
Haage & Schmidt, 312
Habit in diet, 87-88
Hadwen, Dr., 75
Haggard, Mr. Rider, 255
Haig, Dr., 3, 27, 30, 40, 110, 303; on coffee, 4; on asparagus, 5; on diet, 6-8, 17, 24 *n.*, 41-42, 62, 75; on rheumatism, 17-18; 'Uric Acid in Disease,' 23-25; 'Leaflets' on his diet, 24 *n.*, 45, 48-49, 57; 'Diet and Food,' &c., 25, 41, 86; diet for consumption, 41; on uric acid, 44, 46; on the progress of enlightenment, 48; on meal hours for dyspeptics, 57; his diet for fat people, 58; patients' cases, 62-64, 66-71; his ad-

dress, 75; his home for invalids, 76 *seq.*; criticisms on his diet, 78–84; his reply, 84–88

Haldane, R. B., 'Education and Empire,' 36

Hall, W. H. Bullock, 'The Romans on the Riviera,' 166

Hamburg, 357

Hamilton, Major A. C., 340

Hamilton, Sir William and Lady, 185

Hamlet, tomb of, 358

Hampton Court, 236, 273

Hams, curing of, 150

Hanbury, Mr. Thomas, 171

Hanover Road, 403

Hares, 234

Harris, C. J., 43

Harrison, Frederic, quoted, 181–183

Harrod's Stores, 302

Harwich, 329–330, 337

Hatton, Sir Christopher, 309

Hawkweeds, 344

Hawley, 208

Haymarket Stores, 295

Headaches, caused by uric acid, 42; diet for, 45; from indigestion, 120; cure for bilious, 304

Health, 1–88; a stranger's letter on, 2; of the author, 3–5; undermined by adulterated food, 32; where to obtain books on, 32; Natural Health Society, 34; 'Paradise of,' 46; study of, 52; a sure sign of good, 58–59; International Health Association, 112; subjects, legislation on, 158; the houses best for, 277 *seq.*; discussion of, in public, 339–340

Health, books on (*see also under* Cookery): Dewey, Dr. E. H., 'A New Era for Women: Health without Drugs,' 15, 27; — 'True Science of Living,' 32; Haig, Dr., 'Uric Acid as a Factor in the Causation of Disease,' 23; Miles, Eustace, 'Avenues to Health,' 26;

Kellogg, Dr., 'The Stomach,' 32; Pouchet, M., 'Pluralité de la Race Humaine,' 28; Cuvier, 'Le Régne Animal,' 29; Lawrence, Professor, 'Lectures on Physiology,' 29; Bell, Charles, 'Diseases of the Teeth,' 29; Allinson, T. R., 'Medical Essays,' 30; Searle, Henry, 'Tonic System of Treating Affections of the Stomach and Brain,' 31; Fernie, Dr., 'Herbal Simples,' 32; Huxley's 'Elements of Physiology,' 32; 'La Physiologie du Goût' de Brillat Savarin, 32; Lancaster, Owen, 'A Short Account of the Human Body,' 34; 'Water: How it Kills its Thousands,' 35; Maeterlinck, 'The Buried Temple,' 72–74; 'Herald of Health,' 99; Kuhne, Louis, 'New Science of Healing,' 296; Cox, Edwin, 'Degeneracy and Preservation of the Teeth,' 298

'Heart of the Empire, The,' 108

Heath gardens, 208

Heaths, 161–162, 207–210, 233; books on, 322

Hedgerow cuttings for goats, 103

Heidelberg, 349

Helichrysums, 347

Hell, Dante on, 46

Hellebores, 221

Henry IV., 168

Hentzner, Paul, 'Travels in England,' 337

Herb eaters, 29

'Herbal Simples,' 32

Herbs, sweet-smelling, 162

Hertfordshire garden, 236–237

Hervey, Lord Francis, 336

Heywood, Mr., 4

Hidalgood Werokalsi, 320

High feeding, cure, 41; insanity from, 41

— tea, 121

Highland Light Infantry, 403, 405; Brigade, 405

'Hindoo Love Story, A,' 37–39

Holbein's ' Dance of Death,' 357
Holland, third Lord, 242
Holland House, 275, 309 ; rose
    show, 241 ; gardens of, 242-
    243
Holly, 195
Hominy with cheese, 113
— fried, 128
Homœopathy, 30
Horace quoted, 193, 282
Horses, rheumatism in, 18, 92
Horticultural Show, 222
Hospitality, 144
Hotel-keepers and dietists, 64,
    66
Housekeeping in country, 148 seq.
— in town, 135 seq.
Housemaids, 145
Houses, on building, 277-281
Housman, Mr. Laurence, 222
Houston, Miss, 72
Howard, Dr. E., 75
Hyacinths, 154-155, 232, 249
Hyænas, food of, 29
Hydrangeas, 212, 222, 233
Hyères, 161
Hygiene, books on, 32
Hyssop, 327

Ice for thirst, 70
Illness, diet in, 34, 60
Imaginary conversation between
    two doctors, 42-43
Imprisonment, Michael Davitt
    on, 346-347 ; Wilfred Blunt
    on, 347
' In Memoriam,' 101
' In Vinculis,' by Wilfred Blunt,
    347
Income, expenditure of, 138
India and the tea industry, 32,
    51
Indian breakfast dish, 290
Indigestion from tea-drinking,
    31 ; as the cause of constipa-
    tion, 52 ; the cause of diseases,
    60, 120 ; diet in, 61-62, 67
    (see also under Dyspepsia)
Indigofera Gerardiana, 319
' Individualist, The,' 165

Inflammation, gastro-intestinal,
    70
Insanity from high feeding, 41, 48
— from tea-drinking, 142
Intelligence Department, 378,
    380
International Health Association,
    112
Invalids, effect of vegetarian diet
    on, 60, 67-68 ; exercise for, 62,
    86
Ipswich, 295, 337
Ireland, tea-drinking in, 142 ;
    visit to, 204 seq. ; emigration
    from, 213-215
Iris garden, 236
Irises, 154-155, 165, 170, 199,
    235 ; time to plant, 310
Italian Government and Pompeii,
    190-191
Italian may; 200
Italy, diet in, 64
Ivy, 195

Jackman, Messrs., 251, 333
Jam, a new, 285
James I., 275, 307
Jameson, Mrs., ' Diary of an
    Ennuiée,' quoted, 189-191
Jameson Raid, 400
Japan, Bishop of, 121
Japanese decorations, 145, 178,
    192 ; nurseries, 235 ; garden,
    242 ; lilies, 246, 250 ; loniceras,
    282 ; anemones, 319 ; treat-
    ment of wistarias, 330 ; Ginkgo,
    maidenhair-tree, 348
Jasminum nudiflorum, 152
Jeaffreson, J. C., ' Lady Hamilton
    and Lord Nelson,' 185
Jekyll, Miss, 277, 323 ; works on
    gardening by, 244-246
Jersey cows, 103
Jesse, Edward, ' Gleanings from
    Natural History,' 248
Jessop, Miss Florence, 76
Johannesburg, 387
Johnston, Dr., 75
Johnstone, Sir Harry, 256 ; letter
    on monkey's food, 272-273

Jones, Inigo, 309
Jones' ' Views of Seats, Mansions, and Castles of England,' 309
' Journal of a Tour in the North of Europe,' 356–373
' Journal of Horticulture,' 249
Jowett, Benjamin, 3
Julius Cæsar on goat's cheese, 96
July, notes for, 244–265
June, notes for, 224–243, 267–268

KALES, flowering, 197
Karlsbad, 64
Katharine of Arragon, 305
Keats quoted, 147
Keith, Dr., on diet, 8
Kellogg, Dr., 27, 112 ; ' Science in the Kitchen,' 5, 32 ; ' The Stomach,' 32
Kew Bridge, old, 220
Kew Gardens, 155, 209, 238, 242, 318–319, 322 ; of Germany, 350
Khalifa, the, 382
Kid, roast, 93–94
Kidney pudding, illness from eating, 178
Kiew, 370
Kimberley, 379, 382, 392–394, 404–410
Kingsley, Charles, 274–275
Kirby Hall, 308–309
Kitchener, Lord, 382
Kitchens, German, 301–302
Knight, ' Morning Post ' correspondent, 392
Koch, Dr., on tuberculosis, 88, 158 ; antidote for microbes, 188
König, Professor, 112
Kremlin, the, 369–370
Kronstadt, 372
Kruger, ex-President, 388–389, 392
Kugler's ' Handbook of Painting,' 185
Kuhne, Louis, 296

L——, officer, 412
L——, Lord, 381
L——, P., 391, 395, 399
La Cava, station, 200
La Mortola, 171–172
' La Rose de Noël,' poem, 153–154
Laburnum, 231
Lactic acid, 78
Ladies' maids, 140
Ladysmith, 395
Laing's Nek, 382
Lalange, Adolphe, 32
Lancers, 9th, 411, 416–417
' Lancet,' 201
Landcress, as a vegetable, 285
' Landscape Gardening,' 337–338
Lard, substitute for, 117
Larkspurs, 312 ; markings of, 344–345
Las Palmas, 379, 381, 395
Laurels, 236–237
Laurustinus, 170, 225
Lavender, 236, 270 ; oil of, 282
' Lavengro,' 147
Lawless, Miss Emily, ' The Wild Geese,' quoted, 213–216
Lawns, Americans on English, 276
Le Veau, architect, 168
Leghorn, 161, 201
Leichtlin, Mr., 217, 219
Lemann's biscuits, 300
Lemberg, 373
Lemon soufflé, cold, 292
Lemons in Italy, 198
Lentils, 5, 17 ; uric acid in, 18, 48 ; as food for the poor, 108 seq.
Leprosy, 71 ; cause of, 47
Lescot, architect, 167
Letter, an old, 327–329
Lettuce, 261
Liatris, 218
Liddell, Miss, 379
Lilacs, 225, 231
Lilies, 199, 233, 235, 245–246, 323, 332 ; ' for English Gardens,' 245 ; ' of the World,' 249–250
Linden, Mr., 230

Linnæus, on man's frugivorous nature, 29
Lion, sex vigour in the, 55
Lippi, painter, 174
Liverpool, 327–328
Liverpool School of Cookery, 284
'Lives of the Twelve Cæsars,' 183
Lloyd, Mr., 360
Lobelia, 310
Loch family, 391
Loch, Douglas, 402, 405, 412
London, 135 *seq.*; 'Life and Labour of the People in,' 139; flowering plants in, 194–195
London (Bloomfield), Bishop of, 335
London Hospital, 377
Longden, 102
Loniceras, 320˙
Loubet, President, 96
Loudon, Mr. and Mrs., 252
Louis XIV., 168
Louvre, the, 167–168
'Love-in-the-mist,' 323
'Love-lies-bleeding,' 323
'Love's Labour Lost,' by Owen Meredith, 173
Loyal North Lancashire Regiment, 408, 412
Lübeck, 357
Lucerne for goats, 103
Lung Arno, 172
Lyttle, Dr. J. Shaw, 75
Lytton, Bulwer, 191
Lytton, Robert, quoted, 196–197, 317

M——, Lord, 401–404, 407, 411–412
Macaroni, 7; cheese, 113; and cheese patties, 127; à la tripe, 127; and Portugal onion, 287
McI——, storekeeper, 408
Madeira, 162
Maeterlinck quoted, 72–74, 203; his appeal against meat and alcohol, 110
Mafeking, 382, 394, 399, 405–406

Magnolias, 235, 237
Maidenhair tree, 348
Maize, Indian, 125; with cheese, 113; bread, 126
Majuba, 383
Malet, Lucas, 274
Mallock's 'The Individualist,' 165
Malta, 371; goats at, 96
Manchester, International Health Association of, 112
Mangold for goats, 194
Manures, 247
Manuring, 261
Maples, 235
March, notes for, 160–174
Marigolds, 213
Mario, Jesse White, 186
Market gardening, 124
Marrows, 262
Marseilles, 162; 'Flora of,' 165
Martial, quoted, 194
Mary Queen of Scots, 306–308
'Materia Medica,' 9
May, notes for, 203–224, 238
Meals, hours for, 56–57, 116
Meat, one of the causes of cancer, 47; amount of proteid in, 111; the sale of diseased, 177
Meat diet, 6, 10, 62; uric acid in, 16, 18; man's teeth not adapted for, 28; hereditary effect of, 51, 122; in dyspepsia, 61, 67–68
Medical books, old and new, 31 (*see also under* Health books)
Medici, Catherine de, 166–167
Mediterranean heath, 161–162
Medlars, 332
Medstead, 77
Meissonier, Madame, 169
Meissonier's studio, 168–169
Melbourne, 298
Mellin's food, in dyspepsia, 61, 70; as a substitute for tea, &c., 113
Mental irritation bad for digestion, 60
Mentone, 171; 'Flora of,' 161
Menus, vegetarian, 113–115, 117–119

Mercer, Dr. A. B., 75
Mercury in eczema, 46
Meredith, Owen, quoted, 173
'Merrie England,' 249
Mesembryanthemums, 332
Metabolism, 9, 80, 86
Metropolitan Hospital, 77
'Mexican,' R.M.S., 387, 390, 392
Michaelmas daisies, 236, 265, 323
Microbes in dust, 163; antidote for, 188
Miladorovitch, Count, 365
Miles, Eustace, 40, 55, 57; works by, 25-26, 53-54; on diet in uric acid diseases, 45; home recommended by, 77
Milk, 78, 84-85; separated, 4, 7, 90, 93; in diet, 86; in constipation, 87; of goats, 89 seq., 119; ignorance of the properties of, 90-91; of sheep, 96; albumen in, 111 seq.; fat globules in, 93, 119
Milk soup, 288
'Milkwoman, The,' 299
Miller, Dr. E. P., 99
Milton, John, lost poem of, 256-258; 'Lycidas,' 344
Mimosa, 116
Mitchell, Dr., 75
Modder River, 375
Model village, 122
Moggridge's 'Flora of Mentone,' 161
Monkeys, food of, 272-273
Mont St. Michel, 201
Montagu, Lady M. W., 238
Montbretias, 235, 319
Moore, H., & Son, 294
Moore, Mr., 295
Moral self-control, 55
Moravia, 373
'More Pot-Pourri,' 2, 77, 244, 379, 383, 390, 403; American reception of, 354-355
Morley, Mr. John, quoted, 137; his gift to Cambridge, 351-353
'Morning Post' war correspondent, 392, 405
Morrison, Arthur, 107, 115
'Morte d'Arthur,' 101

Moscow, 367-370
Mothers and the training of children, 36
Motley, Mr., 1
Motor-cars, 274-275
Mowat, Dr. Thomas, 75
Mucus colitis, 68, 71
Mulberry-trees, 195, 236
Munby, Mr. A. J., quoted, 298-300
Munthe, Dr. Axel, quoted, 187-189
Murray's 'Guide to Pompeii,' 187; 'Guide to Russia,' 370
Mushrooms, 17, 272; dried, 301; xanthin in, 48
'Mustard' and 'Cress' (two horses), 393, 395
Mustard bath for bilious headaches, 304
Mutton, leg of, mode of using, 151
Myrtle, 162

Naauwpoort, 394-396, 406, 410
Naples, goats in, 89, 185; visit to, 175 seq.; sanitary condition of, 187-189; museum, 103
Narcissi, 154-155
Natal, 381-388, 394, 398-399, 409
National Association for the Prevention of Consumption, 157
'National Review,' 195
Natural Health Society, 34
'Natural History, Gleanings from,' 248
Natural History Museum, 272
Nature in art, 205-206
Neapolitan violets, 170
Nelson, Lord, 185-186, 330
Nemesia strumosa, 316-317
Nene, river, 307-308
Nerves injured by tea drinking, 51; in underfeeding, 61
Nervous diseases, high feeding in, 41
Nervousness, from indigestion, 120; from excessive meat eating, 122
Neuralgia, diet for, 45, 67; intercostal, 63

Neva, river, 363
New English Art Club, 205–206
New South Wales Lancers, 406, 411
' New Trafalgar, A,' 100
Newcastle, 381
Nice, 161
Nicotiana sylvestris alba, 269
Nightingale, Florence, 381
Nightshade, deadly, 48
Nordrach treatment for tuberculosis, 41
Northamptonshire, visit to, 305 *seq.*; maidenhair-tree in, 348
Northumberland Fusiliers, 408
Norvals Pont, 394–398, 408–410
' Nouilles lactées Suisse,' 100
Nüremburg, 161
Nut butter, 112, 117; cutlets 129, 132
Nuts, 43, 111 ; ' Fruits, Nuts, and Vegetables,' 27
Nuts and apples, diet of, 98

Oaks, 236 ; ' Turkey,' 248 ; Government plantation of, 275
Oatmeal, amount of proteid in, 111
Oats, cause of rheumatism in horses, 18 ; bread from, 126
Obesity, cure for, 57–58, 61
Observation, powers of, 149–150
October, notes for, 326–355
Octotis grandis, 332
Odessa, 324, 369–371
Œnotheras, 268, 344
Oil as an aid to digestion, 71, of almonds, 99; cloth, 165 ; olive, 284; stoves, 117
Oïrsted, Professor, 357
Okapi, 272
Old age, diet in, 47, 56
' Old Glory,' 170
Oldfield, Dr., 75
Olive oil, 284
Olive-trees, 172 ; at Cannes, 169
Omdurman, battle of, 382
Onions, 7, 261 ; braised, 131 ; how to serve, 285–286
Open-air treatment of colds, 159

Operations, cinematographs of, 340
Opium, 32
Orange jelly, the best, 293
Orange River, 385, 394, 402–410
Orange River Camp, 375
— trees, 144, 236
Orangeade, as made in Paris, 293
Orchids, 154, 230 ; wild, 232, 348
Ornithogalums, 239
Orwell, river, 337
Ostriches and the soldiers, 413
Owen, Professor, 29, 260
Ox-tongues, to cure, 292

Pæonies, 235
Pæstum, 176, 196
Page-boys, 146
' Palace of Truth,' 36
' Pall Mall Gazette,' 256, 375
' Palmen Garten,' 350
Palms, 170, 272
Parents and the diet of children, 52–53
Parinarium-tree, 272
Paris, 226, 240, 340 ; Opéra Comique, 166 ; restaurants, 64 ; visit to, 168–169
Parkinson, Mr., 251; on cypresstrees, 239–240
Parkyn, receipt for, 127–128
Partridge à la crème, 291
Paths planted with thyme, 163
Paton, Mr. J. B., 247
Patties, macaroni and cheese, 127
Pattison, Mrs. Mark, ' Renaissance of Art in France,' 167
Peace in South Africa, 222, 227
Pear-trees, 212 ; grown from pips, 260, 331 ; prunings of, for goats, 103
Pears, early picking to save, 332
Peas, 4, 17, 261 ; uric acid in, 18, 48, 57
Pegler, Mr., 94
Pelargoniums, 230, 262, 320–321
Penzance, sweet briars, 242, 282
Pergola, at the Capucine convent, 198–200 ; in Suffolk, 333

Pernettias, 233
Peterborough Cathedral, 305, 307
Petersfield, 232
Petrusville, 403
Pheasants, 234
Philanthropic, suggestion for the,
    95–96
Philipstown, 403
Phillips, H., 75
Phlebitis, 63–64
Phloxes, 219, 270, 314
'Picciola,' 345–346
Pigeon-houses in France, 309
Pigs, keeping of, 150
Piles, diet for, 67
Pinus austriaca, 331
Pipes, briarwood, 161–162
Plane-trees, 165
Plants, overfeeding of, 350
'Plasmon,' 13, 54, 56, 110–112;
    biscuits, 8, 18; in rheumatism
    and gout, 55; in mucus colitis,
    68; in gastro-intestinal inflam-
    mation, 70; as a drink, 113
    seq.; in cooking, 127
Platycodon, 218
Playfair, Dr., 41
Pleas for a simple diet, 120 *seq.*,
    204
Pliny on the uses of cork, 165;
    'Letters,' 183
Plumbagos, 220–221, 320, 348
Poems quoted : Keats', 147; Cor-
    revon's 'La Rose de Noël,' 153–
    154; J. Rhoades' poem on
    flowers, 155–156; Owen Mere-
    dith's 'Love's Labour Lost,'
    173; Shelley's 'Ode to Naples,'
    190; Sir Henry Taylor's 'Philip
    van Artevelde,' 194 ; Robert
    Lytton's 'Sorrento Revisited,'
    196–197; Mrs. Browning's
    'Aurora Leigh,' 202; Miss
    Emily Lawless' 'Wild Geese,'
    214–216; Laurence Housman's
    'The Winners,' 222–223 : on a
    pigeon in St. Paul's, 228; 'The
    Settlers,' 229; missing stanza
    of Gray's 'Elegy,' 249; Lord
    Tennyson quoted, 255; lost
    poem by John Milton, 257–

258; R. L. Stevenson's 'The
    Celestial Surgeon,' 258; Miss
    Una van A. Taylor's 'Nous
    n'irons plus au bois,' 276 ; on
    rain, 278; A. J. Munby's 'The
    Milkwoman,' 299–300 ; Robert
    Browning quoted, 324; Crabbe
    quoted, 336; a sailor's street
    song, 339 ; 'The Queen's Cho-
    colate,' 341–343 ; 'Chrysan-
    themums,' 350–351 ; 'Vale,'
    355; J. Rhoades's 'Stars,' 377
Polenta cutlets, 125
Polygonums, 238, 319
Pommes à la caramel, 286
Pompeii, 184; art treasures of,
    178 ; visit to, 186 *seq.*; 'Last
    Days of,' 191 ; books on, 192
Poor, food for the, 108 *seq.*
Portinari family, 174
Post, van de, 340
Postcard box, automatic, 97
'Pot-Pourri from a Surrey
    garden,' 220, 323, 389, 402
Potatoes, poisonous berries of, 48;
    mode of cooking, 71, 288;
    baked, 113; to keep when
    cooked, 116 ; soufflé with nuts,
    130 ; early, 261, 271 ; purée of,
    286
Poultry-keeping, 150
Poussin, artist, 206
Pratt, Anne, 282, 336, 345
Prescriptions for weakness on
    change of diet, 45
Press, and the diet question, 104
— on the training of the eye, 149 ;
    on tuberculosis, 158; and the
    Coronation, 241
Pretoria, 383, 386–387
'Pretty Betty ' (red valerian), 345
'Pretty Polly ' (Pelargonium), 262
Primroses, 232
Privet, 195, 264
Proteid food, 55 ; in obesity, 61 ;
    daily allowance of, 78 *seq.*, 85
    seq., 104, 111, 119–120; in
    separated milk, 90 ; amount of
    in various foods, 111 ; in flour,
    116 (*see also under* Albumen)
Protene, 13, 112

Pruning, necessity of, 211, 237, 319–320, 334
Prunus Pissardii, 152 ; *spinosa*, 153
Prussic acid in almonds, 62, 98–99
Pruth, river, 371, 373
Puddings, fruit, 113 ; raisin and barley, 129 ; semolina, 289
Pullar, Dr., 75
Pythagoreans and the bean, 49

QUACK DRUGS, 32
Quacks, 31
Quadrumana, diet of the, 29
' Queen Mary's Tears ' (thistle), 308
' Queen's Chocolate, The ' (poem), 341–343
' Quo Vadis,' 183

R——, Colonel, 387
Railway carriages, infection in, 156–158
Raisin tea, 112, 131 ; and barley pudding, 129
Raisins, albumen in, 111
Raleigh, Sir Walter, 48
Ramondia pyrenaica, 219
Ranunculus, turban, 172 ; Persian, 210
Raspberries, 260
Récamier family, 32
Receipts, 125–132, 151, 284–297, 302–304
Redruth, 235
Reed, Mr. T. G., 297
Reinholdt, Dr. Chas., 75
Religion, 59
' Renaissance of Art in France,' 167
Rheumatism, 34 ; fish in, 5 ; cause of, 17–18, 86–87 ; in horses, 18, 92 ; water for drinking in, 35 ; diet for, 45–46, 61, 71, 84 ; ' Plasmon ' in, 55
Rhoades, Mr. James, 155 ; poem by, 377
Rhodes, Cecil, 389

Rhododendrons, 231, 233
Rhubarb, 269
Rhus Cotinus and *laciniata*, 319
Ribes sanguinea, 282
Rice, 7 ; with cheese, 113 ; bread from, 126 ; croquettes, 130
Richmond, 318
Rimington's Guides, 396, 399, 403
Rippingille's oil-stoves, 117
' Riviera Notes,' 171
' Riviera, The Romans on the,' 166
Roberts, Captain, 200
— Lord, ' on treating soldiers,' 21–23
Robinson, Mr., 219 ; ' English Flower Garden,' 155, 198 ; on heaths, 208–209 ; on vegetable gardens, 258, 260 ; list of sunflowers, 270
Rockery plants, 310
Rogers, Samuel, 242
Roman ruins at Baiæ, 179
Rome, 175
Romneya Coulteri, 240
Roosevelt, President, quoted, 213
Root crops, 261 ; plants, 270
Rosa rugosa, 319
Rosemary, 236
Roses, 170 ; Christmas, 153, 221 ; in water, 221 ; Holland House show of, 241 ; for English gardens, 246 ; Chinese Guelder, 265 ; soil for, 283 ; on growing, 314–315 ; difficulty of photographing, 335
Rosetta stone, 358
Rossi, Signor, 357
Rotation of crops, 260–262
Royal Engineers, 404
Royal Hospital for Children and Women, 77
Royal Scots Fusiliers, 382
Rubus deliciosus, 222
' Rulers of the South,' 49
Rural districts, depopulation of, 108 *seq.*
Rush-broom, Spanish, 264
Ruskin, John, 247
Russia, tour in, 356 *seq.*

Russia, Alexander I. of, 356 ; death, 364
— Alexander II. of, 357
— Grand Duke Constantine of, 364
— Nicholas I. of, 356 ; accession, 364

S——, Colonel, 410
S——, Sir Leicester, 382
S—— Major, 377–378, 381–382, 388, 390
S——, Sergeant-Major (B.S.A. Police), 399
Sabajone (Venetian dish), 291
Sage, 162
St. Claude, 161
St. Helena, 387
St. Hospice, tower of, 161
St. Paul's Cathedral, 228, 309
St. Petersburg, 356, 360, 363–367
St. Raphael, 160–1
St. Thomas of Canterbury, 101
St. Vincent, 380
Saintine's, ' Picciola,' 345
Salad, cheese, 285 ; potato, 288
Salisbury cure, 6, 46, 50
Salsify, 261
Salt one of the causes of cancer, 47
Salutaris Water, 35
Sande-Torte (cake), 293
Sandhurst, 375
'Sandow' exercises, 378
Sandwiches of almond cream, 70
Sanguisorbas, 265
Sapon (powdered soap), 302
Sardinia, 162
Sauce, green, 286–287 ; Sevillane, 289
Saving, habit of, 133 seq. ; object for, 147
Savoury vegetable stew, 128
Savoys, 252
Saxifrages, 163, 348
School Board curriculum, 120
Schwartzenberg, Prince, 366
Scorzonera, 261
Scots Guards, 406
Scott, Mr., 295

Sea-buckthorn, 238
Seakale beet, 271 ; to cook, 288
Sea Point Horticultural Society, 249
Sebastopol, 370, 372
Second childhood, 346
Sedentary occupation, diet for, 54, 62, 67
Seeds, growing from, 170, 331, 347 ; preservation of, 322
Semolina pudding, 289
Senile decay, 104
Sensuality and diet, 55
September, notes for, 305–325
Servants and diet, 121 ; wages and treatment of, 139, 141
Sewage, dry treatment of, 91–92 ; question 337
Sex vigour in animals, 55
Shanklin, 303
Sheep's milk, 96
Shelley, ' Ode to Naples,' 190 ; cremation of, 200
Shelter for gardens, 234
Shingles, 46, 64
Shopping, 144
Shrubs, treatment of, 225 ; lists of, 237, 253
Shuckburgh, E. S., 184
Sidalcea candida, 218
Sienkiewicz's ' Quo Vadis,' 183
Skin-diseases cured by diet, 72
Slaughter-houses, cruelty in, 177
Slough, 76
Slums, overcrowded condition of, 122–123
Smoke nuisance, 318
Smoking, hereditary effect of, 50–51
Snake's head iris, 155
Snowdrops, 152
Soap, powdered, 302
Social reformers on diet, 108
Society for Prevention of Cruelty to Animals, 177
Socrates quoted, 144
Soda, bicarbonate of, 45 : salicylate of, 45, 63 ; in bread, 65 ; in mucus colitis, 68–69 ; water, 70
Soil, how to tell a good, 231–232

Solanums, 317, 320
Solomon's seals, 225, 232
— wisdom, 326–327
Sorrento, 196 ; poem on, 196–197
Soufflé potatoes with nuts, 130
Soup, vegetable, 284 ; milk, 288
South Africa, deficiency of women in, 272 ; return of troops from, 353 ; Captain Sydney Earle's letters from, 374–417
South Kensington Museum, 308
Southwold, 327
'Spectator' quoted, 222, 229, 323, 341, 344
Spencer, Herbert, on diet, 15
Spikenard of the Bible, 282
Spinach, 261
Spiræas, 200, 225, 238, 265
Spitting, on the habit of, 156–157
Stag's-horn sumach, 318
Standard Bank, 390
Stanford, Bennet, 392
Starch, digestion of, 105 ; in flour, 116
'Stars' (poem), 377
Starvation, cure, 11, 70–71 ; death from, 50 ; from under-feeding, 67
Steamships, food on, 65–66
Stellenbosch, 400
Stevenson, R. L., quoted, 258
Stockholm, 359–360
Stonehenge, 201
Strangford, Lord, 365
Strathfieldsaye, 335
Strawberries, 260
Suffolk, visit to, 326 seq. ; 'Breviary,' 336 ; books on, 337
Sugar in gout and rheumatism, 46 ; in diet, 110
Sugden, Mr. R., 102
'Sun' war correspondent, 405
Sundial, 320–321
Sunflowers, 269–270 ; for goats, 103, 324 ; grown for profit, 324–325
Sunstroke, 255–256
Surrey, gardens in, 232–234
Swanley Horticultural College, 272

Sweden, tour in, 356 seq.
Sweet peas, 316
Swiss Government and the export of goats, 102
Switzerland, goats in, 96 ; analysis of foods in, 100 ; letter from, 344
Synaptase in almonds, 99

T——, A., 409–410
T——, General, 377
Table decoration, 144–145
Tamarisk, 237–238
Taylor, Miss U. van A., quoted, 275
Taylor, Mrs. Fleetwood, M.D., 75
Taylor, Sir Henry, quoted, 194
Tea, 17, 121 ; afternoon, 6–7, 42 ; uric acid in, 18, 48 ; condemnation of, 31–32, 141 ; hereditary effect of, 50 ; excessive drinking of, 51, 142 ; various qualities compared, 51 ; Chinese mode of preparing, 51 ; goat's milk in, 93 ; substitute for, 113
Teeth as an indication of man's proper diet, 25, 28–29 ; books on the, 29, 298 ; early decay of, in the present generation, 89, 297–298
Teetotallers, 83 ; use of tea and coffee promoted by, 50, 142
'Temple Classics,' 101
Tennyson, Lord, 101 ; quoted, 255
Thalictrums, 265
'The Choice' (poem), 214
'The Settlers' (poem), 229
'The Winners' (poem), 222
Theosophy, 20
Thirst, treatment of, in illness, 70
Thistles, 308
Thomas, Mr., 412
Thompson, Dr. Alexander, 183
Thompson, Mr., 345
Thompson, Sir Henry, on diet, 13–14 ; on cremation, 201
Thorne, John, 241, 309

'Three Cups Inn,' The, 330
Thrift, habit of, 133 *seq.*
Thrombosis, 63
Thuyas, 349
Thyme, 162 ; for paths, 163
Tiger, food of the, 29 : sex vigour in the, 55
'Times,' 88, 376 ; war correspondent, 405
Titian's landscapes, 206
Toadflax, wild, 343
Tobacco, to keep moist, 285
Toggenburg goats 91, 96, 102 *seq.*
Tomatoes, stewed, 129
Tool, a new gardening, 321
Toothache among the troops in South Africa, 89
Toulon, 161
Tradescant, Mr., 240
Tragedy caused by eczema, 71-72
Tranquillus, C. Suetonius, 183
Transvaal, 385, 388
'Traveller's joy,' 330
Travelling, diet while, 64-66, 70
'Treating,' Lord Roberts on, 21-23
Tree heaths for briarwood pipes, 161-162
Tuberculosis (*see under* Consumption)
Tuileries, ruins of the, 166-169
Tula, town, 370
Tuli, town, 382
Tulip-trees, 248
Tulips, 207, 213, 222, 225-226
Turkey and Russia, 371-372
Turkeys, 337
Turner, Mr., 400
Turnips, 261

UNDERFEEDING, 19 ; the gums in, 8, 61 ; from non-assimilation, 41 ; Dr. Haig on, 67
'Unit Library,' 101
Upton Church, 248
Uric acid, 17, 18, 42-44, 60-61 ; definition of, 16 ; in eggs, 16, 78 ; as the cause of disease, 23-25, 78 *seq.*, 84 ; as the cause of headaches, 42 ; diet, 45 ; in

flesh and vegetables, 48 ; foods free from, 68-69, 85, 87, 110 ; the cause of gout and rheumatism, 86 (*see also under* Xanthin)

VAAL river, 386
Vaccination, 10
'Vale' (poem), 355
Valescure, 160-161, 169 ; heaths of, 162 ; cork-trees at, 164-165 books on, 165 ; fragments of the Tuileries at, 166-168
Van der Goes, Hugo, 174
Vegetable, stew, 128 ; gardens, 258-262 ; soup, 284
Vegetables, amount of proteid in, 111
'Vegetarian,' newspaper, 43 ; creed, 97, *menus*, 113-115, 117-119 ; Society, 297
Vegetarian diet, 6, 8, 11, 25-27, 43-44 ; books on, 26, 27 ; objections to, 48, 58 *seq.*, 108 ; for children, 52 ; list of doctors who advocate, 75 ; homes for instruction in, 76 *seq.* ; question of quantity in, 78 *seq.*, 81 *seq.* ; constipation in, 82-83 ; as taught in schools, 105 ; on three shillings a week, 109 *seq.* ; plea for a, 120 *seq.* (*see also under* Cookery books and Diet)
Vegetarianism and sociability, 52 ; and sentiment, 60, 177 ; effect on invalids of, 60 ; Maeterlinck on, 72-74 ; and the medical profession, 83 ; Virchow on, 122 ; an American judge on, 284
Vegetarians, unscientific, 41 ; in the Army, 53 ; dining out, 57 ; and cruelty to animals, 177
Veitch & Sons, 210, 213, 222
Venidiums, 268, 332
Veratrum nigrum, 270-271
Verbascum phœniceum, 222
Verbena, 263
Vesuvius, 175-176, 187, 201
Victor Emmanuel, King, of Italy, 174

Vienna, 373; flour, 116
Villiers, Charles, 12
Vilmorin, M., 'Vegetable Gardens,' 227, 258-259
Vine-clad Pergola, 198-200
Vinton & Co., 91
Violets, 171, 232, 249; on growing, 322-323
'Violettes de Parme,' 170
Virchow, Dr., 27, on vegetarianism, 122
Virginia creeper, 271
'Virgin's bower,' 331
Volkousky, Princess Zénéïde, 36

W——, General, 391, 396, 404
W——, sapper, 392
W——, Sir G., 381
W——, F., 379, 395
W——, General F., 395
Wagner, composer, 59
Wales, Henry, Prince of, 275
Wallace, Mrs., 99
Wallflowers, 207, 345-346
Walnuts, 130
Walters, Dr., 75
Wareham, 382
Warsaw, 357, 364
Water, when to drink, 7; for gout and rheumatism, 35; 'How it kills its thousands,' 35
Watercress as a vegetable, 285
Water-lilies, 242
Waterer, Messrs., 231
Waterhouse, Elizabeth, 243
Waters, German, 44, 46
Waterton Collection, 308
Watts, Mr. G. F., 243
Weight in dieting, 58, 61, 78 seq., 85 seq.
Weir-Mitchell, Dr., 41
Wellbank, Mr., 127
Wellington Barracks, 375
Wellingtonias, 349
Welsh polypody, 344
West Surrey Dairy Co., 97

Westminster Abbey, 307
'Westminster Gazette,' 155; letters to the, 156-158, 303; poems from the, 221, 275-276, 350-351, 355
Wheatmeal bread, 126
White, Dr. Charles, 75
Whitmore, Mr., 195
'Wholesome Food for the Poor,' 107-132, 283
Wild flowers, 343-344
— gardening, 198, 208, 233-235, 282
'Wild Geese,' The, 212-216
Wilde, Oscar, 251
Williams, Dr. R., 27
Williams, Mr., 200
Willow herb, 344
Wilson, Dr. Helen, 75
Wilson, Mr. G. F., 233-235, 246
Wimbledon Common, 375
Winchester, 101
Windsor, 290
Wine, the purchase of, 144
Wisley, 225, 233-235, 246
Wistaria creeper, 175, 225, 231, 330
Witteput, 411
Woking, 231; 'Nursery,' 251
Wolf, food of the, 29
Wood gardening, 211, 237, 282
'Workman's budget,' 108
Worthing, 178

XANTHIN, 45, 48, 85; in fish, 49; in eggs, 85; in lentils, 108 (see also under Uric acid)

YEW hedges, 211
Yews, 172, 236, 269
York, 232; Cathedral, 305
Young, Dr. Thomas, 357-358
Yuccas, 207

ZOLA, M., 328

Spottiswoode & Co. Ltd., Printers, New-street Square, London.